LOWLAND GRASSLAN
HEATHLAND HABITATS

Grasslands are everywhere: agricultural land, playing fields and road verges – but, while species-poor, intensively managed grasslands are widespread, colourful semi-natural grasslands and heathlands, buzzing with life, are scarce. These semi-natural habitats are ancient, cultural landscapes, which are of considerable, if not international, importance for biodiversity. Yet despite targets for the conservation and restoration of these valuable grasslands and heathlands, they continue to decline before our eyes.

Lowland Grassland and Heathland Habitats contrasts the uniformity of intensively managed grassland with the diversity of traditionally managed grasslands and heathlands. It examines topics of concern to the ecologist or habitat manager such as causes of the loss and deterioration of these habitats, including inappropriate management, eutrophication and climate change. It evaluates the opportunities for positive change, such as conservation, restoration and creation. A series of case studies illustrate the pressures on some lowland grassland and heathland habitat types and looks at ways to enhance them for biodiversity.

This habitat guide features illustrated species boxes of typical plants and animals, as well as a full species list, a series of projects on the ecology of grassland and heathland species, a colour plate section, up-to-date references and information, and a full glossary. It will provide students and environmentalists with a deeper understanding of the nature and importance of lowland grassland and heathlands.

Liz Price is Senior Lecturer in the Department of Environmental and Geographical Sciences at Manchester Metropolitan University.

HABITAT GUIDES

Series Editor: **C. Philip Wheater**

Other titles in the series:

Upland Habitats
Urban Habitats
Woodland Habitats

Forthcoming titles:

Freshwater Habitats
Marine Habitats
Agricultural Habitats

LOWLAND GRASSLAND AND HEATHLAND HABITATS

•

Elizabeth A. C. Price

Illustrations by Jo Wright

Routledge
Taylor & Francis Group
LONDON AND NEW YORK

First published 2003
by Routledge
11 New Fetter Lane, London EC4P 4EE

Simultaneously published in the USA and Canada
by Routledge
29 West 35th Street, New York, NY 10001

Routledge is an imprint of the Taylor & Francis Group

Typeset in Sabon by RefineCatch Limited, Bungay, Suffolk
Printed and bound in Great Britain by
St Edmundsbury Press, Bury St Edmunds, Suffolk

British Library Cataloguing in Publication Data
A catalogue record for this book is available from the British Library

Library of Congress Cataloging in Publication Data
Price, Elizabeth A. C. (Elizabeth Anne Clewett)
Lowland grassland and heathland habitats / Elizabeth A. C. Price;
illustrations by Jo Wright.
p. cm.
Includes bibliographical references (p.).
1. Grassland ecology. 2. Heathland ecology. I. Title.
QH541.5.P7 P75 2002
577.4–dc21
20020272520

ISBN 0–415–18762–1 (hbk)
ISBN 0–415–18763–X (pbk)

For Jack and Holly Brierley

CONTENTS

ACKNOWLEDGEMENTS

Many people have helped with the production of this book, and I am very grateful for their assistance. In particular, I should like to thank the following: the editor (Phil Wheater) and the publishers, who have been very helpful throughout the process; Jo Wright for her evocative illustrations; two anonymous referees for their positive comments and valuable suggestions; friends and colleagues at Manchester Metropolitan University who contributed in various ways to the production of the book (Jane Boygle, Simon Caporn, Jacky Carroll, Paul Chipman, Rod Cullen, Steve Dalton, Mike Dobson, Becca Gamble, Chrissie Gibson, Sue Hutchinson, Mark Langan, Gregg Paget, Graham Smith); and friends, colleagues and others elsewhere for their input historically or more recently (Mr Barnes, John Day, English Nature Enquiry Service, Dave Fee, Phil Grime, Ben Haines, Paul Hughes, Mike Hutchings, Chris Marshall, Hilary Orrom, Richard Scott, Ken Thompson, Sue Wallis, Tim Walmsley). I should especially like to thank Doris and Peter Price for their constant help and encouragement; Jenny Walsh and John and Cath Brierley for their support and, in particular, Matt Brierley and Jack and Holly for their forbearance, enthusiasm and sense of humour. Any mistakes are mine – I hope you don't find them too irritating!

The author and publishers would like to thank the following for granting permission to reproduce the following illustrations as listed: CEH Monks Wood for Plate 3 (copyright NERC), English Nature for Table 2.6 (English Nature Research Report no. 169), NCC for Table 2.19 (appendix 1, p. 55, of *Focus on Nature Conservation*, No. 17, NCC, 1986) and Hilary Orrom for plate 2.7. Every effort has been made to contact copyright holders for their permission to reprint material in this book. The publishers would be grateful to hear from any copyright holder who is not here acknowledged and will undertake to rectify any errors or omissions in future editions.

SERIES INTRODUCTION

•

The British landscape is semi-natural at best, having been influenced by human activity since the Mesolithic period (*c.* 10,000–4,500 BC). Although these influences are most obvious in urban, agricultural and forestry sites, there has been a major impact on those areas we consider to be our most natural. For example, upland moorland in northern England was covered by wild woodland during Mesolithic times, and at least some was cleared before the Bronze Age (*c.* 2,000–500 BC), possibly to extend pasture land. The remnants of primeval forests surviving today have been heavily influenced by their use over the centuries, and subsequent management as wood pasture and coppice. Even unimproved grassland has been grazed for hundreds of years by rabbits introduced, probably deliberately by the Normans, some time during the twelfth century.

More recent human activity has resulted in the loss of huge areas of a wide range of habitats. Government statistics record a 20 per cent reduction in moorland and a 40 per cent loss of unimproved grassland between 1940 and 1970 (Brown, 1992). In the forty years before 1990 we lost 95 per cent of flower-rich meadows, 60 per cent of lowland heath, 50 per cent of lowland fens and ancient woodland, and our annual loss of hedgerows is about 7,000 km. There has been substantial infilling of ponds, increased levels of afforestation and freshwater pollution, and associated reductions in the populations of some species, especially rarer ones. These losses result from various impacts: habitat removal due to urban, industrial, agricultural or forestry development; extreme damage such as pollution, fire, drainage and erosion (some or all of which are due to human activities); and other types of disturbance which, although less extreme, may still eradicate vulnerable communities. All these impacts are associated with localised extinctions of some species, and lead to the development of very different communities from those originally present. During the twentieth century over 100 species are thought to have become extinct in Britain, including 7 per cent of dragonfly species, 5 per cent of butterfly species and 2 per cent of mammal and fish species. Knowledge of the habitats present in Britain helps us to put these impacts into context and provide a basis for conservation and management.

A habitat is a locality inhabited by living organisms. Habitats are characterised by their physical and biological properties, providing conditions and resources which enable organisms to survive, grow and reproduce. This series of guides covers the range of habitats in Britain, giving an overview of the extent, ecology, fauna, flora, conservation and management issues of specific habitat types. We separate British habitats into seven major types and many more minor divisions. However, do not be misled into thinking that the natural world is easy to place into pigeonholes. Although these are convenient divisions, it is important to recognise that there is considerable commonality between the major habitat types which form the basis of the volumes in this series. Alkali waste tips in urban areas provide similar conditions to calcareous grasslands, lowland heathland requires similar management regimes to some heather moorland, and both estuarine and lake habitats may suffer from similar problems of accretion of sediment. In contrast, within each of the habitat types discussed in the individual volumes there may be great differences: rocky and sandy shores, deciduous and coniferous woodland, calcareous and acid grassland are all typified by different plants and animals exposed to different environmental conditions. It is important not to become restricted in our appreciation of the similarities which exist between apparently very different habitat types and the, often great, differences between superficially similar habitats.

The series covers the whole of Britain, a large geographical range across which plant and animal communities differ, from north to south and east to west. The climate, especially in temperature range and precipitation, varies throughout Britain. The south-east tends to experience a continental type of climate with a large annual temperature range and maximum rainfall in the summer months. The west is influenced by the sea and has a more oceanic climate, with a small annual temperature range and precipitation linked to cyclonic activity. Mean annual rainfall tends to increase both from south to north and with increased elevation. Increased altitude and latitude are associated with a decrease in the length of the growing season. Such climatic variation supports different species to differing extents. For example, the small-leaved lime, a species which is thermophilic (adapted to high temperatures), is found mainly in the south and east, while the cloudberry, which requires lower temperatures, is most frequent on high moorland in the north of England and Scotland. Equivalent situations occur among animals. It is, therefore, not surprising that habitats of the same basic type (such as woodland) will differ in their composition depending upon their geographical location.

In the series we aim to provide a comprehensive approach to the examination of British habitats, whilst increasing the accessibility of such information to those who are interested in a subset of the British fauna and flora. Although the series comprises volumes covering seven broad habitat types, each book is self-contained. However, we remind the reader that the plants and animals discussed in each volume are not unique to, or even necessarily dominant in, the particular habitats but are used to illustrate important features of the habitat under consideration. The use of scientific names for organisms reduces the likelihood of confusing one species with another. However, because several groups (especially birds and to a lesser extent flowering plants) are often referred to by common names, we include common names where possible in the text. We have tried to use standard names, following a recent authority for each taxonomic group (see the species list for further details). In all cases, species mentioned in the text are listed in alphabetical order in the species index and, together with the scientific name, in systematic order in the species list.

1

INTRODUCTION

•

SOME DEFINITIONS

Grasslands and heathlands may be classed together as habitats or ecological communities with several factors in common. Grasslands and heathlands are open habitats, dominated by grasses and dicotyledonous herbs or dwarf shrubs, where trees are usually sparse. In the UK some are natural, occurring where abiotic environmental conditions such as poor soils and harsh climate prevent tree growth, but they can also develop where trees are removed and their regeneration is prevented (Gimingham, 1975; Crofts and Jefferson, 1999). Prehistoric clearance of woodland enabled native grassland and heathland species to spread, and centuries of traditional management allowed the characteristic vegetation of semi-natural grassland and heathland to develop (Duffey *et al.*, 1974; Webb, 1986). More recently the extent of semi-natural grasslands and heathlands has declined dramatically and they are now recognised as priority habitats for conservation (UK Biodiversity Group, 1998a, b).

Grasslands and heathlands occur over a wide altitude range, from sea level to the uplands, but this book examines the ecology and management of heathlands and grasslands in lowland Britain, below about 300 m in altitude and often below the upper level of agricultural enclosure (UK Biodiversity Group, 1998b; Crofts and Jefferson, 1999). The ecology and management of upland heaths and grasslands are described in *Upland Habitats* in this series (Fielding and Haworth, 1999). Lowland grasslands are communities that are dominated by species of grasses and herbs. The vegetation usually forms a closed sward or turf, which may be short or fairly tall and tussocky, but is usually less than 1 m in height (Plate 1). Grasslands can be categorised on the basis of their soil – acidic, mesotrophic or calcareous – and drainage status: wet or dry. The use of terminology is not always consistent, and acidic grasslands are also termed calcifugous grasslands, mesotrophic grasslands may be called neutral grasslands, and calcareous grasslands are also described as calcicolous grasslands. Grasslands can also be grouped according to the intensity of agricultural improvement of the community, i.e. improved, semi-improved or unimproved (semi-natural; Chapter 2) (Rodwell, 1992; Ausden and Treweek, 1995; Crofts and Jefferson, 1999). Areas of good-quality lowland grassland usually support a number of specialist species, and often have high species richness (UK Biodiversity Group, 1998b). Lowland heathlands are communities that are dominated by ericaceous dwarf shrubs such as heather (Plate 1.1, Plate 2) . The terms 'heathland' and 'moorland' are often used interchangeably, but 'heathland' is a broader term that refers to both lowland (i.e. below 300 m) and upland communities, whilst 'moorland' usually refers to particular communities that occur most extensively in the uplands on deep peat substrates. Heath communities may be

Plate 1.1: Dorset heathland

categorised on the basis of their drainage status: dry, humid (moist) or wet (Dolman and Land, 1995; Thompson *et al.*, 1995). Good-quality lowland heathland typically consists of areas of dry and wet heath, bogs, open water, areas of bare ground and some scattered trees and scrub (UK Biodiversity Steering Group, 1995).

EXTENT AND LOCATION OF LOWLAND GRASSLANDS AND HEATHLANDS IN THE UK

The areas of the UK covered by grassland and dwarf shrub heath habitats (upland and lowland) in 1998, based on field survey data from the Countryside Survey 2000 (CS2000), are shown in Table 1.1. It is estimated that dwarf shrub heath occupies 6.1 per cent of the UK, while grassland of different types occupies 33.9 per cent.

The extent and location of UK grasslands and heathlands in 1998 are shown in Plate 3. Comparison of Plate 3 with the altitude map, Fig. 1.1, indicates that large blocks of semi-natural grassland (i.e., neutral, calcareous, acid) and heathland are concentrated in the uplands, and in the lowland fringes, below 300 m, bordering the uplands. Elsewhere in the lowlands, surviving semi-natural grasslands and heathlands have become increasingly scarce and fragmented. In contrast, improved grasslands are generally restricted to the lowlands. Grasslands are the most widespread habitats in the UK (Haines-Young *et al.*, 2000) and even in towns and cities large proportions of urban land may be grassland (Carr and Lane, 1993; Baines, 1995) (Plate 1.2). However, the majority of UK grassland is species-poor improved mesotrophic grassland (24.7 per cent of the UK). In contrast, semi-improved and unimproved neutral (mesotrophic) grasslands combined occupy 3.5 per cent (Table 1.1) and the actual value for unimproved lowland mesotrophic grassland is really much less (11,600 ha, 0.05 per cent of the UK) (Jefferson and Robertson, 1996) (Table 1.2). Acid grasslands are an extensive semi-natural habitat in Britain (Table 1.1), although the majority (over 1.2 million ha) are in the uplands. Acid grassland

Table 1.1: Estimated stock of grassland and heathland (upland and lowland) in the UK, 1998 (area, 000 ha)

Habitat (upland and lowland)	England and Wales		Scotland		GB		Northern Ireland		UK	
	000 ha	*%*	*000 ha*	*%*	*000 ha*	*%*	*000 ha*	*%*	*000 ha*	*%*
Improved grassland	4,431	29.1	1,051	13.1	5,482	23.7	568	41.0	6,050	24.7
Neutral grassland	444	2.9	168	2.1	613	2.7	254	18.3	867	3.5
Acid grassland	547	3.6	748	9.3	1,295	5.6	28	2.0	1,324	5.4
Calcareous grassland	38	0.2	27	0.3	65	0.3	1	0.1	66	0.3
Dwarf shrub heath	485	3.2	1,002	12.5	1,487	6.4	13	0.9	1,500	6.1

Source: Haines-Young *et al.* (2000). Based on field survey data from Countryside Survey 2000 and the Northern Ireland Countryside Survey 2000.

Note: GB Great Britain, *UK* United Kingdom.

Figure 1.1: Approximate distribution of land over 300 m in altitude in the UK, indicated by dark shading

in the lowlands probably covers less than 30,000 ha (0.12 per cent of the UK). Calcareous grassland in the lowlands covers between 33,000 and 41,000 ha of the UK (0.14–0.17 per cent of the UK) (Rodwell, 1992; UK Biodiversity Group, 1998b). The area of lowland acid grassland is intermediate between that of lowland unimproved mesotrophic grassland and lowland calcareous grassland (UK Biodiversity Group, 1998b) (Table 1.2). The UK has approximately 58,000 ha of lowland heathland (UK Biodiversity Steering Group, 1995), which is about 0.24 per cent of the UK (Table 1.2). These values are rather crude estimates, but they clearly show that, whilst improved grasslands are widespread, semi-natural lowland heaths and grasslands are scarce.

WHY LOWLAND GRASSLANDS AND HEATHLANDS?

If the majority of lowland grasslands and heathlands are man-made habitats, what is their value? Many are ancient habitats, composed of native species, which are of considerable, and in many cases international, ecological importance. For example, despite

Plate 1.2: Urban grassland, Manchester

Table 1.2: Estimates of the stock of lowland grassland and heathland in the UK (area in 000 ha)

Lowland habitat type	*000 ha*	*% of UK*
Lowland heathland	58	0.24
Improved grassland	6,050	24.7
(Semi-improved and unimproved neutral grassland combined – above and below 300 m)	(867)	(3.5)
Lowland unimproved mesotrophic grassland	11.6	0.05
Lowland acid grassland	30	0.12
Lowland calcareous grassland	33–41	0.14–0.17

Sources: UK Biodiversity Steering Group (1995); Jefferson and Robertson (1996); UK Biodiversity Group (1998b).

lowland heathland losses of about 80 per cent since 1800, 20 per cent of the international total survives in the UK (UK Biodiversity Steering Group, 1995; English Nature, 2000). In addition, some lowland grassland and heathland types, such as the crested dog's-tail – common knapweed grassland community, are largely confined to the British Isles, and the UK therefore has a special responsibility for their conservation (Jefferson *et al.*, 1999). In the absence of trees, grassland and heathland species flourish in the climate of the UK, and regional differences in the climate and in soils create conditions suitable for a range of distinctive heathland and grassland community types (Rodwell, 1991, 1992, 2000). Maritime conditions (in particular the effects of salt spray) can also influence the composition of coastal grassland and heathland (Mitchley and Malloch, 1991). Superimposed on this pattern, caused by variation in soils and climate, is the influence of different management practices, which can further affect the character of these communities. Thus, although grasslands and heathlands share some fundamental characteristics, the differences within and between these habitat types are probably just as significant as their similarities. Within the large range of lowland grassland and heathland community types found in the UK, some are now extremely restricted in extent, for example unimproved lowland mesotrophic grasslands, which total about 11,6000 ha (Table 1.2), can be divided into a number of types, some of which cover less than 100 ha in total (Jefferson and Robertson, 1996).

Following the Earth Summit in Rio de Janeiro in 1992, the UK government published its Strategy for Sustainable Development and pledged to conserve and enhance biological diversity in the UK (DOE, 1996a). One of the main approaches to this was the launch of 'Biodiversity: the UK Action Plan' in 1994, which involved the preparation of action plans for key or priority habitats and species. Priority habitats include lowland heathlands and semi-natural grasslands, because the UK has an international obligation to protect them, they are at risk or rare, and they contain important species (DOE, 1996a; UK Biodiversity Group, 1998a, b). UK Biodiversity Action Plan targets for semi-natural grasslands and heathlands include habitat conservation, restoration and expansion (UK Biodiversity Group, 1998b) but, despite these targets, the extent and quality of semi-natural grasslands continue to decline. For example, the stock of calcareous grassland in the UK declined in the 1990s by 18 per cent (Haines-Young *et al.*, 2000). Similarly, while the majority of lowland heathland is now notified as SSSI (Sites of Special Scientific Interest, DOE, 1996a), the habitat is still affected by factors that lead to a decline in heathland quality, such as lack of appropriate management, nutrient enrichment, fragmentation and disturbance from development (UK Biodiversity Steering Group, 1995).

At a time when the role of land management practices in rural and urban areas is being increasingly questioned (e.g. Hindmarch and Pienkowski, 2000), lowland grasslands and heathlands provide ideal systems to examine the significance of human activity in shaping habitats. What is the range of grassland and heathland types in the UK? What are the factors that account for the differences? What are the interactions between organisms in these communities? What are the impacts of natural factors such as climate and substrate, and anthropogenic influences such as disturbance, agricultural improvement, fragmentation, pollution and neglect on grasslands and heathlands? Why is the information important? Can economic and environmental objectives be reconciled in managing these habitats? These are some of the questions this book addresses.

THE STRUCTURE OF THE BOOK

This volume has the same structure as other books in the series, first presenting background material that supports information in later chapters. Chapter 2 describes the ecology of UK lowland grasslands and lowland heathlands and examines the range of communities that exist today. The variety of grasslands and heathlands is enormous and they are described on the basis of their vegetation characteristics (Rodwell, 1991, 1992, 2000) because this is probably the most widely adopted method of habitat classification in the UK. Characteristic animals are also described, and rare species associated with each habitat are used to illustrate the range of species present and threats to biodiversity. Chapter 3 evaluates the management of grasslands and heathlands, including maintenance, improvement, restoration and creation, in relation to problems and opportunities that are especially relevant to these habitat types. Considerable attention is paid to past management and economic factors and protection measures because these have a huge bearing on the management of grasslands and heathlands today. In Chapter 4, a number of current issues such as climate change, nitrogen deposition and ecological mitigation are illustrated in greater detail through case studies. Finally, in Chapter 5, some practical projects are suggested to encourage readers to become involved in examining ecological responses to environmental factors for themselves. Having read the book, I hope that you will be able to understand the important issues facing grasslands and heathlands. Habitat ecology is a broad topic, and it has been difficult to decide what should be included. On the whole, I have selected information that students have found particularly interesting or useful. All the topics covered in this book are interrelated, and sub-headings allow you to read selectively and cross-reference with other relevant sections. Some aspects are covered only briefly, but references and sources of additional information are given at the end of the book.

2

LOWLAND GRASSLAND AND HEATHLAND HABITATS

•

The combined effects of different environmental conditions and management practices have created a wide range of grasslands and heathlands in the UK. This chapter explains the importance of climate and soils, outlines general characteristics of grasslands and heathlands, and examines the range of grassland and heathland communities that exists today. Relevant management is mentioned where appropriate, but the role of historical and current management practices is examined in detail in Chapter 3.

THE IMPORTANCE OF CLIMATE AND SOILS

Climate

Heathland

There are several parts of the world where the climate is suitable for the development of dwarf shrub communities, and in north-west Europe heathland communities occur primarily in the areas bordering the North Sea, the English Channel and the Atlantic. The distribution of heathland in north-western Europe coincides with the area that experiences a temperate, oceanic (or maritime) climate, lacking temperature extremes, but with abundant rainfall throughout the year (Gimingham, 1972; Webb, 1986). Although the whole of the British Isles falls within the climatic region suitable for heathland formation, around 6,000 years ago most of lowland UK was covered by woodland (Godwin, 1975). Heathland occurred naturally only in a few locations where poor soils, climate and wind exposure prevented tree growth (Putwain and Rae, 1988). The expansion of heathlands began in the Atlantic period, about 5,000 years ago, when the climate of the British Isles became cooler and wetter (more oceanic) and less favourable for tree growth, and woodland clearance began (Chapter 3) (Gimingham, 1975; Webb, 1986; Read and Frater, 1999). Although the shift in climate may have facilitated heathland expansion, the majority of lowland heathland is predominantly a man-made landscape (Gimingham, 1975; Webb, 1986).

Grassland

Similarly, there are several parts of the world where the climate is suitable for the development of temperate grassland, but grassland usually occurs as the natural vegetation type where the seasonal climate favours the dominance of perennial grasses and is too dry or cold to support woodland (Bell and Walker, 1992; Archibold, 1995). In the UK natural grassland, like naturally occurring heathland, is restricted to areas where environmental conditions prevent tree growth (Ausden and Treweek, 1995). Elsewhere, where woodland has been cleared and prevented from regenerating, the temperate, oceanic climate and high rainfall of the UK favour the growth of grass

(Smith, 1980; Lane, 1992). In some regions, grass has a growing season of over nine months a year (Moore, 1966; Fielding and Haworth, 1999).

Regional differences

Whilst the UK experiences a generally oceanic type of climate, regionally the climate is not uniform. For example, the climate in the east of the UK is relatively continental, with greater annual extremes of temperature and less rainfall than towards the more oceanic (or maritime) west (Fielding and Haworth, 1999). Also, northern areas of the UK tend to be cooler, wetter and less sunny than southern parts (Smith, 1980). Differences in climate influence species distribution patterns. For example, for some species, the UK is at the northern or southern geographical limit of their natural range (DOE, 1996a). For this reason, grassland and heathland communities differ regionally according to climate (Rodwell, 1991, 1992). Altitude also has a significant effect on climate and communities (Fielding and Haworth, 1999). Climatic changes at smaller spatial scales, e.g. in relation to topography, aspect (north or south-facing) or microclimate at the soil surface, can also influence the character of a community. For example, an increase in turf height following the removal of grazing animals has marked effects on microclimate, such as changes in temperature and humidity at ground level (Smith, 1980).

Other environmental conditions

Coastal grasslands and heathlands are influenced by maritime conditions, in particular the effects of salt spray deposition (UK Biodiversity Group, 1999c), and in exposed situations management may not be required to maintain these communities (Mitchley and Malloch, 1991).

Geology and soils

Grassland

Grasslands can be categorised on the basis of their soil, acidic (calcifugous), mesotrophic (neutral) or calcareous (calcicolous), and drainage status (wet or dry) (Rodwell, 1992; Ausden and Treweek, 1995). Across this range, species-rich grasslands are generally associated with low levels of plant-available soil phosphorus and potassium (Critchley *et al.*, 2002). Lowland acid or calcifugous grassland typically occurs on nutrient-poor (oligotrophic), generally free-draining soils with pH ranging from 4 to 6 (Table 2.1). These soils overlie acid rocks such as sandstones and acid igneous rocks or superficial deposits such as sands and gravels (Rodwell, 1992; UK Biodiversity Group, 1998b). The absence of calcium carbonate results in acidic soils with different chemical characteristics from calcareous soils (Rorison, 1990). For example, as pH falls the solubility of several toxic metal ions (aluminium, manganese and iron) increases. This increased solubility can also lead to nutrient deficiencies (for example, of calcium, magnesium and phosphorus) because soluble ions may be leached or precipitated as metal salts (Table 2.1). In acid soils, ammonium nitrogen is the main source of available inorganic nitrogen, whilst nitrate is the main source in most other soils (Rorison, 1986).

Mesotrophic grassland is often referred to as neutral grassland. This broad type includes all improved, semi-improved and unimproved grassland on mesotrophic (moderately fertile) to nutrient-rich mineral soils of pH 4.5 to 6.5. They may occur on moist brown soils, periodically flooded gleyed soils or permanently

Table 2.1: Mean values of selected soil properties for different grassland types

Grassland type	*pH*	*Extractable P (mg l^{-1})*	*Extractable K (mg l^{-1})*	*Total N (%)*	*AWC (%)*
Mesotrophic	6.3	13.6	171	1.04	25.8
Calcareous	7.8	7.9	155	0.98	26.6
Acidic	5.8	20.7	168	0.56	19.6
Mires	6.5	5.8	76	1.93	32.0

Source: Critchley *et al.* (2002).

Notes: Data were collected from a range of mesotrophic, calcareous and acid grassland types and averaged, e.g. mesotrophic grassland includes unimproved, semi-improved and improved community types. For values for individual community types refer to Critchley *et al.* (2002). *Extractable* Plant-available. *P* Phosphorus (Olsen method of extraction). *K* Potassium. *N* Nitrogen. *AWC* Available water capacity.

moist sites (Rodwell, 1992; UK Biodiversity Group, 1998b). Brown soils are fertile, often deep, soils that are favoured for agriculture, and large areas of mesotrophic grassland have been agriculturally improved (Chapter 1: 'Extent of lowland grasslands and heathlands in the UK'). The natural vegetation cover is deciduous woodland, and so they may be termed brown forest soils. Gley soils are characterised by permanent or temporary waterlogging (Scottish Natural Heritage, 1996). Unimproved mesotrophic grasslands require relatively restricted soil conditions and are potentially sensitive to altered soil properties, including soil pH. Waterlogging or periodic inundation is essential to maintain species-rich mire and wet grassland communities (Table 2.1) (Critchley *et al.*, 2002).

Lowland calcareous or calcicolous grasslands occur on shallow, lime-rich soils, generally overlying limestone rocks, including chalk, typically of pH 7.0 to 8.4 (Table 2.1). The limestone rocks underlying calcareous grassland soil are usually porous, and calcareous grassland soils are very free-draining and well aerated. Some may occur on excessively draining rendzina soils (Rodwell, 1992; Ausden and Treweek, 1995). Rendzinas are shallow soils, with an organic layer overlying calcareous bedrock, sometimes with a rubbly layer between (Smith, 1980; White, 1997). A general feature of calcareous soils is the lack of available nitrogen, potassium and phosphorus (although total amounts of nitrogen and phosphorus, mainly in organic forms, may be high), and they are characteristically rich in calcium as calcium carbonate ($CaCO_3$) which results in a pH at or just greater than neutral. At this pH range, metal availability is low, and metal toxicity is rarely a problem in uncontaminated calcareous soil. However, the metals required by plants in trace amounts, particularly iron, are in low supply (Rorison, 1990). Calcareous grasslands as a group require restricted soil conditions, but are generally not well differentiated from one another on the basis of their soil properties. Factors such as management and climate apparently have a greater effect on calcareous grassland species composition (Critchley *et al.*, 2002).

Heathland

Lowland heaths are usually characterised by acidic (pH within the range 3.4–6.5) and often sandy soils (Webb, 1986). In particular, they are normally deficient in available nitrogen and calcium. In some cases they are also particularly low in phosphorus (Gimingham, 1972, 1992). Heathland generally forms in regions where the underlying rock strata result in the formation of poor, acidic soils, but the presence of superficial layers of acidic or base-poor, leached deposits over basic rocks such as limestone may also support heathland vegetation (Gimingham, 1972; Rackham, 1986; Webb, 1986). For example, chalk heath has a mixture of chalk grassland species with heather or bell-heather (Grubb *et al.*, 1969; Chapman and Reiss, 1992).

Beneath most heathlands, podsolic (or podzolic) soils are found. A vertical cross-section of the soil is known as a soil profile. The profile of a podsol reveals well defined layers, or horizons (Plate 2.1). The surface of a podsol is covered by plant litter at various stages of decomposition. Heathland generates plant litter that is resistant to decomposition, and does not support many detritivores and decomposers. Beneath the litter is the uppermost horizon of the soil, the A horizon, a mixture of minerals and organic matter. Beneath the A horizon is the ash-grey E horizon, from which humus and oxides of iron are leached by percolating water. These are redeposited lower down the profile in the B horizon, where they form hard pan layers. The lower C horizon resembles the parent rock beneath. The ash-grey layer and the hard pan layers are characteristic features of a podsol, although they may be modified under certain environmental conditions (Gimingham, 1972, 1992; Webb, 1986).

Podsol formation depends on a number of factors, including rock type, climate and vegetation. Podsols usually form in acid, coarse-textured, well drained materials (Scottish Natural Heritage, 1996). Climate is an important factor because podsols form in regions where the amount of rainfall exceeds the amount of evapotranspiration (water evaporated from the soil surface or transpired by plants). Rain percolating through the soil transports humus particles and fine clay from the upper to the lower horizons. Bases such as calcium, magnesium, sodium and potassium are leached by chemical processes and are replaced by hydrogen ions. This causes acidification of the upper soil layers (Webb, 1986). Peat formation is associated with areas of even

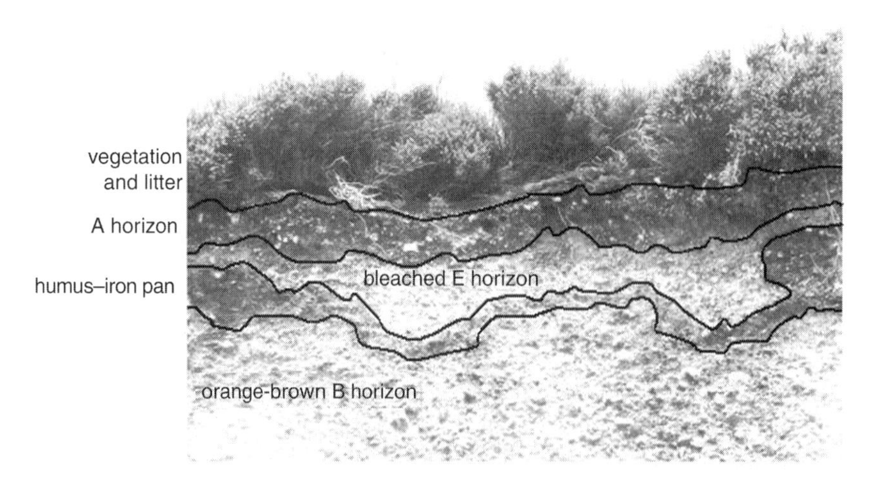

Plate 2.1: Profile of a heathland podsol, showing the bleached E horizon and the humus–iron pan

higher rainfall (Dimbleby, 1984). Heathland vegetation also modifies soil formation processes. Ericaceous dwarf shrub vegetation facilitates the development of podsols through the accumulation of acidic plant litter (Gimingham, 1972; Scottish Natural Heritage, 1996). Heathland podsols may also be derived from former acid brown forest soils through the formation of the ash-grey E horizon as iron oxides are leached from the upper layers and the development of the B horizon (with hard pan layers) as they are redeposited. It is thought that the removal of woodland cover by Neolithic people 4000–5000 years ago (Chapter 3) accelerated the rate at which acid brown soils were converted to podsols (Gimingham, 1972; Webb, 1986).

Lowland heathland varies according to environmental gradients such as increasing soil wetness, resulting in dry, damp (humid) and wet types of heathland (Rodwell, 1991). Wet heathlands grade into acidic mires and blanket peats, whereas dry heathlands often grade into grassland.

Soil bacteria and fungi

The characteristics of grassland and heathland soils can also influence soil microbial activity. Bacteria and fungi are important soil organisms and play a significant role in nutrient cycling (Wood, 1995) (Species box 2.1). Soil biological activity is usually greater in neutral or slightly alkaline soils. In general, bacteria have a relatively narrow pH tolerance, typically in the alkaline range, while fungi have a broad range of tolerance but tend to be more active in acid soils (Curry, 1994). Management also influences soil microbial communities, for example grassland management

Species Box 2.1: Field mushroom *Agaricus campestris*

The field mushroom is the common wild edible mushroom. It consists of a feeding system of colourless threads (the mycelium) that derives nourishment from organic matter in the soil, and specialist structures, or sporophores, which liberate spores. The sporophore consists of a cap bearing vertical gills, supported on a central stalk. The cap is 3–10 cm across, domed before expanding fully, and is usually white, scaly or smooth. The gills are deep pink, maturing to purplish brown. Field mushrooms are most likely to be found in old pasture or lawns in late summer to autumn, often growing in 'fairy rings' marked by grass of a darker green. The leading edge of the mycelium expands outwards, while the trailing edge dies. Grass growth is stimulated at the outer margin of a fairy ring, because there is rapid decomposition of organic matter in the mycelial zone.

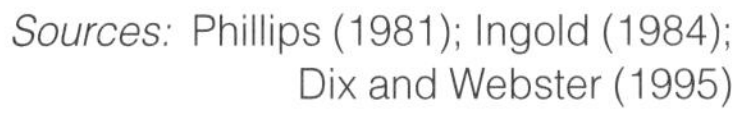

Sources: Phillips (1981); Ingold (1984); Dix and Webster (1995)

intensification generally leads to a significant decline in the proportion of fungi relative to bacteria in the soil community (Bardgett and McAlister, 1999; Donnison *et al.*, 2000; Grayston *et al.*, 2001). Some microbes form symbioses with plant roots, for example mycorrhizas (intimate symbiotic associations of fungi and roots) and nitrogen-fixing symbioses (e.g. between rhizobia bacteria and legume plants) (Wood, 1995).

Mycorrhizas

The roots of most plant species under natural conditions form symbiotic associations, called mycorrhizas, with specialised fungi (Wood, 1995). There are several types of mycorrhizal associations, each involving different groups of plants and fungi. The most widespread are the arbuscular (or vesicular-arbuscular, VA) mycorrhizas, in which the fungus penetrates the root cells. Vesicular-arbuscular mycorrhizas are formed between about two-thirds of plant species and a restricted group of fungi, and most herb and grass species of temperate grassland have VA mycorrhizas. Ectomycorrhizas are sheathing forms that cover the outside of roots and spread between cells. Ectomycorrhizas are formed between about 10 per cent of plant species and fungi, and are characteristic of roots of conifers and a number of broad-leaved trees (Dix and Webster, 1995). The other main mycorrhizal types are restricted in one case to orchids (orchid mycorrhizas) and in the other to ericaceous species, such as heather (ericoid mycorrhizas) (Hadley and Pegg, 1989; Fitter, 1997). The mycorrhizal association enhances nutrient uptake by plants. In particular, the hyphae of the fungus transport phosphate ions from the soil back to the root (Fitter, 1997), although in ericaceous species, such as heather, the association also increases nitrogen uptake (Bradley *et al.*, 1982; Strandberg and Johansson, 1999). The fungus obtains carbohydrate from the plant, and is dependent on the plant for its energy requirements. Transfer of nutrients and carbohydrates has also been shown to occur between plants connected by hyphal links, but the significance of this is not fully established (Allen, 1992). Addition of fertiliser can reduce the mycorrhizal infection of grasses (e.g. mat grass) and ericaceous species (e.g. heather) (Hartley and Amos, 1999). Mycorrhizas may also have other roles; for example, in soils with high, potentially toxic, metal concentrations, ericoid mycorrhizas can accumulate metals and significantly reduce the accumulation of metals in heather shoots (Bradley *et al.*, 1982). There is also evidence that VA mycorrhizas can protect plants against pathogen attack (Newsham *et al.*, 1994). In addition, diversity of VA mycorrhizal fungi appears to be a major factor contributing to plant biodiversity in calcareous grassland, since plant species composition and relative abundance change when the VA mycorrhizal fungi taxa that are present are varied experimentally (van der Heijden *et al.*, 1998). Practices such as ploughing, application of fertilisers and community translocation are likely to be detrimental to VA mycorrhizal communities (Crofts and Jefferson, 1999; Merryweather (2001).

Nitrogen fixation

Symbioses between plants and nitrogen-fixing bacteria are of great agricultural importance and are of ecological significance because nitrogen is in limited supply in semi-natural grasslands and heathlands. A number of symbioses exist, but the most widely studied is that of rhizobia with legumes (Begon *et al.*, 1996; Fitter, 1997). Most members of the pea family (Fabaceae) (legumes) form nodules on their roots that contain a group of bacteria known as rhizobia that can fix atmospheric nitrogen. Nitrogen fixation in nodulated legumes declines when nitrate fertilisers are added (Begon *et al.*, 1996) and the abundance of legumes may also decline in response to the addition of nitrogen. For example, experimental addition of nitrogen to grassland can lead to a fivefold reduction in the abundance of pea species and bird's-foot trefoils (Tilman *et al.*, 1994). The tissues of legumes such as white clover usually have a high nitrogen content and increase the local level of soil nitrogen as they decompose (Begon *et al.*, 1996) (Species box 2.2).

Climate and soils summary

Before prehistoric forest clearance, grasslands and heathlands would have been confined to areas where climatic or maritime conditions prevented tree growth, or to thin infertile soils, or to clearings maintained by wild herbivores (Jeffrey and Pigott, 1973; Duffey *et al.*, 1974). Today grasslands occur on a range of soils,

---

Species Box 2.2: White clover *Trifolium repens*

White clover is a common native clonal perennial species found on grassy and rough ground, most abundant in moist, fertile habitats. Clonal plants grow by producing shoot and root units, called ramets, which can potentially survive on their own. White clover is a low-growing, stoloniferous legume, rooting at nodes. Flowers are usually white, sometimes pale pink, rarely red or mauve, and are insect-pollinated, but recruitment from seed in established populations is rare, and clonal growth of individuals is important for propagation. White clover has been an important fodder crop since the seventeenth century. It is the most important pasture legume in Britain, and seventy-five cultivars are recorded in Europe. White clover is valuable to agriculture and for amenity planting because its symbiont *Rhizobium* can fix atmospheric nitrogen in habitats where other major nutrients, such as phosphorus, are not limiting. Nitrogen fixation by white clover often creates fertile soil that encourages invasion and temporary dominance by grasses because, under high nitrogen conditions, grasses outcompete clover. Under grass, soil nitrogen levels decline, which favours reinvasion by white clover, because it is competitively superior to grasses under low soil nitrogen conditions. The result is that clover and grass patches cycle over time. *T. repens* is generally intolerant of shade, but is tolerant of grazing, trampling and cutting. In part this is because plants of white clover are highly physiologically integrated, which allows resources to be shared between connected ramets. Although the plant is palatable to stock, some genotypes release hydrogen cyanide when tissues are damaged, which gives some protection against invertebrate and small mammal herbivores.

Sources: Grime *et al.* (1988); Cain *et al.* (1995); Marshall and Price (1997); Alpert and Stuefer (1997); Stace (1997)

whereas heathlands have a more restricted distribution. Heathlands and acid grasslands typically occur on acid soils, and one may grade into the other. Heathland may be replaced by acid grassland as a result of nutrient enrichment or intense grazing or burning (Gimingham, 1992) (Chapter 3, 'Problems' and 'Habitat maintenance and enhancement'). Semi-natural grasslands and heathlands are generally chartacterised by low nutrient availability, while improved grasslands are associated with more fertile conditions. Climatic variation and drainage patterns are superimposed on these soil-related differences.

SOME GENERAL CHARACTERISTICS OF GRASSLANDS AND HEATHLANDS

Ecological succession

Most lowland grasslands and heathlands, in the absence of grazing, cutting or burning management, will undergo changes in their species composition and structure over time (ecological succession). In general, the change is towards scrub and woodland communities (the climax community over most of the UK is some type of woodland). As succession progresses, there is an associated decline in species richness and an increase in soil nutrients (Webb, 1986; Crofts and Jefferson, 1999).

Most species-rich grassland and heathland communities occur in areas of intermediate soil fertility and disturbance, where competitive species are unable to thrive (Table 2.2). Management, such as cutting, grazing or burning prevents a build-up of soil nutrients and prevents fast growing, competitive species from achieving dominance by repeatedly removing new growth (Plate 2.2). In contrast, application of fertiliser increases productivity and allows competitive species to dominate. Disturbance created by livestock trampling and rabbit activity also creates gaps in the vegetation which allow establishment of plant species from seed and which provide conditions suitable for animals such as invertebrates and reptiles (Grime *et al.*, 1988; Tilman, 1997; Crofts and Jefferson, 1999).

Table 2.2: Plant strategies (Grime *et al.*, 1988)

Plant species can be characterised on the basis of their 'strategy' or functional type (the C–S–R model). The three strategies of competitor (C), stress tolerator (S) and ruderal (R) are extremes; intermediate strategies (e.g. stress-tolerant competitor, S–C) exploit intermediate conditions

C	*S*	*R*
Low-stress, low-disturbance conditions	High-stress, low-disturbance conditions	Low-stress, high-disturbance conditions
e.g stinging nettle (*Urtica dioica*)	e.g sheep's fescue (*Festuca ovina*)	e.g annual meadow grass (*Poa annua*)

The C–S–R model proposes that vegetation that develops in a particular situation is the result of an equilibrium between the intensities of stress (factors that constrain the rate of growth, such as shortages of mineral nutrients), disturbance (factors that result in the destruction of biomass, such as grazing or mowing) and competition (for light, soil resources, space). Competitors are associated with low stress, low disturbance conditions and have a high potential growth rate

Source: Hodgson *et al.* (1995).

Note: Stress Factors that constrain rate of growth. *Disturbance* Factors that result in destruction of biomass.

Plate 2.2: Cattle grazing semi-natural grassland, Side Farm Meadows, a reserve owned by Plantlife. Cattle are often preferred animals in conservation grazing schemes, as they create a variable sward

General characteristics of grassland plants

Some grassland plants avoid grazing through mechanisms such as mechanical or chemical defences; for instrance, aromatic herbs such as wild thyme are unpalatable to most herbivores (Crofts and Jefferson, 1999). Others are able to tolerate grazing or cutting; for example, grasses are kept in check, but are able to regenerate because their leaves grow continually from the base (they have intercalary meristems), so that, as they are defoliated, they are able to produce new tissue (Fitter *et al.*, 1984). New shoots, or tillers, are also produced from the base of the plants, but grasses differ in their tillering capacity, and hence their response to defoliation. Perennial rye grass produces many tillers, while crested dog's-tail produces relatively few (Crofts and Jefferson, 1999). Other species that are defoliated survive because they have a creeping (e.g. white clover) or rosette growth form (e.g. common knapweed). They produce shoots close to the soil surface, so cutting or grazing does not remove their apical meristems, although flowers and seed heads may be lost (Thompson, 1994). Many grassland perennials are long-lived species that do not need to reproduce by seed annually (Species box 2.3). In contrast, short-lived perennials and annuals such as yellow rattle will need to set seed each year (Grime *et al.*, 1988), unless they have a persistent seed bank (Hutchings, 1997). Some species, such as meadowsweet, will germinate in deep shade under a tall sward; others, such as harebell, require light to germinate (Grime *et al.*, 1988; Crofts and Jefferson, 1999).

General characteristics of heathland plants, heather in particular

Heathland plants grow in nutrient-poor conditions. Their nitrogen content is low, and falls from a spring peak as the growing season progresses (Webb, 1986). Heather can thrive on nutrient-poor acid soils, probably because it has a relatively slow growth rate (Species box 2.4). It is a woody dwarf shrub with small leaves, and once past the seedling stage, it can tolerate freely drained sands and gravels that dry out readily, although it can be killed by

Species Box 2.3: Creeping buttercup *Ranunculus repens*

Creeping buttercup is a clonal perennial stoloniferous herb that occurs on fertile, moist soils throughout the British Isles. Yellow, glossy, insect-pollinated flowers are produced mainly from May to June, and seed is set in June and July. Clonal growth of individuals is important for propagation. Clonal plants grow by producing shoot and root units, called ramets, which can potentially survive on their own. In clonal plants, daughter ramets receive support in terms of resources from the parent plant. Clonal growth is a relatively low-risk method of proliferation, but sexual reproduction by seed produces new genetic recombinations. In creeping buttercup, stolons bearing ramets are produced in early summer. Stolons wither in late summer or autumn and detached ramets overwinter as rosetts of leaves. The species is highly variable. Creeping buttercup, bulbous buttercup (*R. bulbosus*) and meadow buttercup (*R. acris*) are frequent in meadows and pastures, but *R. repens* has a wider habitat range than *R. bulbosus* or *R. acris*. All three species are dependent on bare patches for regeneration by seed, but the distribution of creeping buttercup is particularly associated with disturbed habitats. The three species differ in a number of ways, including the relative importance of recruitment from seed (see table). In some ridge-and-furrow grassland the bulbous buttercup is found on the fairly well drained ridge tops, the meadow buttercup occurs on the moister flanks of ridges, and creeping buttercup occupies the wetter furrows.

Sources: Fitter *et al.* (1985); Grime *et al.* (1988): Waite (1994); Lynn and Waldren (2001)

Ranunculus bulbosus Bulbous buttercup	*Ranunculus acris* Meadow buttercup	*Ranunculus repens* Creeping buttercup
Short, stem bases swollen	Medium/tall	Short/medium, creeping
Lobed leaves, the end lobe stalked	Deeply cut leaf lobes, the end lobe unstalked	Lobed, rather triangular leaves, the end lobe stalked
Recruitment mainly from seed, rarely by clonal growth (daughter 'corms')	Recruitment mainly from seed, with some clonal growth (daughter rosettes)	Recruitment from seed and clonal growth important (widely spaced ramets on stolons)
Sepals turned downwards	Sepals erect	Sepals erect

Species Box 2.4: Heather *Calluna vulgaris*

Heather is an evergreen dwarf shrub with upright or trailing stems. It produces large, spreading plants that may reach 80 cm in height and 50 cm in diameter. Leaves are 2–3.5 mm long and triangular in cross-section. Roots have ericoid mycorrhizas. Purplish pink (or white), insect or wind-pollinated flowers are produced from August to September. Seed is shed from September onwards, and some may overwinter on the plant. Seeds vary in germination characteristics, and may be incorportated into a persistent seed bank. Seed germinates between spring and autumn, and establishment appears to depend on the creation of bare ground. Heather is a relatively short-lived (normally less than thirty years) shrub, but plants can be rejuvenated by appropriate management. Heather is a native species that is abundant in suitable places such as heaths, moors, rocky places, bogs and open woodland. It is most frequent and abundant on nutrient-poor, acid (usually pH less than 5), sandy or peaty substrates, where the presence of mycorrhizas increases nutrient uptake and growth rate of plants. Because of the destruction of lowland heaths, heather is predominantly found in the uplands, but it has also decreased in some upland areas owing to conversion to forestry and overgrazing. *C. vulgaris* is also recorded from acidic man-made habitats such as toxic metal spoil. Survival on contaminated soils depends on the ability of the mycorrhizal associate to bind toxic metals. A study of thirty-four heather populations from different locations in Great Britain revealed considerable genetic variability within populations. Genetic variability between populations was related to site latitude and longitude and not to management history.

Sources: Bradley *et al.* (1982); Grime *et al.* (1988); Stace (1997); Meikle *et al.* (1999); Strandberg and Johansson (1999)

prolonged drought. It also occurs on soils that retain water, but is killed by waterlogging and lack of aeration around the roots (Gimingham, 1975). At the seedling stage, plants are pyramid-shaped, but the leading shoot usually dies during winter, and within a few seasons the plant becomes dome-shaped. Two types of heather shoots have been distinguished: long shoots, which are the actively growing terminal shoots, and short shoots, which are small lateral shoots and are the main leaf-bearing photosynthetic part of the plant. The terminal bud of a long shoot typically dies during the winter, and in spring some short shoots grow out to form new long shoots.

This growth form is important because it means that the plant can tolerate grazing, cutting and burning. The edible parts of the plant are generally non-woody long and short shoots formed during the current growing season. Following grazing, sufficient short shoots generally remain to grow and replace the long shoots. If all the current season's growth is removed, the plant can regenerate from reserve buds in the axils of leaf clusters. If grazing is fairly intense, the plant responds by producing new shoots at the expense of woody tissues, but if grazing is very heavy, shoot replacement cannot match removal and the plant does not survive. Heather regenerates after fire because there are reserve buds at the base of the stem, which is often insulated from the heat of the fire by litter or soil (Gimingham, 1975).

Cyclical change

Cyclical change is particularly important in heathlands because the dominant species, heather, has a life cycle consisting of four phases: pioneer (three to ten years), building (seven to thirteen years), mature (twelve to thirty years) and degenerate (over thirty years) (Gimingham, 1972; Webb, 1986). The pioneer phase is the phase of early establishment, whether from seedlings, or following burning. In the building phase, the canopy reaches its maximum density. In the mature phase, vigour is reduced and the canopy density declines. Finally, plants become degenerate and the central branches die back, outer stems lie flat on the ground and canopy cover is low. In the degenerate phase, bare patches of ground and the low canopy cover allow other species, such as bell heather, grasses (e.g. wavy-hair grass), or tree seedlings to colonise, and establishment of grasses or scrub can prevent the regeneration of heather. Grazing can delay the onset of mature and degenerate phases of heather, but it is also usually necessary to rejuvinate the vegetation periodically by cutting or burning (Gimingham, 1992) (Chapter 3, 'Habitat maintenance and enhancement').

THE PLANT SPECIES COMPOSITION AND CLASSIFICATION OF MAJOR GRASSLAND AND HEATHLAND HABITATS

Classification of habitat types

A community may be defined as an assemblage of species that occur together in a common environment, although communities are rarely discrete entities with sharp boundaries. However, community description approaches to habitat classification are valuable because they allow identification of vegetation types and communication at the national and international level. They can also be used to monitor changes in habitat extent and condition, and are used in the implementation of legislation. There have been several attempts to categorise these habitat types based on recognisable plant community descriptions or vegetation characteristics, for example the NCC Phase 1 Habitat Survey (Nature Conservancy Council, 1990), the European-based CORINE Biotopes Habitat Classification (e.g. Benstead *et al.*, 1999), the National Vegetation Classification (NVC), for British plant communities (e.g. Rodwell, 1991, 1992, 2000), Biodiversity Action Plan (BAP) definitions (e.g. UK Biodiversity Steering Group, 1995; UK Biodiversity Group, 1998b, 1999c), the Countryside Surveys of 1978, 1990 (CS1990) and 2000 (CS2000) and Land Cover Maps 1990 and 2000 (LCM 1990 and LCM 2000) (Bunce *et al.*, 1999; Firbank *et al.*, 2000; Haines-Young *et al.*, 2000). Different approaches often use slightly different definitions and subdivisions, for example Table 2.3 shows Countryside Survey 2000 (CS2000) /Biodiversity Action Plan Habitat Definitions and National Vegetation Classification equivalents (Rodwell 1991, 1992; Haines-Young *et al.*, 2000).

Probably the most detailed system for the British Isles (i.e. excluding Ireland) is the NVC. The NVC recognizes a large number of relatively distinct types of plant community and sub-communities that are divisions of a continuum of vegetation types (Rodwell, 1992; Sanderson, 1998). NVC communities and sub-communities are dis-

Table 2.3: Countryside Survey 2000/Biodiversity Action Plan habitat definitions and National Vegetation Classification equivalents

CS2000 habitat	*BAP broad habitat*	*CS2000/BAP definition*	*NVC equivalents*
Improved grassland	Enclosed farmland	Improved grassland occurs on fertile soils and is characterised by the dominance of a few fast growing species (e.g. rye grass and white clover). Typically used for grazing and silage, improved grassland can also be managed for recreational purposes. They are often intensively managed using fertilizer and weed control treatments. They may also be ploughed and sown as part of the rotation of arable crops (leys)	Mesotrophic grasslands
Neutral grassland	Enclosed farmland	Neutral grasslands are found on soils that are neither very acid nor alkaline. They do not contain calcifuge (lime-avoiding) plants, which are found on acid soils, or calcicole (lime-loving) plants which are found on calcareous soils. Unimproved or semi-improved neutral grasslands may be managed as hay meadows, pastures or for silage. They differ from improved grassland in that they are less fertile and contain a wider range of herb and grass species. Usually the cover of rye grass is less than about 25%	Mesotrophic grasslands
Acid grassland	Mountain, moor, heath and down	Acid grassland vegetation is dominated by grasses and herbs. These grasslands occur on a range of lime-deficient soils that have been derived from acid bedrock or from superficial deposits such as sands and gravels. They characteristically include a range of calcifuge (lime-avoiding) plants	Calcifugous grasslands
Calcareous grassland	Mountain, moor, heath and down	Calcareous grassland vegetation is dominated by grasses and herbs. These grasslands occur on shallow, well drained soils that are alkaline as a result of the weathering of chalk, limestone or other types of base-rich rock. They characteristically include a range of calcicoles (lime-loving) plants	Calcicolous grasslands
Linear grassland features	Boundary and linear features	This habitat includes a diverse range of linear landscape features including grass strips and semi-natural vegetation along road verges	Mesotrophic grasslands and others
Dwarf shrub heath	Mountain, moor, heath and down	Dwarf shrub heath comprises vegetation that has a greater than 25% cover of plant species from the heath family or dwarf gorse species. It generally occurs on well drained, nutrient-poor acid soils	Lowland dry heaths, transition to lowland wet heaths, maritime heaths

Sources: Rodwell (1991, 1992); Haines-Young *et al.* (2000).

Notes: Broad habitat categories were developed for the UK Biodiversity Action Plan (BAP), and these were used in the Countryside Survey 2000 (CS 2000) and the Northern Ireland Countryside Survey 2000. NVC National Vegetation Classification (Rodwell, 1991, 1992).

tinguished by the species composition and character of the vegetation (e.g. the frequency and abundance of different species). Individual NVC community types can be related to the European CORINE system (Hill, 1996) (Chapter 5).

The following sections describe the range of British lowland grasslands and heathlands based on vegetation type, and illustrate some of the factors that determine their distribution. The descriptions are based on the NVC because it is probably the most widely adopted method of habitat classification in the UK. Sub-communities are generally excluded for clarity. The range of descriptions is fairly comprehensive and is intended to be a source of reference for particular communities of interest. Grasslands are divided into mesotrophic, calcareous, acid and coastal types, and heathlands into dry heaths, wet heaths and maritime heaths. An overview of each type is given at the beginning of the appropriate section.

Mesotrophic grasslands

Mesotrophic (or neutral) grasslands are grasslands on moderately fertile to nutrient-rich soils that are neither strongly acid nor too basic. Because they often occur on the more fertile soils throughout the UK lowlands, a large proportion have been converted to arable or agriculturally improved (Ausden and Treweek, 1995). They may be defined as NVC communities MG1 to MG13 (Table 2.4). Mesotrophic grasslands are a particularly broad category, with variations generally dependent on the intensity of management (unimproved, semi-improved and improved) and drainage conditions (free-draining, periodically inundated with water, permanently damp or waterlogged).

Improved mesotrophic grasslands

The majority of all grassland in rural and urban parts of the UK is improved. Unfertilised semi-natural grasslands tend to have a relatively low and rather variable growth rate, particularly in the early part of the growing season, and semi-natural grassland soils are usually too low in nutrients such as nitrogen to support highly productive grass growth (Tallowin, 1997). Agricultural management of improved grasslands aims to maintain the dominance of productive species and suppress others. Improved grasslands are often sown or created by modification of unimproved grasslands by the use of drainage, inorganic fertilisers, slurry or high doses of manure and selective herbicides. Inorganic fertilisers tend to be most effective when applied to free-draining, neutral soils. Fertiliser application is therefore often accompanied by drainage of wet soils and application of lime to neutralise acid soils (Rodwell, 1992; Ausden and Treweek, 1995). Improved grasslands are species-poor communities characterised by the abundance of rye grass and white clover (e.g. perennial rye grass leys and related grasslands, MG7).

MG6 grassland (Table 2.4) is the major permanent pasture type on moist but freely draining brown soils in lowland Britain. It occurs where there has been intensive improvement for pasture. MG6 grassland may be derived from a wide range of vegetation types, including well drained and wet mesotrophic grasslands, acid and calcareous grasslands and fixed sand dunes. Species characteristic of these vegetation types can persist at low frequency. Where past improvements have been uneven, transitions to original unimproved vegetation may remain, and if the origin of the grassland is clearly recognisable, e.g. calcareous or acid, the vegetation may be classed as semi-

Table 2.4: Characteristics of NVC mesotrophic (neutral) grassland communities (excluding sub-communities)

NVC code	*Community title*	*Location*	*Comments*
Mown and ungrazed grasslands			
MG1	*Arrhenatherum elatius* grassland (false oat-grass grassland)	Common on road verges and in neglected agricultural and industrial habitats	Rather variable community, characteristic of approximately neutral soils throughout the British lowlands. Mown once or twice a year for amenity. Ungrazed and unmanured or sporadic grazing
MG2	*Arrhenatherum elatius–Filipendula ulmaria* tall-herb grassland (false oat-grass –meadowsweet tall-herb grassland)	Localised in northern Britain	Confined to steep slopes on rendziniform soils over carboniferous limestone in cool, damp situations. The vegetation would probably not withstand heavy grazing. Lack of grazing by large mammals results in a build-up of organic matter and there is a large annual nutrient turnover (Ausden and Treweek, 1995). A small number of sites, e.g. Lathkill Dale, Derbyshire, support the national rarity Jacob's ladder (*Polemonium caeruleum*)
Well drained permanent pastures and meadows			
MG3	*Anthoxanthum odoratum–Geranium sylvaticum* grassland (sweet vernal-grass –wood crane's-bill grassland)	Cool and wet upland fringes in northern England	Mown July–September and autumn and winter-grazed until April–May (normally sheep or cattle). Occasional dressings of manure and lime. Traditionally treated meadows characteristically on approximately neutral brown soils. Influenced by higher rainfall and lower temperatures than MG5
MG4	*Alopercus pratensis–Sanguisorba officinalis* grassland (meadow foxtail–great burnet grassland)	Flood plain meadow, restricted occurrence in the lowlands	Characteristic of periodically flooded meadows on alluvial soils. Mown in July and autumn and winter-grazed from August (e.g. Lammas Day) until February (normally sheep, cattle and horses). Often survives where common rights have maintained traditional management practices, e.g. North Meadow, Cricklade
MG5	*Cynosurus cristatus–Centaurea nigra* grassland (crested dog's tail–common knapweed grassland)	Throughout the British lowlands, becoming increasingly rare	Meadows mown annually for hay in June and autumn and winter-grazed until April, by cattle, sheep and horses, occasional dressings of manure and lime. Pastures – low-intensity grazing and manured by stock. Optimum sward height 5–10 cm. Occurs on approximately neutral brown soils. Presence of ridge-and-furrow may indicate absence of ploughing since medieval times. Road verges are an important refuge for this community (Firbank *et al.* 2000)

Table 2.4: contd

NVC code	*Community title*	*Location*	*Comments*
MG6	*Lolium perenne–Cynosurus cristatus* grassland (perennial rye grass–crested dog's tail grassland)	Common and widespread in the British lowlands, occurring wherever there has been intensive improvement for pasture	Moist approximately neutral brown soils. Often resown, application of chemical fertiliser. Pasture grazed in rotation through the year, used for occasional crops of hay or silage. Coarse recreational sward, road verges and lawns. *Senecio jacobaea* and *Cirsium arvense* may be abundant weeds
Long-term leys and related grasslands			
MG7	*Lolium perenne* leys (perennial rye grass leys and related grasslands)	Throughout lowland Britain	Intensively treated, species-poor sown swards (specially sown on prepared ground or by seeding into meadows and pastures), chemically fertilised, grazed through the year, or cut for silage. Sometimes long-term leys (grown in rotation with arable crops). Permanent recreational and amenity swards
Ill drained permanent pastures			
MG8	*Cynosurus cristatus–Caltha palustris* grassland (crested dog's tail–marsh marigold grassland)	Localised in lowland Britain	Localised community, surviving as traditionally managed pasture on seasonally flooded ground by rivers. Also occurs below springs, flushes and seepage lines. Typically occurs on gleyed soils. May be the vegetation that was managed in the past as water meadow to provide a supplement to spring grazing. Most water meadow stands have been neglected or drained and improved. Sometimes termed flood pasture
MG9	*Holcus lanatus–Deschampsia cespitosa* grassland (Yorkshire fog–tufted hair grass grassland)	Widespread in suitable sites throughout the British lowlands	Moist or periodically inundated, approximately neutral gleyed soils. May be grazed or ungrazed. Unimproved or neglected agricultural land on heavy soils, clearings and road verges
MG10	*Holcus lanatus–Juncus effusus* rush-pasture (Yorkshire fog–soft rush rush-pasture)	Widespread in suitable sites throughout the British lowlands and upland fringes	Characteristic of permanently moist sites over a wide range of soils of varying pH. Generally grazed, but also common on unimproved or neglected agricultural land, damp verges and ditches

Table 2.4: contd

NVC code	*Community title*	*Location*	*Comments*
Inundation grasslands			
MG11	*Festuca rubra–Agrostis stolonifera–Potentilla anserina* grassland (red fescue–creeping bent–silverweed grassland)	Frequent near sea level on flood plains and upper salt marsh	Frequently inundated with fresh or brackish water. May be the natural vegetation on light-textured brown earths and alluvial soils that experience frequent superficial inundation. Extensive on flood plains of major rivers and on upper salt marsh, where it is often used as pasture for low-intensity livestock grazing from May to October (sheep, cattle and horses). Optimum sward height, 5–10 cm. Fragmentary stands occur elsewhere, e.g. on road verges
MG12	*Festuca aruninacea* grassland (tall fescue grassland)	Exclusively a coastal community, along the banks of tidal rivers, upper salt marsh and clay cliffs	Frequently inundated with brackish or salt water. Generally ungrazed
MG13	*Agrostis stolonifera–Alopercus geniculatus* grassland (creeping bent–marsh foxtail grassland)	Widely distributed throughout the British lowlands, with the most extensive stands in eastern England	Silty, approximately neutral soils kept moist and sometimes waterlogged at fluctuating margins of fresh waters. Often winter-flooded and grazed from May to October (low intensity, with sheep, cattle and horses). Optimum sward height 5–10 cm. On the Ouse Washes areas of washland are managed for feeding birds

Sources: Rodwell (1992); Crofts and Jefferson (1999).

Table 2.5: Characteristics of unimproved, semi-improved and improved mesotrophic grassland

Characteristics	*Unimproved*	*Semi-improved*	*Improved*
Description	Species diversity often high, with a low proportion of agricultural species	A transition category. Less diverse than unimproved grasslands, but wider range of species than improved grassland	Limited range of species. More than 50 % perennial rye grass, white clover and other agricultural species
Intensity of management	Traditional practices	Fertiliser application, ploughing and reseeding	Fertiliser application, ploughing and reseeding
Productivity	Low	Intermediate	High
Floristic diversity	High	Intermediate	Low
Examples	*Anthoxanthum odoratum–Geranium sylvaticum* grassland, *Briza* sub-community (MG3b) (Rodwell, 1992)	*Anthoxanthum odoratum–Geranium sylvaticum* grassland, *Bromus* sub-community (MG3a); transitions to *Lolium perenne–Cynosurus cristatus* grassland (MG6) (Rodwell, 1992)	Examples of *Lolium perenne–Cynosurus cristatus* grassland (MG6), *Lolium perenne* leys (MG7) (Rodwell, 1992)

Sources: Nature Conservancy Council (1990); Rodwell (1992).

improved (Nature Conservancy Council, 1990) (Table 2.5).

In the past fifty years or so, improved grasslands have increased by approximately 90 per cent in area (Table 3.3), although the rate of increase has apparently slowed in some regions since 1990 (Chapter 3, 'The history of grasslands and heathlands'; Haines-Young *et al.*, 2000). Improved swards may be permanent (pasture) or short-term agricultural grassland sown as part of an arable crop rotation (ley). Improved grasslands are used for grazing and silage. They are also widespread as recreational swards (Plate 2.3) or as amenity swards on greens, road verges and lawns. Bryophytes (mosses and liverworts) generally have a low cover in MG6 and MG7 swards and usually become established on areas of bare soil (Rodwell, 1992; Ausden and Treweek, 1995; Haines-Young *et al.*, 2000).

Unimproved mesotrophic grasslands

Unimproved mesotrophic grasslands have not been reseeded, nor treated with chemical fertilisers or pesticides. Their sward composition has not been modified by agricultural treatments, although they may have been treated with low levels of farmyard manure. Species diversity is usually high, with species characteristic of the area and soils (Nature Conservancy Council, 1990). Unimproved mesotrophic (neutral) grasslands are rare and fragmented (Jefferson and Robertson, 1996) (Table 2.6). They have traditionally been managed as meadow (mown and autumn- and winter-grazed) or pasture (grazed, used for occasional crops of hay). The traditional management of hay meadows (e.g. MG4 grassland, Table 2.4), and water meadows (e.g. MG8 grassland, Table 2.4) is described in Chapter 3. Outside

Plate 2.3: Amenity grassland (*Matt Brierley*)

Table 2.6: Estimates of the extent of mesotrophic grassland in England (excluding MG1, MG6 and MG7)

NVC community type (code)	*Area (ha)*
MG2	<100
MG3	<1,500
MG4	<1,500
MG5	<5,000
MG8	<500
MG11	<1,000?
MG13	<2,000? [a]
Approximate total	<11,600

Source: Jefferson and Robertson (1996).

Note: [a] 800 ha in Ouse Washes.

the agricultural setting, swards may be mown for amenity, and commons, greens and churchyards are often remnants of old, unimproved grassland (Rackham, 1986).

Coarse, ungrazed grassland. Coarse, ungrazed grasslands tend to be tall and tussocky (e.g. MG1 and MG2, Table 2.4). They are rather variable, but are characteristically dominated by false oat grass. MG1 grassland is widespread throughout the lowlands, for example many roadside verges, and neglected agricultural and industrial habitats. MG2 tall-herb grassland is a type of species-rich mesotrophic grassland that occurs locally on steep slopes in northern Britain. Bryophytes have a limited occurrence in MG1, but may form a lush ground cover in MG2 grassland (Rodwell, 1992).

Well-drained permanent pastures and meadows. Well-drained permanent pastures and meadows include grasslands that are often termed hay meadows, e.g. MG3, MG4 and MG5 (Table 2.4). MG4 and MG5 are lowland vegetation types, while MG3 occurs in the upland fringes, characteristically between 200 m and 400 m altitude (UK Biodiversity Group, 1998b) but is included in the lowland category by some authorities (Crofts and Jefferson, 1999). Unimproved hay meadows usually have a diverse flora with a large number of grass species and often a high proportion of herbaceous dicotyledons (herbs) (Plate 5). In the UK, the main community of unimproved hay meadows and pastures is MG5, but this is now highly fragmented. Herbs make up a large percentage of MG5 vegetation, and exceptionally may account for 95 per cent of the cover. MG4 seasonally flooded grassland is less widely distributed. It occurs in scattered sites from the Thames valley, through the Midlands and Welsh borders, to Yorkshire. MG4 has a species-rich and rather varied sward of grasses and herbs, including snake's head fritillary (Species box 2.5) (Rodwell, 1992; UK Biodiversity Group, 1998b). The main concentrations of MG3 grassland are in the northern Pennines of North Yorkshire, Durham and east Cumbria. The vegetation has a dense growth of grasses and herbs, including wood cranesbill and great burnet (UK Biodiversity Group, 1998b). Yellow rattle, a semi-parasitic annual species, is a characteristic species of MG3, although it also commonly occurs in a number of mesotrophic and calcareous grassland types (Rodwell, 1992).

Species Box 2.5: Snake's head fritillary *Fritillaria meleagris*

The snake's head fritillary occurs in damp meadows on alluvial soil, typically of pH 6.0–7.5, and is particularly associated with MG4 grassland. It has declined dramatically since the 1930s owing to agricultural intensification and conversion to arable. It is now frequent only in Suffolk and the Thames valley, although it has been planted and naturalised in places. The largest natural colony of snake's head fritillary is at North Meadow NNR, Wiltshire, where continuity of centuries-old management, involving maintenance of periodic flooding, hay cutting after 1 July and aftermath grazing, has allowed a very old meadow community to survive. *F. meleagris* takes several years to reach maturity. On mature plants, stems with three to six leaves emerge from a small bulb, and cup-shaped flowers are produced in April or May. Flowers are chequered light and dark purple and cream, or are sometimes white. Hay making in July ensures that plants set seed.

Sources: Trist (1981); Stace (1997); Treweek *et al.* (1997)

Hay meadows are grazed in winter (except in unfavourable weather), and then shut up for hay (i.e. not grazed) in spring (February to March in the lowlands, April to May in the upland fringes) so the vegetation can grow and be mown for hay between June and August or September. The regrowth following cutting may be grazed. The timing of cutting and subsequent grazing varies between sites and community type (Table 2.4). Loss of nutrients through removal of hay is balanced by a light application of manure in early spring and, in some cases (e.g. MG4), deposition of silt during winter floods (Rodwell, 1992; Ausden and Treweek, 1995). Application of fertilisers is incompatible with preserving the high plant species diversity of hay meadows. For example, low phosphorus availability and low overall nutrient availability in spring and early summer appear to be important factors in maintaining plant species richness (Tallowin, 1997), and the species richness of hay meadows can be significantly reduced even by low to moderate inputs of nitrogen fertiliser (25 kg N ha^{-1}, Tallowin and Smith, 1994; 50 kg N ha^{-1}, Mountford *et al.*, 1993). The absence of grazing during spring and summer allows a succession of plants to flower. Nationally rare (occurrence in fifteen or fewer 10 km squares) and nationally scarce (occurrence in sixteen to 100 10 km squares) vascular plant species occurring in hay meadows include lady's-mantle species (MG3); downy-fruited sedge (MG4); dwarf mouse-ear (MG5); tuberous thistle (MG5); snake's head fritillary (MG4, MG5); spignel (MG3); narrow-leaved waterdropwort (MG4); greater yellow rattle (MG5) and sulphur clover (MG5) (Jefferson and Robertson, 1996). Other specialist declining plant species include dyer's greenweed (MG5), green-winged orchid (MG5), greater butterfly orchid (MG5) (Plate 2.4) and pepper saxifrage (MG4, MG5, MG9) (UK Biodiversity Group, 1998b). Bryophytes tend to be frequent in MG3, MG4 and MG5 grassland, although their cover may be variable (Rodwell, 1992). The conservation status of lower plants (mosses, liverworts and lichens) of mesotrophic grasslands is not well known (Jefferson and Robertson, 1996).

Plate 2.4: Greater butterfly orchid in a hay meadow

Wet grassland. There is a range of habitats that can be defined as lowland wet grassland (Rodwell, 1992; Dargie, 1993; Ausden and Treweek, 1995; Benstead *et al.*, 1999), but the terminology is often not used consistently (Dargie, 1993). Most definitions of wet grassland embrace a range of inland and coastal grassland types occurring in areas with a high water table (e.g. ill-drained permanent pastures, Table 2.4), or subject to periodic flooding (e.g. inundation grasslands, Table 2.4) (Jefferson and Grice, 1998; Benstead *et al.*, 1999). Wet grasslands therefore include river floodplain meadows and pastures, lakeside meadows and pastures, washlands (extensive semi-natural areas adjacent to rivers created as flood storage), floodplain grazing marshes or coastal grazing marshes behind a sea wall (claimed from salt marshes) (UK Biodiversity Steering Group, 1995; Jefferson and Robertson, 1996; Benstead *et al.*, 1999). Seasonally flooded meadow on well drained soils (well drained permanent pasture and meadow) such as MG4 hay meadow (Table 2.4), is included in the wet grassland category by some authorities, e.g. Benstead *et al.* (1999), but not by others, e.g. Rodwell (1992); Ausden and Treweek (1995). The definition of wet grassland also includes types of swamp and mire (Jefferson and Robertson, 1996). Most wet grasslands of conservation value have many or all of the following attributes: they have an appropriate flooding or water regime, are regularly mown or grazed, are agriculturally unimproved and are associated with drainage channels or ditches containing standing water, temporary pools or oxbows (Joyce and Wade, 1998; Benstead *et al.*, 1999). They support many indigenous species of plants, invertebrates and birds, and are important for flood alleviation, groundwater replenishment and water quality improvement (Jefferson and Grice, 1998; Benstead *et al.*, 1999). There is no accurate figure for the extent of lowland wet grassland in England, and much has been agriculturally improved. The largest extent of wet grassland is the Somerset Levels (Plate 2.5).

As noted above, wet grasslands can be conveniently divided into ill drained permanent pastures (with a high water table), inundation grasslands (that are periodically flooded) and fen meadow or rush pasture communities (Tables 2.4 and 2.7) (Rodwell, 1992; Dargie, 1993; Ausden and Treweek, 1995; Jefferson and Grice, 1998). Fen meadow and rush pasture (NVC communities M22–M24 and MG10) occur where the ground is moist or waterlogged for most of the year. They occur on base-rich or moderately acid soils and

Plate 2.5: The largest extent of wet grassland in Britain is the Somerset Levels

Table 2.7: Characteristics of NVC rush-pasture and fen-meadow communities (excluding sub-communities)

NVC code	*Community title*	*Location*	*Comments*
M22	*Juncus subnodulosus–Cirsium palustre* fen meadow (blunt-flowered rush–marsh thistle fen-meadow)	Wide distribution on suitable soils throughout the southern British lowlands	Occurs on suitably wet and base-rich soils. Derived from other sorts of wetland vegetation by mowing and/or grazing. Extent has been reduced by abandonment and intensive land improvement. Usually grazed by cattle from April to November. A few sites cut for hay with aftermath grazing. Optimum sward height 5–40 cm
M23	*Juncus effusus/acutiflorus–Galium palustre* rush pasture (soft rush/sharp-flowered rush–common marsh bedstraw rush pasture)	Widespread through the west of Britain	Occurs on a variety of moist, moderately acid to neutral soils in the cool and rainy lowlands of western Britain. Grazing, and occasionally mowing, maintain this vegetation and prevent succession to woodland. Agricultural improvement has reduced the extent of M23. Usually grazed by cattle from April to November. A few sites cut for hay with aftermath grazing. Optimum sward height 5–40 cm
M24	*Molinia caerulea–Cirsium dissectum* fen meadow (purple moor grass–meadow thistle fen-meadow)	Widespread through the lowland south of Britain	Occurs on moist to fairly dry peaty soils in the warmer lowlands. Maintained by mowing or grazing. Reduced and fragmented by reclamation and neglect. Usually grazed by cattle from April to November. A few sites cut for hay with aftermath grazing. Optimum sward height, 5–40 cm. Periodic localised winter burning may also be practised

Sources: Rodwell (1991); Crofts and Jefferson (1999)

support a high proportion of rare and scarce vascular plants (Jefferson and Grice, 1998). They are difficult to define, but, with the exception of MG10 (Rodwell, 1992), are described as mire communities by Rodwell (1991) on the basis of their species composition.

Ill-drained permanent pastures such as MG8, MG9 and MG10 are characterised by a range of moisture-loving or moisture-tolerant species. For example, tufted hair grass is tolerant of high levels of soil moisture, perhaps partly owing to its well developed root aerenchyma (tissue with large, air-filled intercellular spaces) (Davy, 1980; Rodwell, 1992). The MG9 grassland community occurs in a wide range of habitats with wet, anaerobic (lacking in oxygen), nutrient-poor soils that are unsuitable for many other mesotrophic grassland species. MG10 rush pasture is rather similar to MG9, but soft rush is more frequent (Rodwell, 1992). MG8 grassland survives as traditionally treated pasture on seasonally flooded ground. MG8 flood pasture is scarce and localised, with less than 1,000 ha cover in England and Wales. Scotland is estimated to have about 600–800 ha of this community (UK Biodiversity Group, 1998b). The MG8 community is species-rich, with no single species consistently dominant, although some species, such as marsh marigold, which is unpalatable to stock, and meadowsweet, with its frothy cream flowers, are sometimes noticeably abundant (Rodwell, 1992). Bryophytes have a patchy, rather infrequent occurrence in MG8, MG9 and MG10 (Rodwell, 1992).

Inundation grasslands tend to contain species that can establish on periodically flooded and disturbed areas. MG11 and MG12 grasslands are frequently inundated with fresh or brackish water (Table 2.4). They share a number of salt-tolerant species such as sea mayweed and sea milkwort. MG11 occurs as extensive stands on the flood plains of major rivers and on the upper salt marsh where the community is frequently grazed. The perennial rye grass sub-community (MG11a) includes stands that have been improved by inorganic fertilisers, and sometimes by ploughing and reseeding, for intensive use as pasture. MG12 is generally ungrazed. MG13 grassland is associated with fluctuating margins of fresh waters (Table 2.4). It occurs on the seasonally inundated flood plains of large rivers and is particularly associated with the washlands, which were built in the seventeenth century to control flooding in the reclaimed Fens (Rodwell, 1992; Ausden and Treweek, 1995). Coastal and floodplain grazing marsh is a term that is used to describe periodically inundated pasture or meadow with ditches that maintain the water level. The exact extent of grazing marsh in the UK is not known, but estimates suggest that there may be a total of 300,000 ha. However, only a small proportion (10,000 ha) of this grassland is unimproved (UK Biodiversity Steering Group, 1995). Bryophytes are usually sparse or infrequent in MG11, MG12 and MG13 (Rodwell, 1992).

Lowland calcareous grasslands

Calcareous (or calcicolous) grasslands are developed on shallow, lime-rich, oligotrophic (nutrient-poor) soils most often derived from limestone rocks, including chalk, with a pH typically between 7.0 and 8.4, although some examples occur on soils with a pH of 5.5 (Rodwell, 1992) (Table 2.8). Lowland calcareous grassland communities are confined to the warmer and drier climates of the southern and eastern parts of the UK (although coastal calcareous grassland (machair, Table 2.12) may also develop in the north and west of Scotland and in western Ireland on wind-blown shell sand). Lowland types are normally defined as communities CG1 to CG8 of the NVC which are restricted to the lowlands, and CG9, which occurs in both upland and lowland situations: two sub-communities of CG9 are principally lowland types (Table 2.8) (Rodwell, 1992; Jefferson and Robertson, 1996; UK Biodiversity Group, 1998b). Calcareous grasslands are a nationally and internationally rare habitat, but their extent in England exceeds that of unimproved mesotrophic grassland. Estimates indicate that the amount of lowland calcareous grassland remaining in the UK is around 33,000 ha to 41,000 ha, with less than 1,000 ha of this in Wales. The majority is found on chalk (25,000 ha to 32,000 ha) (UK Biodiversity Group, 1998b).

Lowland calcareous grasslands largely survive on escarpments or dry valley slopes (Plate 2.6) but, more rarely, remnants survive on flatter sites such as in Breckland and military training areas such as Salisbury Plain. They include a range of plant communities in which calcicoles (lime-loving plants) are characteristic. Calcareous grassland species that survive the low nutrient availability and high concentrations of calcium carbonate are either slow-growing species with low nutrient demands or

Table 2.8: Characteristics of NVC calcicolous (calcareous) grassland communities (excluding sub-communities)

NVC code	*Community Title*	*Location*	*Comments*
Calcicolous grasslands of the south-east lowlands			
CG1	*Festuca ovina–Carlina vulgaris* grassland (sheep's fescue–carline thistle grassland)	Scattered sites on harder limestones (carboniferous limestone and chalk) around the southern and western coast of England and parts of the Welsh coast	Largely confined to steep and rocky, warm and dry slopes with a generally southerly aspect and excessively draining rendzina soils. Heavy grazing, usually by sheep and rabbits, maintains the vegetation in its characteristically open state. Close to coastal resorts, trampling is considerable. These sites may have been important refuges for grassland species such as hoary rock-rose, white rock-rose and honewort during the period of post-glacial forest advance. Optimum sward height 2–10 cm with > 10% bare ground
CG2	*Festuca ovina–Avenula pratensis* (*Helictotrichon pratense*, Stace, 1997) grassland (sheep's fescue–meadow oat-grass grassland)	Widely distributed over a variety of southern lowland limestones, including the chalk of the South Downs and the carboniferous limestone in Derbyshire	Characteristic of free-draining, calcareous soils in a relatively warm and dry lowland climate. Maintenance of the community depends on a certain balance of grazing, traditionally by sheep and rabbits, although also by cattle and horses. Much reduced in extent because of changes in pastoral practice and the conversion of land to arable. Optimum sward height 2–10 cm with up to 10% bare ground
CG3	*Bromus erectus* (*Bromopsis erecta*, Stace, 1997) grassland (upright brome grassland)	Especially frequent over the chalk in south-east England. *Bromus erectus* reaches an approximate north-western climatic limit around the Humber–Severn line	Lightly grazed or ungrazed grassland (sheep, cattle, horses and rabbits), mostly on calcareous soils. Many stands of CG3 may have originated as grazing of CG2 grassland was relaxed. The decline of many typical CG2 species may be a response to the shading effect of *B. erectus* foliage and accumulation of its litter. Optimum sward height 2–15 cm
CG4	*Brachypodium pinnatum* grassland (tor-grass grassland)	Characteristic of cooler and damper sites than CG3, e.g. towards the western and northern fringe of the chalk. *Bracypodium pinnatum* has a similar, but less restricted, climatic distributional limit to *Bromus erectus*	Like *Bromus erectus*, *Brachypodium pinnatum* is associated with an absence or relaxation of grazing (sheep, cattle, horses, rabbits) in predominantly calcicolous swards such as CG2 grassland. As in CG3, the decline of many typical CG2 species in CG4 grassland may be a response to the shading effect of *B. pinnatum* foliage and accumulation of its litter. Optimum sward height 2–15 cm
CG5	*Bromus erectus* (*Bromopsis erecta*, Stace, 1997)–*Brachypodium pinnatum* grassland (upright brome–tor-grass grassland)	Calcareous soils along the north-western fringe of the southern lowland limestones	Associated with the absence or reduction of grazing where conditions suitable for the vigorous growth of both *B. erectus* and *B. pinnatum* coincide. Burning has had an important influence on this vegetation, resulting in a moderately rich associated flora. Some sites are subject to heavy trampling pressure. Optimum sward height 2–15 cm

Table 2.8: contd

NVC code	*Community Title*	*Location*	*Comments*
CG6	*Avenula pubescens (Helictotrichon pubescens*, Stace, 1997) grassland (downy oat-grass grassland)	Scattered locations over a variety of lowland limestones	Moister, more mesotrophic, calcareous soils on flat or gently sloping sites, where there is sometimes a history of disturbance. Grazing by sheep, cattle, horses or rabbits, although CG6b is associated with little or no grazing, except by rabbits. Optimum sward height 2–15 cm
CG7	*Festuca ovina–Hieracium pilosella (Pilosella officinarum*, Stace 1997)–*Thymus praecox (polytrichus*, Stace, 1997)/*pulegioides* grassland (sheep's fescue–mouse-ear hawkweed–wild/large thyme grassland)	Breckland and scattered localities on chalk and southerly exposures of carboniferous limestone	Associated with very thin, free draining oligotrophic calcareous soils, more continental climatic conditions and livestock or heavy rabbit grazing. In some kinds of CG7 grassland, the effects of disturbance such as ploughing or the mound-building activities of ants may be important. Optimum sward height 1–5 cm and >5% bare ground
Sesleria albicans grassland			
CG8	*Sesleria albicans (Sesleria caerulea*, Stace, 1997)–*Scabiosa columbaria* grassland (blue moor grass–small scabious grassland)	Lowland Durham, largely confined to slopes, abandoned quarries and road verges	Cool, dry climate, free-draining calcareous soils over magnesian limestone. Maintained by grazing of stock (sheep, cattle and horses) and rabbits. Reduced in extent by changes in agricultural practice, myxomatosis and quarrying of the bedrock. Optimum sward height 2–15 cm
CG9	*Sesleria albicans (Sesleria caerulea*, Stace, 1997)–*Galium sterneri* grassland (blue moor grass–limestone bedstraw grassland)	Morecambe Bay, North Yorkshire, Upper Teesdale	Free-draining moist soils over carboniferous limestone. From below 250 m altitude (Morecambe Bay) to parts of upland Britain. Two sub-communities, the typical sub-community (CG9b) and the *Helianthemum canum (oleandicum*, Stace, 1997)–*Asperula cynanchia* sub-community (CG9a), are principally lowland types. The *Carex capillaris–Kobresia simpliciuscula* sub-community (CG9d) is confined to the sugar limestone in Upper Teesdale, where climatic and substrate conditions keep the vegetation sparse and open. In other sub-communites, grazing (sheep and cattle) prevents scrub invasion. Optimum sward height 2–15 cm

Sources: Rodwell (1992); Crofts and Jefferson (1999).

Notes: Communities CG10–CG14 are calcicolous grassland of the north-west uplands (Rodwell, 1992), and are not included in this volume (see Fielding and Haworth, 1999).

Plate 2.6: Calcareous grassland, Derbyshire Dales

species that are potentially fast growing but which can persist under low nutrient conditions (Rorison, 1990). These grasslands are typically managed as part of pastoral or mixed farming systems, supporting sheep, cattle or occasionally horses. A few examples are cut for hay (UK Biodiversity Group, 1998b). The character of lowland calcareous grasslands is influenced by factors such as climate, soil characteristics and the balance of grazing.

CG1 grassland (Table 2.8) occurs on warm, dry slopes with excessive drainage at locations around southern and western coasts of England and Wales. At some sites CG1 may be climax vegetation, whereas at others it is maintained by grazing (Rodwell, 1992).

CG2 grassland is widespread, and characteristically occurs south of the Humber–Severn line on chalk, although it also occurs on the carboniferous limestone in Wales and Derbyshire. At sites in the warmest part of its range species with a continental distribution in Europe are found, such as horseshoe vetch, dwarf thistle and early spider orchid (Species box 2.6). Although this CG2a sub-community is not strictly confined to the chalk, it is commonly described as chalk grassland (e.g. Castle Hill National Nature Reserve (NNR), Sussex, Plate 2.7). It typically forms a short, even, closed turf and is characteristically species-rich with up to twenty-five species, and sometimes more than forty species, per square metre (Rodwell, 1992). Towards the north, with cooler, moister conditions, this sort of sward becomes confined to south-facing slopes, and there is a change in species composition to the *Dicranum scoparum* (a moss) sub-community (CG2d), in which continental species are usually absent and species associated with calcareous swards in the northern and western uplands, such as sweet vernal grass, are occasionally found. Limestone bedstraw occurs in this vegetation at the southern limit of its English distribution in Derbyshire. Bryophyte occurrence varies in the different CG2 sub-communities, while lichens are typically rare (Rodwell, 1992). This transition is a good example of the influence of climate on the species composition of vegetation. Other transitions in calcareous

Species Box 2.6: Early spider orchid *Ophrys sphegodes*

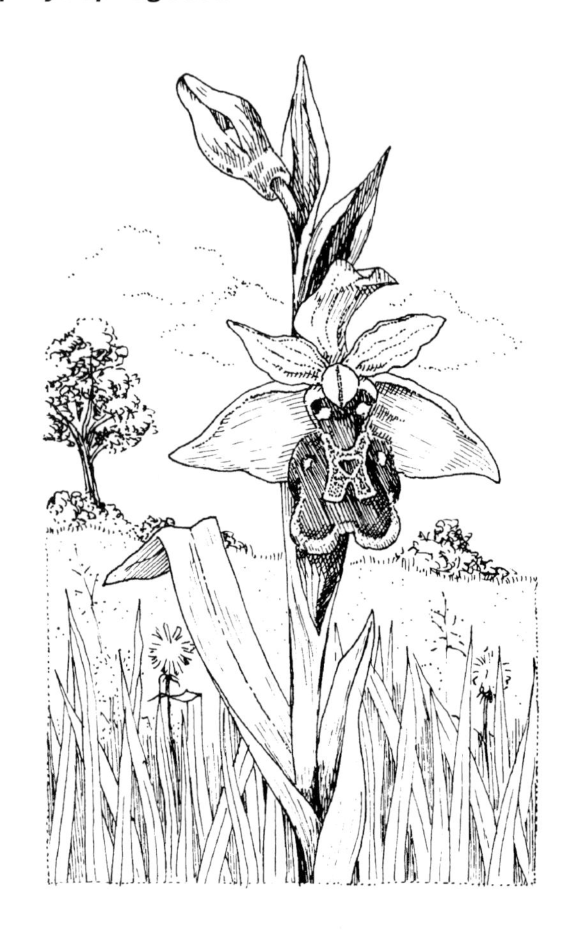

The early spider orchid is native, occurring very locally on grassland or spoil heaps on chalk or limestone. Between the 1930s and 1980s the species declined, and it was feared that it would disappear from the British Isles by 2000. Since then, *O. sphegodes* has increased in abundance at many of its known sites and has appeared at new locations. The increase has been attributed to appropriate management of nature reserves, and to climate change, which may favour a number of southern species. The early spider orchid produces tubers from which rosettes of leaves emerge above ground in autumn. Flower spikes are produced early in the year. Flower spikes are usually 10–20 cm tall in the UK and most flowering plants bear from two to four flowers. On the Continent, flower spikes may be as tall as 45–60 cm, and may bear up to ten flowers per spike. The petals of *O. sphegodes* are yellowish, the lip is velvety in texture, and is dark purplish brown in colour, with variable blue markings. On the Continent up to 45 per cent of seed capsules ripen, but in the UK the figure is about 10 per cent. In the UK there also appears to be little recruitment from vegetative propagation. Pollination of the early spider orchid is usually by pseudocopulating males of the solitary bee species *Andrena nigroaena*. Bees are lured to the flowers by visual cues and volatile semiochemicals. A number of subtle variations in the odour of flowers act as signals to guide bees to unpollinated flowers. This sophisticated strategy should increase pollination, and therefore the number of seeds produced. Although self-fertilisation appears to occur, the poor rate of seed set in the UK suggests that a lack of pollinators may be a major limitation on seed production. Trampling damage, and consumption of flower spikes by grazing stock, further limit seed production, although light grazing out of the flowering season is usually necessary to maintain a short sward. The species is relatively short-lived, and unless regular recruitment is achieved through seeds, populations of *O. sphegodes* are likely decline rapidly.

Sources: Hutchings (1987, 1989); Stace (1997); Ayasse *et al.* (2000); Schiestl *et al.* (2000); Schiestl and Ayasse (2001); Rose (2002)

grassland vegetation are related to climate. For example, in the southern grasslands towards the east, particularly over the chalk of East Anglia (e.g. Breckland), there is a marked shift away from dominance by grasses, and lichens, especially *Cladonia* spp., can be very prominent in some types of sward (Rodwell, 1992; Jefferson and Robertson, 1996). The

Plate 2.7: The short turf in the background is chalk grassland, Castle Hill National Nature Reserve, Sussex (*Hilary Orrom*)

vegetation here is CG7 grassland (Table 2.8). This shift is associated with the transition to the most extreme kind of continental climate in Britain. In Breckland there are frequent summer droughts and cold winters. The grasslands here are the closest British approximation of the steppe grasslands of eastern Europe (Ausden and Treweek, 1995). In certain situations, CG7 may be the climax vegetation, but where conditions are less harsh the community is maintained by grazing (Rodwell, 1992).

The balance of grazing is a significant factor in determining the character of calcareous grasslands. The most important grazers on lowland calcareous grassland have traditionally been sheep and rabbits. Large areas of calcareous grassland over southern chalk (e.g. CG2a) were traditionally managed as sheepwalk (Smith, 1980) where flocks of sheep grazed the grassland during the day tended by a shepherd and were folded in pens on arable land at night. Manure deposited in the pens transferred nutrients from the already nutrient-poor grassland to the arable land (Ausden and Treweek, 1995). Stands of the CG2d sub-community in areas like Derbyshire and the Mendips also owe their character and diversity to the close, even grazing of sheep. However, historically these grasslands were grazed by different breeds of sheep and also by dairy cattle. Folding pastoralism, characteristic of the chalk, was probably not practised here (Rodwell, 1992).

The post-war policy of converting suitable grassland to arable meant that many surviving areas of calcareous grassland were often too small for sheep grazing to be profitable (Chapter 3, 'Grasslands in historical times'). Many grasslands were maintained by increasing numbers of rabbits until myxomatosis spread in 1954. Responses of different species to abandonment and a decline in rabbit grazing are complex, but the most obvious is the spread of coarse grasses, such as tor grass, previously controlled by grazing, and the rapid decline of many low-growing, light-demanding species of the close-cropped sward (Bobbink and Willems, 1987; Willems, 1990). Litter produced by these coarse grasses also seems to inhibit the growth of other chalk grassland species, although tor grass and upright brome can germinate and grow up through litter and vegetation (Crofts and Jefferson, 1999; Hurst and John, 1999). An increase in the abundance of tor grass has also been attributed to elevated nitrogen deposition (Bobbink and Willems, 1987; de Kroon and Bobbink, 1997) (Chapter 3, 'Problems') (Species box 2.7). Communities of such rank, tussocky swards, dominated by different combinations of upright brome, tor grass, red

Species Box 2.7: Tor grass *Brachypodium pinnatum*

Tor grass is a clonal perennial, deciduous grass species, with erect shoots, rhizomes (underground stems) and an extensive root system. The majority of tor-grass rhizomes are very short, which gives rise to small clumps of shoots, typically 2–8 cm in diameter. All individuals beyond a threshold size produce small numbers of longer rhizomes and stands may become extensive. Shoots may reach up to 120 cm in height. The species flowers in June–August, and seed is shed from September onwards. Tor grass is native to Britain and occurs in grassland typically on chalk and limestone. The species is found mainly in the south and east of Britain. In northern and upland sites seed is rarely produced, and initial establishment appears to be restricted, although, once established, the species is frost-resistant. Elsewhere in northern Europe, isozyme studies suggest that individual tor-grass clones can achieve considerable age (possibly 245 years) and size (5.73 m^2). Under the influence of atmospheric nitrogen deposition, and a decline in appropriate management practices, such as grazing, the abundance of tor grass has increased in existing sites in the UK and elsewhere. In experiments to investigate the effects of increased nitrogen inputs on chalk grassland plots, the main above-ground response of tor grass was an increase in shoot size (height and biomass), and light transmission at 5 cm above the soil surface was reduced from 32 per cent in control plots to 4 per cent in nitrogen-fertilised plots. Below-ground, total rhizome length was greater in fertilised plots than in control plots. Tor grass recycles nitrogen on an annual basis, and the amount of

nitrogen stored in shoots tripled in response to fertilisation. Effective nitrogen acquisition and recycling, expansion of the rhizome system and production of tall shoots that overtop and shade other species are features that help the species to become dominant under elevated levels of nitrogen. Cattle grazing is particularly effective at controlling tor grass where it has become dominant, but other measures may be employed if cattle grazing is not appropriate.

Sources: Grime *et al.* (1988); Baxter and Farmer (1993); de Kroon and Bobbink (1997); Stace (1997); Shlapfer and Fischer (1998); Shaw (2000)

fescue and downy oat grass, are CG3, CG4, CG5 and CG6 grassland (Table 2.8) (Rodwell, 1992). CG5 swards are sometimes quite short or open, and bryophytes are frequent. In other coarse grass community types, such as CG6, bryophytes are rather uncommon (Rodwell, 1992).

Two distinctive, more northern species-rich calcareous grassland communities occur on magnesian limestone in lowland Durham, CG8, and carboniferous limestone in northern England, CG9 (Table 2.8). Two sub-communities of CG9, the hoary rock rose – squinancywort sub-community (CG9a) and the typical sub-community (CG9b) are principally lowland types (Rodwell, 1992; Jefferson and Robertson, 1996). Another distinct vegetation type that occurs on calcareous bedrocks towards the upland fringes is the OV37 community (Table 2.9). This is a local, open turf community that is restricted to the spoil heaps of lead mines or outcrops of veins of heavy metals. The large amounts of heavy metals, especially zinc, present in the spoil strongly influence the species composition of the community. Spring sandwort, sometimes known as leadwort, and Alpine pennycress are the most distinctive components of this community, and can accumulate zinc. The conspicuous mountain pansy occasionally occurs in lead spoil vegetation (Plate 2.8), for example at Priestcliffe Lees Nature Reserve, owned by the Derbyshire Wildlife Trust. The typical sub-community (OV37a) is found on spoil with poorly developed soils, while the yarrow–eyebright sub-community (OV37b)

Table 2.9: Characteristics of NVC metal mining spoil grassland communities (excluding sub-communities)

NVC code	*Community Title*	*Location*	*Comments*
OV37	*Festuca ovina–Minuartia verna* community (sheep's fescue–spring sandwort community)	The community occurs locally in the Mendips, Derbyshire dales, Yorkshire dales and north Pennines	OV37 is a local community restricted to the spoil heaps of lead mines or outcrops of veins of heavy metals among calcareous bedrocks. It occurs down to altitudes of 150 m, but is mostly characteristic of the upland fringes of northern and western Britain. The character of the vegetation is strongly influenced by the mineralogy of the soil parent material. In particular, colonisation is limited by the large amounts of zinc present in the spoil. Characteristic species include spring sandwort and Alpine pennycress, which can accumulate zinc. Sheep's fescue, common bent and common sorrel have strains tolerant of heavy metals. Other species may be able to colonise only spoil that has been ameliorated with less toxic parent material. Vegetation change in the absence of management is likely to be slow, owing to the extreme nature of the environment. Long-term management requirements are not fully understood, but the vegetation is usually open to grazing stock

Sources: Crofts and Jefferson (1999); Rodwell (2000).

Plate 2.8: Mountain pansy, a species that occasionally occurs in lead spoil vegetation

occurs on more developed soils and in transition to less contaminated ground around the spoil (Rodwell, 2000).

In addition to the influences of climate, soils and grazing, agricultural improvement has a significant impact on lowland calcareous grasslands. The application of inorganic fertilizers and re-seeding has converted many calcareous swards to MG6 grassland (Table 2.4). Transitions through various grades of improved sward (i.e. the semi-improved–improved continuum) are a common feature of much lowland calcareous grassland (Rodwell, 1992).

Lowland calcareous grasslands support an extremely rich flora, including a very large number (seventy-four) of nationally rare and nationally scarce species, including monkey orchid and pasque flower (Jefferson and Robertson, 1996; UK Biodiversty Group, 1998b).

Lowland acid grasslands

Acid grasslands are probably the most extensive semi-natural habitats in Britain, yet little is known about their true extent or conservation management requirements, especially in the lowlands. There are probably over 1.2 million ha of acid grasslands in the uplands, where they form the majority of rough grazing land, but in the lowlands the extent is much less, and is unlikely to be more than 30,000 ha (UK Biodiversity Steering Group, 1995) and more recent estimates suggest that the area of lowland dry grassland is in the approximate range 15,000 ha to 22,000 ha (Sanderson, 1998) (Table 2.10). The extent of lowland acid grassland is therefore intermediate between that of lowland unimproved mesotrophic grasslands (approximately 11,600 ha, Table 2.6) and lowland calcareous grasslands (33,000 ha to 41,000 ha, UK Biodiversity Group, 1998b).

Lowland acid grassland can be defined as enclosed and unenclosed acid grassland throughout the UK lowlands, normally below 300 m (UK Biodiversity Group, 1998b), although the definition is not used consistently (Sanderson, 1998; Crofts and Jefferson, 1999) (Table 2.11). Important concentrations of lowland acid grassland occur in southern and

Table 2.10: Estimates of the extent of lowland dry acid grassland in England

NVC community type (code)	*Area (ha)*
U1	8,000–12,500
U2	3,000–5,500
U3	2,000–3,000
U4	3,000–5,000
Inland SD10	150–450
Inland SD11	60–300

Source: Sanderson (1998).

Notes: for definitions of U1–U4, SD10 and SD11, see Tables 2.11 and 2.12.

south-western England and the Welsh and English border hills. Scotland is estimated to have less than 5,000 ha, mostly on the upland fringes. Lowland acid grasslands usually occur on generally free-draining, nutrient-poor soils with pH 4 to 5.5, overlying acid rocks or superficial deposits such as sands and gravels (UK Biodiversity Group, 1998b). Lowland acid grassland is often associated with heathland and in some circumstances may be derived from dwarf shrub heathland communities as a result of heavy grazing and burning (Crofts and Jefferson, 1999). Acid grasslands also occur in parklands and locally on coastal cliffs and shingle (Table 2.12). Acid grassland is normally managed as pasture (UK Biodiversity Group, 1998b). Lowland types are defined as NVC communities U1, U2 and U3. U4 occurs in the upland fringes, although the U4b sub-community is common towards lower altitudes (Table 2.11). Localised inland stands of SD10b and SD11b vegetation may also be included (Table 2.12) (Rodwell, 1992; 2000; UK Biodiversity Group, 1998b). Acid grasslands are generally less species-rich than mesotrophic and calcareous grasslands, although lowland acid grasslands are often richer than upland acid grasslands (Ausden and Treweek, 1995).

U1 grassland (Table 2.11) is the most widespread type of acidic grassland in southern Britain. The terms grass heath and lichen heath are often used to describe this vegetation type, but it is probably best described as parched acid grassland (Gimingham, 1992; Rodwell, 1992; Ausden and Treweek, 1995; Sanderson, 1998). The vegetation of U1 is distinctive, with an open sward of small, tussocky grasses, ephemerals (species that germinate, set seed and die in a short period within a year) and sometimes an extensive cover of lichens and/or mosses. Scattered dwarf shrubs such as heather are usually kept in check by grazing. Some ephemerals are continental in their range, e.g. Spanish catchfly, others are southern lowland species, e.g. common storksbill, while others occur widely in the British lowlands, e.g. dovesfoot cranesbill. The dry conditions, particularly in spring and summer, can restrict growth, and this is one of the reasons why the grass sward is open. This allows other species, such as spiked speedwell, to persist in the vegetation (Rodwell, 1992). In the more extreme situations the grassland may be maintained by climatic and soil conditions, but in many cases grazing and disturbance are very important. Sheep grazing, from the Neolithic period onwards, has probably played a crucial role in maintaining the open landscape, and prior to myxomatosis, many stands of acid

Table 2.11: Characteristics of NVC calcifugous (acid) grassland communities (excluding sub-communities)

NVC code	*Community Title*	*Location*	*Comments*
U1	*Festuca ovina–Agrostis capillaris–Rumex acetosella* grassland (sheep's fescue–common bent–sheep's sorrel grassland)	Occurs widely over suitable substrates throughout the warm and dry lowlands of England and Wales	Characteristic of base-poor, oligotrophic and summer parched soils. Grazing (sheep, cattle, horses and rabbits) and disturbance often very important. The most extensive stands are found in Breckland. The decline of grazing by stock and rabbits has led to the loss of many stands. Optimum sward height, 1–5 cm with >15% bare ground
U2	*Deschampsia flexuosa* grassland (wavy-hair grass grassland)	Widespread but local distribution in moderately oceanic parts of the lowlands but more common towards upland fringes	Characteristic of base-poor, free-draining or moist soils. Grazing important (sheep, cattle, horses and rabbits). Most stands probably derived from woodland, heath and mire. Optimum sward height 5–10 cm
U3	*Agrostis curtisii* grassland (bristle bent grassland)	Suitable soils in warm, oceanic parts of central and south-west Britain	Develops over moist, base-poor soils, in conditions intermediate between those suitable for dry and wet heath, in response to burning and moderate to heavy grazing (sheep, cattle, horses and rabbits) or disturbance. It can represent a temporary phase in heath regeneration, but too frequent burning and heavy grazing can establish it more permanently. Optimum sward height 1–5 cm with >15% bare ground
U4	*Festuca ovina–Agrostis capillaris–Galium saxatile* grassland (sheep's fescue –common bent–heath bedstraw grassland)	Widespread through the cool, wet sub-montane zone of north-west Britain	Extensive pasture on better drained, base-poor mineral soils. Character determined by climate and soils but the community is dependent upon grazing by sheep, cattle, horses and rabbits. Occurs on rough grazing land towards the limits of enclosure and over unenclosed slopes at moderate altitude. The Yorkshire fog–white clover sub-community (U4b) is common towards lower altitudes. Optimum sward height 5–10 cm

Sources: Rodwell (1992); Crofts and Jefferson, (1999).

Notes: Communities U5 and U6 are calcifugous grassland of the uplands (Rodwell, 1992), and are not included in this volume (see Fielding and Haworth, 1999).

grassland were associated with warrens. Declines in heath grazing and rabbit populations have led to the loss of many examples of the community (Rodwell, 1992).

In cool, wet regions of western Britain, on permeable soils, the U1 community is replaced by swards of U4 moist acid grassland (Sanderson, 1998) (Table 2.11). The maintenance and

Table 2.12: Characteristics of NVC maritime grassland communities (excluding sub-communities)

NVC code	*Community Title*	*Location*	*Comments*
Sand dune grassland			
SD6	*Ammophila arenaria* mobile dune community (marram mobile dune community)	Occurs widely on suitably mobile dune systems all around the British Isles	SD6 is the most widespread and extensive colonising vegetation of mobile sands above the tidal limit. Dominates young dunes and blow-outs until reduced accretion limits the vigour of marram and allows an increase in other species
SD7	*Ammophila arenaria–Festuca rubra* semi-fixed dune community (marram –red fescue semi-fixed dune community)	Occurs all round the British coast where there are suitable semi-mobile sands	Found on semi-mobile substrates that are generally drought-prone and impoverished. In the SD7 community, where deposition of sand is probably less than 5 cm annually, the vigour of marram grass declines
SD8	*Festuca rubra–Galium verum* fixed dune grassland (red fescue–lady's bedstraw fixed dune grassland)	This grassland can be found on suitable more calcareous, stable dunes and sand plains around the British coast	Typically found where distance or shelter prevents fresh deposition of beach material. The substrate is usually lime-rich (pH 6.5–8.5). Grazing and differences in regional climate affect the character of the community. The machair landscape of north-west Scotland is an important example of this community
SD9	*Ammophila arenaria–Arrhenatherum elatius* dune grassland (marram–false-oat grass dune grassland)	This community occurs locally on suitably stable dunes around many parts of the British coast (commoner in north-east England)	Typically confined to stretches of more stabilised, calcareous coastal sands where there is little or no grazing. In many places the community could probably progress readily to scrub
SD10	*Carex arenaria* dune community (sand sedge dune community)	Occurs in suitable situations on dunes all around the British coast and inland in Breckland and Lincolnshire	Pioneer vegetation among coastal dunes and at a few inland sites where species have invaded from surrounding grass swards and heaths. Many sites are subject to grazing. Sub-community SD10b occurs locally on inland sands
SD11	*Carex arenaria–Cornicularia aculeata* dune community (sand sedge–*Cornicularia aculeata* dune community)	Mainly on the east coast of Britain and inland on the sands of Breckland	Lichen-rich swards, characteristic of fixed, rather acid sands. Heavy grazing by rabbits has maintained the vegetation in the past, although greater disturbance may erode the surface. Sub-community SD11b occurs locally on inland sands
SD12	*Carex arenaria–Festuca ovina–Agrostis capillaris* dune grassland (sand sedge–sheep's fescue–common bent dune grassland)	Scattered localities around the coasts of south-east England, commoner towards the north, where older, leached sands have been grazed	Characteristic of grazed stretches of fixed, acidic sands, developing on quartz sand, or after prolonged leaching of calcareous wind-blown sediments. Grazing by cattle, sheep and rabbits is important in preventing progression to heath or woodland

Table 2.12: contd

NVC code	*Community Title*	*Location*	*Comments*
Maritime grassland			
MC8	*Festuca rubra–Armeria maritmia* maritime grassland (red fescue–thrift maritime grassland)	Occurs around the whole of the cliffed coastline of Britain	Generally occurs on steep to moderate slopes up to about 50 m above sea level, receiving large amounts of sea spray. MC8 is found on a variety of rock types on soils of neutral pH. MC8 is usually inaccessible to stock. Where grazing occurs, generally by sheep, it influences the character of the community. It is best to maintain grazing where it exists, but not to extend it to long-ungrazed areas
MC9	*Festuca rubra–Holcus lanatus* maritime grassland (red fescue–Yorkshire fog maritime grassland)	Occurs widely on British sea cliffs, except along the south coast	Characteristic of less maritime (i.e. receives less salt spray), somewhat sheltered situations, either on the top of cliffs or on lee slopes. The character of the community is influenced by soil moisture and nutrient status. MC9 is usually ungrazed or lightly grazed
MC10	*Festuca rubra–Plantago* spp. maritime grassland (red fescue–plantain spp. maritime grassland)	Predominantly a northern community, reaching its most pronounced development where heavy sheep grazing occurs in regions of high rainfall	Characteristic of less maritime situations (i.e. receives less salt spray), differing from MC9 largely because it is consistently and heavily grazed, generally by sheep. Heavy grazing obscures the boundaries between maritime and inland plant communities
MC11	*Festuca rubra–Daucus carota* ssp. *gummifer* maritime grassland (red fescue–sea carrot maritime grassland)	Most common on the chalk and limestone of the south coast of England and south Wales	Characteristic of less maritime situations (i.e. receives less salt spray), virtually confined to cliffs of calcareous rocks. Ungrazed or occasionally grazed

Sources: Mitchley and Malloch (1991); Rodwell (2000).

character of the community are determined by grazing, and it is a grassland type of major agricultural importance in the upland fringes, where sheep show a preference for these swards (Rodwell, 1992; Fielding and Haworth, 1999). The Yorkshire fog–white clover sub-community (U4b) is very common towards lower altitudes, generally between 100 m and 250 m. U4b swards are not as productive as MG6 swards, but they are of importance in the dairying areas of western Britain and are also used for winter pasturing of hill sheep. Bryophytes are not very abundant in the closed turf.

Species-poor grasslands dominated by wavy-hair grass (U2 grassland, Table 2.11) have a widespread but local distribution throughout the relatively oceanic parts of the lowlands and upland fringes (Species box 2.8). They typically occur on free-draining,

Species Box 2.8: Wavy-hair grass *Deschampsia flexuosa*

Wavy-hair grass is a native species which occurs throughout Britain, especially in the north and west, in unproductive, relatively undisturbed base-poor habitats such as acid heaths, moors and open woods and drier parts of bogs. It is a winter green, densely tufted perennial, sometimes spreading by rhizomes, with narrow, mid- to dark-green leaves. Wavy-hair grass flowers in June and July and seed is shed in August or September. The delicate flowerheads resemble 'a pink mist on the ground', and the fine flowerhead branches are strongly wavy. The species is shade tolerant and is also tolerant of low pH, low nutrient supply and high substrate concentrations of aluminium and manganese. However, it shows strong positive growth responses to fertiliser addition, particularly in high light conditions, and high rates of nitrogen deposition in the Netherlands have contributed to the replacement of heather-dominated dry heaths by grassy vegetation dominated by wavy-hair grass. Although rabbits and sheep eat wavy-hair grass, new shoots of heather are preferred, and grazing can lead to an increase in *Deschampsia* in moorland and heathland habitats.

Sources: Phillips (1980); Grime *et al.* (1988); Hester *et al.* (1991); de Kroon and Bobbink (1997); Stace (1997)

moderately moist soils. Wavy-hair grass occurs in a number of acid grassland types, but U2 grassland, in which wavy-hair grass is obviously dominant, is dependent on a climate that is fairly moist. Heather is also a constant species in the vegetation, and this community is also described as grass heath; the low cover of heather and the presence of a greater number of grassland species distinguishes U2 grassland from heaths such as H9 (heather–wavy-hair grass) heath (Sanderson, 1998). Treatments such as burning and grazing are important in maintaining the community, and most stands have probably been derived from woodlands, heaths and mires. For example, when heather–dwarf gorse (H2) and H9 lowland heaths (Table 2.15) are burned, wavy-hair grass often becomes dominant in the early phases of regrowth, and this dominance may be prolonged by grazing (Rodwell, 1992; Ausden and Treweek, 1995).

Grasslands dominated by bristle bent, U3 (Table 2.11) occur on moister acid soils, in warm, oceanic parts of central southern and south-west England. In the lowlands, there is a gradation from dry stands with early hair grass to damper stands with purple moor grass (Sanderson, 1998). They are usually associated with humid heathland (i.e. intermediate between wet and dry heathland), and transitions between U3 and H3 and H4 vegetation (Table 2.16) are likely to be frequent (Sanderson, 1998). These grasslands are often derived from heathland in response to excessive burning (Rodwell, 1992; Ausden and Treweek, 1995; Jefferson and Robertson, 1996). The cover of bryophtes may be locally high, but lichens are sparse (Rodwell, 1992).

Lowland acid grasslands are associated with a considerable number of rare (seventeen) and scarce (twenty-two) vascular plants, including chives, small alison, mossy stonecrop, Deptford pink, maiden pink, eyebright, blue fescue, smooth cat's ear, sickle medick, grape hyacinth, perennial knawel, sand catchfly, Spanish catchfly, Breckland thyme, suffocated clover and spiked speedwell. The majority of rare and scarce plants associated with lowland acid grasslands occur in U1. A number of species that occur in U1 also occur in CG7 grassland (Jefferson and Robertson, 1996; Sanderson, 1998). Breckland retains the greatest extent of both these vegetation types (Rodwell, 1992). U1 grasslands are also important for lower plants, in particular lichens, and U4 grasslands are of particular value for fungi (Sanderson, 1998).

Coastal grasslands

Coastal habitats are diverse and include rocky shores, shingle, sea cliffs, dunes and salt marshes (Frid and Evans, 1995). Coastal grasslands can be broadly divided into two categories: sand dune grassland and maritime grassland that is typically associated with sea cliffs and slopes (Table 2.12).

Sand dune grassland normally occurs on more stabilised dunes, for example SD8 fixed dune grassland and SD9 dune grassland, although semi-mobile and mobile dune vegetation may be dominated by grasses such as marram, for example the marram–red fescue semi-fixed dune community (SD7) and the marram mobile dune community (SD6) (Table 2.12) (Rodwell, 2000) (Species box 2.9).

The SD8 grassland community of fixed dunes consists of a closed sward of red fescue and other grasses, herbs and mosses in which marram is common but no longer dominant. Grazing and differences in regional climate affect the character of the community (Rodwell, 2000). The substrate is usually lime-rich and low in nitrogen and phosphorus, and these calcareous fixed dunes support a particularly wide range of plant species (UK Biodiversity Group, 1999c). The machair grassland of north-west Scotland and the Outer Isles is an important example of the SD8 community. Machair grassland is a type of dune pasture that has typically developed on calcareous sand, with a high shell content, sometimes 80–90 per cent, blown inland by very strong prevailing winds. Machair grassland plains contain a mosaic of wet and dry grassland communities, related to management patterns and substrate characteristics. Machair grassland forms part of the wider machair landscape, which includes the beach, hay meadows and areas of blown sand overlying peat. Traditionally, moderate numbers of sheep and cattle have been pastured on SD8, and enrichment from manure has encouraged the development of the daisy-meadow buttercup (SD8d) sub-community of dry machair and the self-heal (SD8e) sub-community of wet machair. Machair stands have also been influenced by rotational arable cultivation that began with Viking occupation. Traditionally, seaweed has been spread on these fields to add organic matter and nutrients to the infertile sands. It is estimated that machair grassland is restricted to about 25,000 ha worldwide, with

Species Box 2.9: Marram grass *Ammophila arenaria*

Marram is a coarse perennial grass with stems up to 120 cm in height. Leaves are greyish-green, with tightly inrolled margins; flowers are produced from June to August. The species is native to the UK and occurs on mobile sand dunes. Marram grass plays an important role in the dune-building process, and binds younger dune systems together with its long, tough rhizomes. The species is often planted to stabilise dunes. In actively accreting sand, marram grass growth can keep pace with a burial rate of up to 1 m a year, and where sand remains very mobile, marram is sometimes the only species in the vegetation (SD6 community). As the substrate becomes more fixed the vigour of marram grass declines and conditions allow the development of a fairly rich associated flora (SD7 community). The vigour of marram grass declines where sand is not actively accreting because the species actually benefits from regular burial by windblown sand. As new sand is deposited it creates a surface rooting zone of sand that is free of soil-borne pathogens (plant parasitic nematodes and fungi). Frequent deposition of sand is necessary for the vigorous growth of marram, because fungal and nematode species are able to colonise the deposited sand layer within one year.

Sources: Fitter *et al.* (1984); Stace (1997); de Rooij-van der Goes *et al.* (1998); Rodwell (2000)

17,500 ha in Scotland and the rest in western Ireland (UK Biodiversity Group, 1999c; Rodwell, 2000). SD9 dune grassland is confined to stretches of fixed calcareous coastal sands, including dunes, where there is little or no grazing (Rodwell, 2000).

On dunes that are initially acid, or have become acidified by leaching, acid dune grasslands or dune heaths develop (UK Biodiversity Group, 1999c). The sand sedge dune community SD10 occurs on freshly deposited calcareous and acid sands in more sheltered places on dunes all around the British coast, and the sheep's fescue sub-community (SD10b) is found inland on the highly acidic sands in Breckland (Table 2.12). The sand sedge–*Cornicularia aculeata* (a lichen) dune community (SD11) is characteristic of fixed and rather acid sands. It is found mainly on the east coat of Britain, but the sheep's fescue sub-community (SD11b) also occurs inland on the sands of Breckland (Rodwell, 2000). Inland examples of SD10 and SD11 are included in the lowland dry acid grassland Habitat Action Plan (UK Biodiversity Group, 1998b). Dune slack (or dune valley) communities are similar to the vegetation of mires and other periodically flooded habitats inland (Rodwell, 2000), and are not described in this book.

Fixed dune communities are generally maintained by grazing by livestock or rabbits, and undergrazing often leads to invasion by coarse grasses and scrub. Overgrazing can also have damaging effects, and parts of some stabilised dunes have been converted to arable use (UK

Biodiversity Group, 1999c). Fixed dune habitats are particularly threatened and are included in the EC Habitats Directive and the coastal sand dunes Habitat Action Plan (UK Biodiversity Group, 1999c). There is a separate Habitat Action Plan for machair (UK Biodiversity Group, 1999c).

Maritime grasslands occur in less exposed locations (beyond the most spray-splashed zone) on sea cliffs. They include MC8, MC9, MC10 and MC11 (Table 2.12; Rodwell, 2000). A maritime form of red fescue is a constant component of the vegetation. Other maritime species include thrift, sea plantain and sea carrot. The character of the vegetation is influenced by salt input from sea spray, the type of soil and the impact of grazing (Table 2.12). Traditionally, livestock were grazed on cliff grasslands, where they maintained the open maritime grassland vegetation. Post-war intensification of agriculture led to maritime grassland on level ground being converted to arable, while that on sloping ground has been abandoned. Where abandoned grassland is not maintained by exposure, it is often overgrown by scrub (Oates, 1999; UK Biodiversity Group, 1999c). Maritime grasslands are included in the Habitat Action Plan for maritime cliffs and slopes (UK Biodiversity Group, 1999c).

Open communities

A number of open communities have been recognised by Rodwell (2000). Open communities form a broad category that includes arable weed communities and spoil communities (e.g. OV37, earlier). The category also includes coarse weedy grassland vegetation such as the OV23 perennial rye grass–cock's-foot community, which is characteristc of resown recreation areas like verges and playing fields, where there is occasional mowing and regular disturbance or some neglect (Table 2.13).

Grassland community summary

The previous sections illustrated the broad range of mesotrophic, calcareous, acid and coastal lowland grassland vegetation types that have been identified in the UK, and outlined some of the factors that influence their distribution. A number of rare plants, lichens and fungi are associated with grasslands, and

Table 2.13: Characteristics of NVC open communities (excluding sub-communities)

NVC code	*Community title*	*Location*	*Comments*
OV23	*Lolium perenne–Dactylis glomerata* community (perennial rye grass–cock's-foot community)	The community is ubiquitous throughout the British lowlands	Charateristic of resown recreation areas around residential areas, playing fields and urban road verges, where there is occasional mowing, regular disturbance or some neglect. This community is commonly found in zonations and mosaics with other grassland and weed communities, patterns depend on the frequency of trampling and mowing. Refer to Rodwell (2000) for further information

Source: Rodwell (2000).

several of these are featured in Species Action Plans, which set targets for the conservation of priority species (UK Biodiversity Steering Group, 1995; UK Biodiversity Group, 1998a; UK Biodiversity Group, 1999a) (Table 2.14) (Chapter 3, 'Current problems and opportunities'). Table 2.14 serves to emphasise that species from a wide range of grasslands are declining, but that the causes of decline are generally similar, in particular agricultural improvement or a decline in appropriate management.

Table 2.14: Grassland vascular plants, lichens and fungi featured in Species Action Plans

Name	*Type of grassland and other requirements*	*Factors causing loss or decline in relation to grassland*
Vascular plants		
Creeping marshwort *Apium repens*	Wet, usually base-rich permanent pasture subject to winter flooding	Agricultural intensification, including the use of herbicides, control of winter flooding, overgrazing and ploughing
Wild asparagus *Asparagus oficinalis* ssp. *prostratus*	Coastal dunes and cliff tops	Lack of grazing, gradual loss of genetic variation, the small number of plants, sparse flower production and naturally low proportion of female plants result in very low levels of seed production
Prickly sedge *Carex muricata* ssp. *muricata*	Bushy and grassy areas, mostly on limestone	Inappropriate grazing: intolerant of excessive grazing and of competition when grazing is absent. Erosion from recreation
Scottish scurvygrass *Cochlearia scotica*	Cliff-top grassland, mature dune grassland	Coastal pollution and development, possibly effects of climate change
Lady's slipper orchid *Cypripedium calceolus*	Moderately grazed species-rich limestone grassland in the north Pennines	Uprooting and picking, habitat destruction due to increased grazing pressure
Deptford pink *Dianthus armeria*	Dry pastures	Conversion of pasture to arable and building land, cessation of grazing, successional changes
Eyebrights *Euphrasia* spp. endemic to the UK	Maritime grassland, dune grass[a]	Loss of habitats, lack of grazing
Early gentian *Gentianella anglica*	Limestone and chalk grassland, dunes, cliffs	Loss of suitable habitat, inappropriate management, particularly reduction in grazing
Hawkweeds *Hieracium* spp. (sect. *Alpestria*)	Pastures, meadows[a]	Pasture improvement – cultivation and reseeding, or application of lime and fertilisers, intense grazing leads to flower head removal and low seed production, development
Cut-grass *Leersia oryzoides*	Wet meadows	Cessation of traditional watercourse management, cessation of grazing to reduce competition and poaching of ground to provide suitable germination conditions, possibly water pollution

Table 2.14: contd

Name	*Type of grassland and other requirements*	*Factors causing loss or decline in relation to grassland*
Pennyroyal *Mentha pulegium*	Very short turf, subjected to grazing and trampling, traditionally managed lowland village greens, trackways. Trampling enables stems to root in the soil, little recruitment from seed	Loss of seasonally wet habitats, abandonment of disturbed habitats, agricultural intensification, development
Irish lady's tresses *Spiranthes romanzoffiana*	Marshy meadows	Fertiliser and herbicide use, inappropriate stock densities, timing and type of grazing, change from extensive cattle grazing to intensive sheep grazing
Lichens		
Starry breck-lichen *Buellia asterella*	Calcareous, sandy, lichen dominated turf grazed by rabbits	Lack of grazing, scrub encroachment, soil acidification by conifer seedlings, possibly agricultural chemical spray drift and nitrogen deposition
Scaly breck-lichen *Squamarina lentigera*	Sandy, calcareous open sward	Loss of bare ground due to rabbit decline, possibly agricultural chemical spray drift
Fungi		
Date-coloured wax cap *Hygrocybe spadicea*	Neutral grassland, mown parkland, dunes, limestone pasture, mainly in the uplands	Grassland improvement through ploughing or fertiliser addition, succession on dune sites, trampling
Earth-tongue *Microglossum olivaceum*	Grassland habitats, such as on limestone sea cliffs, believed to be saprophytic on mosses	Agricultural improvement through ploughing or application of fertilisers and herbicides, scrub encroachment
Nail fungus *Poronia punctata*	Possibly the rarest fungus in Europe, it occurs in the dung of horses and ponies that have fed on unimproved pasture or hay. Generally confined to New Forest, although recently recorded from Dorset sites following reinstatement of grazing (J. Day, pers. comm.)	Changes in agricultural practices, particularly the decline in the use of horses and the loss of unimproved grassland (particularly hay meadows)

Sources: UK Biodiversity Steering Group (1995); UK Biodiversity Group (1998a, 1999a).

Notes: [a] Also associated with heathland. Species may also be associated with other habitats.

Lowland dry heath

Lowland dry heath communities occur on generally free-draining soils. They are generally relatively species-poor, but the character of these communities is influenced by variations in regional climate, soils and management such as grazing and burning. They include heather–sheep's fescue heath (H1), heather–dwarf gorse heath (H2), Cornish heath–common gorse heath (H6), heather–western gorse heath (H8) and heather–wavy-hair grass heath (H9) (Table 2.15; Rodwell, 1991).

H1 occurs in parts of eastern England such as Breckland, and is the most continental of the southern British dwarf shrub communities. Here heather is often the only dwarf shrub and occurs with short tussocks of sheep's fescue and mosses such as *Dicranum scoparum* and *Hypnum cupressiforme* and lichens, with patches of bare ground. Locally, sand sedge and bracken are common. Grazing by sheep and rabbits is important for maintaining the community, which now has a very fragmented distribution (Rodwell, 1991; Gimingham, 1992).

The H2 community is found to the south and west, from the Weald (e.g. Ashdown Forest) in the east, through the New Forest, to the Poole Harbour area (Dorset heathland) (Plate 2), where the temperature range is less extreme and the rainfall is higher (Rodwell, 1991). It is transitional between the continental heathland in eastern England and the more oceanic heathland of the south-west peninsula, and the Dorset heathlands are associated with a mixture of species at the limits of their geographical distribution (Webb, 1986)

H2 is characterised by mixtures of heather, bell heather, dwarf gorse and wavy-hair grass. In some stands, seedlings of oak, birch and Scots pine are common. The distinctive species common dodder also occurs in this community. It is a slender, climbing annual plant that is parasitic principally on gorse and heather (Webb, 1986). The incidence of bryophytes and lichens in H2 may be patchy. Maintenance of the community is dependent upon a grazing and/or burning regime. Heathland reclamation to arable and neglect mean that H2 generally occurs as fragmented stands, although larger areas survive in the New Forest (Rodwell, 1991). Historical information suggests that the floras of the Dorset heaths and the heaths of the New Forest were similar in the early 1900s. Today the two areas are very different. The New Forest remains largely intact, and traditional management survives. In contrast, the Dorset heaths have become fragmented and traditional management has declined. These changes appear to have led to a marked decline in the rarer plants of the Dorset heaths, while many of these species continue to flourish in the New Forest (Byfield and Pearman, 1996).

Farther west, in the more equable parts of the southern and western lowlands, such as the South West Peninsula and Wales, and round to the southern Pennine fringes, dwarf gorse is replaced in dry heath vegetation by western gorse, H8 heath. H8 also occurs on the mildly oceanic coast of north Norfolk. The variable pattern of rainfall in this large H8 zone and management practices influence the character of the community, although constant species are heather, bell heather and western gorse. Where examples of H8 occur on superficial acid deposits over calcareous bedrock, a number of calcicoles (lime-loving species) may occur in the vegetation (limestone heath). In general, bryophytes and lichens are not very diverse. Locally, on the Lizard peninsula in Cornwall, the distinctive H6 heath community is found. The occurrence of the nationally rare

Table 2.15: Characteristics of NVC lowland dry heath communities (excluding sub-communities)

NVC code	*Community title*	*Location*	*Comments*
H1	*Calluna vulgaris–Festuca ovina* heath (heather–sheep's fescue heath)	East Anglia	Confined to base-poor and oligotrophic sandy soils in the more continental lowlands of eastern England (i.e. low rainfall and a wide annual range of temperature). Reduced and fragmented as a result of a decline in traditional management and improvement for agriculture and forestry, but some large stands remain in Breckland. Occurrence of intricate grass/heath mosaics at some sites
H2	*Calluna vulgaris–Ulex minor* heath (heather–dwarf gorse heath)	Weald and Hampshire basin, including the New Forest	Impoverished acid soils, predominantly free-draining, in south-east and central southern England. These areas are wetter and have a less extreme temperature range than East Anglia (i.e. slight tendency towards an oceanic climate). Character influenced by climate and traditional burning and grazing treatments. In some sites, neglect has led to the progression to woodland, in others small fragments have become isolated by agriculture or forestry. Some large areas exist in better condition, e.g. in the New Forest
H6	*Erica vagans–Ulex europaeus* heath (Cornish heath–common gorse heath)	Lizard, Cornwall	Character of these heaths influenced by the warm oceanic climate, in combination with the fairly base-rich but quite oligotrophic free-draining brown earths of the Lizard peninsula. The community survives most extensively over serpentine. Character influenced by soil conditions, burning and, to a lesser extent, grazing
H8	*Calluna vulgaris–Ulex gallii* heath (heather–western gorse heath)	South West Peninsula, Wales, southern Pennine fringes, north Norfolk	A community of free-draining, generally acid to approximately neutral soils in the more oceanic regions of southern and western lowlands. Grazing and burning, and in some situations exposure to wind, prevent succession to woodland. This kind of heath is now usually restricted to marginal grazing land. On limestone covered by suitable soils, limestone heath occurs
H9	*Calluna vulgaris–Deschamsia flexuosa* heath (heather–wavy-hair grass heath)	Southern Pennines, North York Moors, Midland plain	Characteristic sub-shrub vegetation of acid and impoverished soils at low to moderate altitudes through the Midlands and northern England. Cooler and wetter climate. Community modified by burning, grazing and atmospheric pollution

Source: Rodwell (1991).

Notes: H10 and H12–H22 are heath communities of the uplands (Rodwell, 1991) and are not included in this volume (see Fielding and Haworth, 1999).

Cornish heath and common gorse in this vegetation type is associated with the warm oceanic climate and free-draining brown earth soils (Species box 2.10). The character of this plant community is also influenced by burning and grazing practices. H9 heaths are drier heaths of the lowlands and upland fringes in the Midlands and northern England. They are associated with a cooler and consistently wetter climate than H8 heaths. Characteristic species are heather, wavy-hair grass and the moss *Pohlia nutans*, although bryophytes and lichens are rarely abundant. Western gorse and bell heather are less frequent, bilberry becomes more common, and cowberry and crowberry occur locally. Some areas are frequently burned, but this practice is less common on lowland sites (Rodwell, 1991).

Lowland wet heath

Dry heaths generally occur on higher, free-draining sandy areas which grade into valleys, where wet heath and valley mire communities occur. In wet heaths, heather is generally replaced by cross-leaved heath (Dolman and Land, 1995). Transitions to lowland wet heath communities are shown in Table 2.16. They include the humid or transitional communities

Species Box 2.10: Common gorse *Ulex europaeus*

Common gorse is a densely spiny, spreading, evergreen leguminous shrub, up to 2.5 m in height. It is found in grassy places and heathland, mostly on sandy or peaty soil. It is native to central and western Europe, but has been introduced to other regions. Leaves are present during the seedling stage, but are subsequently found in a reduced form as very strong, deeply grooved spines. Hermaphrodite golden yellow flowers are produced mainly between January and May and are usually insect-pollinated. Gorse is able to grow on relatively infertile soils, and is more tolerant of soil acidity than most legumes. (It is most frequent within the pH range 4–6.) However, it also invades more fertile sites. Regeneration is by seed, and seeds may remain dormant yet viable in the soil for up to thirty years. The species generally requires disturbance for establishment and can be very common in infrequently disturbed sites on relatively infertile soils, such as roadsides and sea cliffs. On the infertile, highly acidic soils of heathland the presence of common gorse usually indicates where some form of disturbance has ameliorated highly infertile soils, although common gorse is also characteristic of a distinct heathland community on the brown earths of the

Lizard in Cornwall. More typically, western gorse (*U. gallii*) occurs in the heaths of the west and north, and dwarf gorse (*U. minor*) occurs mainly in the south-east. Hybrids between common and western gorse occur in the south-west. Common gorse forms extensive impenetrable thickets, with persistent, nitrogen-rich litter. Individual plants degenerate after about twenty years, and bushes become leggy, creating open patches where species such as birch can colonise. However, gorse can regenerate from stumps and can be maintained by cutting or burning.

Sources: Grime *et al.* (1988); Rodwell (1991); Stace (1997)

Table 2.16: Characteristics of NVC transitions to lowland wet heath communities (excluding sub-communities)

NVC code	*Community title*	*Location*	*Comments*
H3	*Ulex minor–Agrostis curtisii* heath (Dwarf gorse–bristle bent heath)	South Dorset and Hampshire	Characteristic of impoverished acid soils with reasonably high levels of moisture and a moderately oceanic climate. Appearance also often affected by burning and grazing
H4	*Ulex gallii–Agrostis curtisii* heath (Western gorse–bristle bent heath)	South West Peninsula, South Wales	Confined to warm, oceanic parts of Britain, on a variety of moist, acid soils. Burning and grazing are important in maintaining the community
H5	*Erica vagans–Schoenus nigricans* heath (Cornish heath–black bogrush heath)	Lizard, Cornwall	Character of these heaths influenced by the warm, wet climate. Confined to the wet, base-rich but calcium-poor soils on the Lizard. Influenced by fire and to a lesser extent grazing. Effects of grazing difficult to separate from those of burning. Seasonal waterlogging of the ground may have protected these heaths from widespread cultivation
M15	*Scirpus cespitosus (Trichophorum cespitosum,* Stace, 1997)*–Erica tetralix* wet heath (deer grass–cross-leaved heath wet heath)	Widespread occurrence at low altitudes in wetter north-west Britain	Character influenced by the wetter climate of north-west Britain. Moist, acid, shallow peat or peaty mineral soils. Drainage and peat cutting have extended its coverage on to once deeper and wetter peats. Grazing and burning have important effects on the vegetation
M16	*Erica tetralix–Sphagnum compactum* wet heath (cross-leaved heath–*Sphagnum compactum* wet heath)	Largely confined to south-east Britain but widely distributed, extending to north-east Scotland	Community of seasonally waterlogged, acid shallow peat or oligotrophic mineral soils. A distinctive feature of mire/heath zonations in southern England. Grazing and burning are important in maintaining the vegetation. At some sites, drainage has allowed M16 migration on to once wetter peats, at others drainage has destroyed M16 by lowering the water table. Distribution in lowland England is much fragmented by heathland reclamation. Grazing and burning are important

Sources: Mitchley and Malloch (1991); Rodwell (1991)

Notes: H10 and H12–H22 are heath communities of the uplands (Rodwell, 1991) and are not included in this volume (see Fielding and Haworth, 1999).

dwarf gorse–bristle bent heath (H3), western gorse–bristle bent heath (H4) and Cornish heath–black bogrush heath (H5) and the wet heath communities deer grass–cross-leaved heath wet heath (M15) and cross-leaved heath–*Sphagnum compactum* (a moss) wet heath (M16). M15 and M16 are classified as mire communities by Rodwell (1991), but wet heaths are an integral part of lowland heathland (UK Biodiversity Steering Group, 1995), and so they are included here.

In eastern England, where the climate is continental, there is a clear difference between the wet heaths on wetter soils (e.g. M16) and dry heaths on free-draining soils (e.g. H1). Towards the south-west, conditions become more oceanic and the boundary between wet and dry heaths becomes less distinct, because wet-heath species can grow on free-draining soils in the wetter climate and transitional types of heath vegetation (often called humid heaths) occur between drier and wetter heaths in this region.

H3 is the humid community that occurs with H2 dry heath in the moderately oceanic area around the Hampshire basin. Characteristic species include bristle bent, heather, bell heather, cross-leaved heath, purple moor grass and dwarf gorse. Rare species include Dorset heath. A number of distinctive bryophytes and lichens occur. H4 is the humid alternative to H8 dry heath, and is found in south-west England. Characteristic species include bristle bent, heather, bell heather, cross-leaved heath, purple moor grass, tormentil (Plate 4) and western gorse. Rare species include Dorset heath and Cornish heath. Bryophtes and lichens are not widespread. Common dodder occurs in both H3 and H4. H5 occurs on the distinctive soils of the very oceanic Lizard in Cornwall. The Lizard heaths (H5 and H6) are unusual heathland communities that exist nowhere else in Britain (Webb, 1986).

Wet heaths, such as M15 and M16, are characteristic of periodically waterlogged shallow peat and peaty mineral soils (Table 2.16). Some wet heaths require very little management because the wetness of the soil prevents tree encroachment. In other cases, particularly where the vegetation extends on to drier ground, light grazing and burning are important for maintaining the community (Rodwell, 1991; Gimingham, 1992).

M16 extends to the north-east of Scotland, but it is a distinctive feature of transitions between mire and heath in south-east England. Characteristic species include cross-leaved heath, heather, purple moor grass and the moss *Sphagnum compactum*. It is associated with a number of rare species, including Dorset heath and marsh gentian. Wet heaths are acidic and low in nutrients, particularly nitrogen, and several plant species obtain a supplementary source of nitrogen by trapping and digesting insects (Webb, 1986). The insectivorous plants common or round-leaved sundew and oblong-leaved sundew are characteristic of the open, muddy patches associated with wetter areas of heathland and are found in the M16 community (Gimingham, 1992) (Species box 2.11). M16 can support a diverse bryophyte community and lichens such as *Cladonia* spp. quite often occur. M15 is a similar vegetation type that occurs in wetter north-west Britain. Characteristic species include purple moor grass, tormentil, cross-leaved heath, heather and deer grass (a sedge). Common cotton grass (a sedge), common sundew and great sundew are distinctive species that occur in this community (Plate 2.9). A number of bryophytes are common, such as *Sphagnum capillifolium*, which is usually crimson or pinkish, while others occur fairly frequently, such as *Hypnum cupressiforme*. Lichens do not occur consistently, but *Cladonia* spp. can be locally abundant (Rodwell, 1991).

Maritime heath

Two heathland communities are associated with maritime conditions. Heather–spring squill heath (H7) occurs on less exposed parts of sea cliffs (Plate 2.10), while heather sand

Species Box 2.11: Round-leaved sundew *Drosera rotundifolia*

Round leaved or common sundew is a low, perennial, insectivorous plant, with a rosette of round, long-stalked leaves, up to 5 cm in length. The leaves are reddish and covered in sticky hairs, which curve inwards to trap insects. A spike of white flowers is produced from the rosette. Round-leaved sundew is a native species, found in wet, acid, peaty places with little shade, such as wet heaths, moors and sphagnum bogs. Plant carnivory is an adaptation to the low-nutrient, bright, waterlogged habitats in which these species occur. *D. rotundifolia* supplements nutrients taken up by the roots with nutrients such as nitrogen and phosphorus absorbed from prey trapped by the leaves. In many carnivorous species, traps may also absorb organic substances derived from prey, but most potassium, calcium and magnesium is taken up from the soil by the roots. Carnivorous species differ in their reliance on nutrients derived from insects, for example insect nitrogen contributes about 26.5 per cent of the total nitrogen content of *D. rotundifolia*, although some studies suggest higher values. In this species, carnivory can increase leaf growth and enhance flowering and seed set, and nutrients derived from carnivory may be stored in winter buds and used for growth the following season. The flowers of *D. rotundifolia* are held on a tall stem, and it is possible that the spatial separation of flowers and traps promotes segregation of prey and insect pollinators. Carnivorous species may be particularly responsive to atmospheric deposition of nitrogen, but four years of experimental addition of nitrogen (above 10 kg N ha^{-1} yr^{-1}) led to a significant decrease in establishment and survival of *D. rotundifolia*, caused by increased competition for light with taller species that also responded positively to increased nitrogen inputs.

Sources: Adamec (1997); Stace (1997); Bobbink *et al.* (1998); Ellison and Gotelli (2001)

sedge heath (H11) is characteristic of sand dunes (Table 2.17; Rodwell, 1991). The most obvious difference between H7 maritime cliff heathland and other heathland communities is the influence of sea spray. This community is probably a climax community, although grazing contributes to its maintenance in more sheltered locations. It occurs all around the coast of mainland Britain and offshore islands, except in the east and south. Regional differences occur due to climate and soil factors, and grazing also affects the character of the vegetation. The community is quite species-rich, and includes a number of rare species, such as spring squill and Scottish primrose. Mosses and lichens are not abundant. H11 is found on older stretches of coastal dunes, grading into grassland or scrub. It occurs on dunes along the coasts of western England and Wales, and is commoner in Scotland. This community is generally confined to sands with a pH of less than 5 that are base-poor or leached. The character of the community, and reversion to grassland or progression to scrub, are dependent on the level of grazing. Mosses and lichens are a

Plate 2.9: Common cotton grass (a sedge)

Plate 2.10: Maritime heath, Anglesey

distinctive feature of this community (Rodwell, 1991). As with coastal grasslands, a decline in coastal slope grazing and scrub management has led to the invasion of bramble, bracken, gorse and blackthorn and a decline in heathland plants and invertebrates (Oates, 1999).

Heathland community summary

The previous sections illustrated the range of dry, wet and maritime heathland vegetation types that have been identified in the UK, and outlined some of the factors that influence their distribution. A number of rare plants and

Table 2.17: Characteristics of NVC maritime heath communities (excluding sub-communities)

NVC code	*Community title*	*Location*	*Comments*
H7	*Calluna vulgaris–Scilla verna* heath (heather–spring squill heath)	Less exposed parts of sea cliffs around the coast of Britain except the more sheltered east and south between Durham and Dorset	Occurs over a wide range of moderately base-poor soils. Influenced by salt spray deposition on exposed sea cliffs. H7 can probably be seen as a climatic climax community, although grazing probably contributes to its maintenance, particularly in more sheltered places. In the most exposed situations there is no requirement for any management, but in more sheltered areas some light grazing is useful
H11	*Calluna vulgaris–Carex arenaria* heath (heather–sand sedge heath)	Local community on dunes around the coast of Britain, mainly in the north	Characteristic sub-shrub vegetation of stabilised base-poor sands or leached lime-rich sediments. Develops by the colonisation of dune grassland. Maintenance, reversion to grassland or progression to scrub affected by the level of grazing

Sources: Mitchley and Malloch (1991); Rodwell (1991).

lichens are associated with heathlands, and several of these are featured in UK Species Action Plans, which set targets for the conservation of priority species (UK Biodiversity Steering Group 1995; UK Biodiversity Group 1998a, 1999a) (Table 2.18) (Chapter 3, 'Current problems and opportunities'). Table 2.18 illustrates that species from different heathland types are declining, often owing to habitat loss (e.g. caused by development) and inappropriate habitat management.

FAUNA OF LOWLAND GRASSLANDS AND HEATHLANDS

This section describes the characteristic animals that are associated with the range of lowland grassland and heathland habitats described in this chapter. The invertebrates, amphibians and reptiles, mammals and birds of grasslands and heathlands are examined separately, although a number of species occur in both habitat types. Rare or declining species associated with each habitat are used to illustrate the range of species that are present and the threats to biodiversity.

Fauna of lowland grasslands

Invertebrates

Many semi-natural grasslands support large and very interesting invertebrate communities, but grassland types vary in the richness of the invertebrate fauna they contain, and in the number of rare (occurs in one to fifteen 10 × 10 km squares), scarce (occurs in sixteen to 100 10 × 10 km squares) and declining species they support. These differences often reflect variation in such factors as climate, soil type, vegetation composition and management. For example, sward structural diversity (including the presence of bare ground) has an important influence on surface dwelling invertebrates, with architecturally varied swards tending to

Table 2.18: Heathland plants and lichens featured in Species Action Plans

Name	*Type of heathland and other requirements*	*Factors causing loss or decline in relation to heathlands*
Vascular plants		
Eyebrights *Euphrasia* spp. endemic to the UK	Lowland heaths, maritime heaths [a]	Loss of habitat, particularly inland heath in Cornwall, lack of grazing
Hawkweeds *Hieracium* spp. (sect. *Alpestira*)	Dry heaths [a]	Intensive grazing, development
Marsh clubmoss *Lycopodiella inundata*	Wet heaths	Habitat loss, drainage, cessation of traditional management and associated successional changes, nitrate and phosphate pollution and associated increase in growth of competitive species, possibly atmospheric pollution, including heavy metals, nitrogen and sulphur dioxide
Perennial knawel *Scleranthus perennis* ssp. *prostratus*	Semi-open, very short grassy heaths, generally on acidic soil. A poor competitor and open conditions are required for seedling establishment	Increased use of herbicides and fertilisers, heathland abandonment, afforestation and development, habitat deterioration due to inappropriate grazing
Lichens		
A reindeer lichen *Cladonia mediterranea*	Serpentine heath	Small population size, trampling, fires
A lichen *Cladonia peziziformis*	Coastal or montane heathland, up to 400 m	Inappropriate use of burning, succession, inappropriate levels of grazing, trampling

Sources: UK Biodiversity Steering Group (1995); UK Biodiversity Group (1998a, 1999a).

Notes: [a] Also associated with grassland. Species may also be associated with other habitats.

support more diverse and abundant invertebrate communities (McLean, 1990; Curry, 1994; Key, 2000; Morris, 2000; Kirby 2001) (Species box 2.12). Ground nesting ants tend to be most abundant in unmanaged or lightly managed grasslands, and increased disturbance associated with intensive management is accompanied by a decline in their importance (Curry, 1994) (Species box 2.13). In general, the invertebrate species richness and nature conservation interest of calcareous grassland is high, and some species are confined to calcareous grasslands. Wet grassland and dry grassland on sandy soil (acid grassland) are moderately rich in invertebrates, and neutral and hay-cut grassland are rather poor, at least in scarce species (Kirby, 1994). The areas of greatest importance for invertebrates in wet grasslands are the ditches and their margins and fens, which may support large numbers of nationally rare invertebrate species. The surrounding fields are often of relatively low value, particularly if they are managed as hay meadows (Drake, 1998;

Species Box 2.12: White lip banded snail *Cepaea hortensis*

The white lip banded snail (*C. hortensis*) is widespread in the UK and is found in grassland, woods, hedges and dunes, but is absent from very acid sites. *C. hortensis* and the brown lip banded snail (*C. nemoralis*) may occur together, but *C. hortensis* has a more northerly range and usually occupies wetter or cooler habitats than *C. nemoralis*. The colour of the lip of the shell cannot always be used reliably to discriminate between the species, but the shell of *C. nemoralis* is usually larger than that of *C. hortensis*. Both species show shell polymorphism, which involves variation in colour and banding. Shells may be yellow, pink or brown in colour, and may be plain or have up to five bands, fused together in various combinations. Factors that influence the frequency of different shell morphs include visual selection by predators and climatic selection for paler morphs in hotter areas. A study of the two species on Marlborough Downs, southern England, has shown that between 1960 and 1985 *C. hortensis*, the species with the more northerly range, had spread at the expense of *C. nemoralis*. In contrast, in both species there was an increase in the frequency of paler shells, favoured in hotter climates. The changes in species distribution may reflect an increase in vegetation height following a decline in grazing due to myxomatosis: as turf thickness increases, the microclimate at the base of the sward becomes cooler and more humid. The change to paler-colour morphs in both species may be due to climatic effects, such as particularly hot summers, or to camouflage effects because paler shells are better camouflaged against vegetation.

Sources: Cameron (1976); Cowie and Jones (1998)

Jefferson and Grice, 1998). It is unclear what factors contribute to these variations, but differences in management are probably important. For example, hay meadows are structurally uniform and subject to periodic sudden disturbance (Kirby, 1992).

Structural diversity is very important for butterflies on chalk grassland, and different species have preferences for different turf heights (BUTT, 1986) (Table 2.19). Sward height can influence the microclimate at the soil surface, for example the soil beneath 1 cm tall horseshoe vetch plants can be 6°C warmer than under 6 cm tall plants (Thomas, 1990). In some cases there are precariously balanced relationships between invertebrates that are strongly influenced by vegetation structure. For example, the caterpillars of the large blue butterfly feed on the flower heads of wild thyme before being adopted into the nests of a species of red ant (*Myrmica sabuleti*) (Species box 2.14). The large blue can survive to adulthood only with colonies of this species. Between 1955 and 1963 the large blue started to disappear, while its habitat of warm, south-facing grassland on well drained acidic and calcareous soils apparently remained unchanged. In fact *M. sabuleti* had also largely

Species Box 2.13: Yellow meadow ant *Lasius flavus*

The yellow meadow ant is fairly common in grassland throughout the British Isles, where it forages mostly underground, but forms conspicuous mounds. It is also found on heathland. The 30 cm high, vegetation-covered mounds of yellow meadow ants are built up year by year, and large mounds on the Porton Ranges, Wiltshire, may be eighty to 150 years old. Hundreds of mounds can be found in undisturbed grassland. Anthill soil is fine and friable, differs from the surrounding soil in terms of nutrient levels, contains fewer plant-parasitic nematodes, and has a lower water content and higher pH than the surrounding soil. In calcareous grassland in the UK and across a large part of Europe, *L. flavus* mounds can have a pronounced effect on vegetation. Some plant species may be largely confined to ant hills (e.g. thyme-leaved sandwort), some occur rarely on ant hills (e.g. salad burnet), while others are found both on ant hills and in the surrounding turf (e.g. eyebrights). There may also be a significant concentration of myrmecochrous plants in turf immediately surrounding the mounds, probably due to seeds deposited there by workers. *L. flavus* responds to an increase in vegetation height by making taller nest mounds, but eventually hills may be deserted. One of the most interesting aspects of ant ecology is their association with other animals. These include insects that provide ants with honeydew or glandular secretions and are in turn nurtured by the ants – for example, caterpillars of blue butterflies – and 'guests' that are accepted by the ants as if they were part of the colony, for example guest beetles. The yellow meadow ant is host to the guest beetle *Claviger testaceus*.

Sources: Smith (1980); Chinery (1986); Skinner and Allen (1996); Blomqvist *et al.* (2000); Kovarova *et al.* (2001)

disappeared, to be replaced by other, unsuitable species of red ant. *M. sabuleti* declined because it needs longer periods of warmth to rear its brood than most other red ants, and it depends on the warm soil of grassland that is cropped very short. With the decline of rabbits following myxomatosis the vegetation grew too tall, the soil microclimate became unsuitable for *M. sabuleti*, and the large blue butterfly declined markedly and finally became extinct in the UK. However, a site supporting *M. sabuleti* on the edge of Dartmoor in Devon was actively managed to create the right sward conditions, and in 1983 and 1986 caterpillars were reintroduced using a race from Sweden and the large blue became re-established in Britain. The large blue has also been successfully reintroduced to sites in Somerset, but introduction to sites in the Cotswolds has been less successful (Thomas, 1999).

Management of the habitat of the large blue benefited a range of other butterfly species that also require warm conditions. Although the red ant, the large blue and the butterfly larval food plant, wild thyme, are all suited to exceptionally warm grassland in Britain, they are susceptible to severe drought. Management therefore encourages site patchiness, with pockets that are too moist, cool or hot during normal years but may provide refuge

Table 2.19: Turf height preference of chalk grassland butterflies

	Preferred turf height on typical sites (cm)														
Species	*0.5*	*1*	*2*	*3*	*4*	*5*	*6*	*7*	*8*	*9*	*10*	*15*	*20*	*30*	*>30*
Adonis blue	◆	◆	♦	♦											
Silver-spotted skipper [a]	♦	◆	◆	◆	◆	♦									
Chalkhill blue	♦	♦	♦	◆	◆	◆	♦	♦	♦	♦	♦				
Silver-studded blue [a]			◆	◆	◆	◆									
Brown argus			◆	◆	◆	◆									
Common blue	♦	♦	♦	♦	◆	◆	◆	◆	◆	◆	♦				
Dingy skipper			◆	◆	◆	◆									
Grayling [a]			◆	◆	◆	◆	◆								
Grizzled skipper		◆	◆	◆	◆	◆	◆	◆							
Small heath			◆	◆	◆	◆									
Small copper-brood 1						♦	♦	♦	♦	♦	◆	◆	◆		
Small copper-brood 2		◆	◆	◆	◆	◆	◆	◆	◆	◆					
Wall [a]			◆	◆	◆	◆	◆	◆	◆	◆	◆				
Small blue			♦	♦	♦	◆	◆	◆	◆	◆	◆	◆			
Meadow brown				♦	♦	♦	◆	◆	◆	◆	♦	♦	♦		
Green hairstreak [b]					◆	◆	◆	◆	◆	◆	◆				
Marsh fritillary			♦	♦	◆	◆	◆	◆	◆	◆	◆	◆			
Marbled white						◆	◆	◆	◆	◆	◆	◆	◆	♦	
Dark green fritillary					♦	♦	♦	♦	◆	◆	◆	◆			
Gatekeeper [b]											◆	◆	◆		
Large skipper									◆	◆	◆	◆	◆		
Duke of Burgundy [b]				♦	♦	◆	◆	◆	◆	◆	◆	◆	◆		
Ringlet												◆	◆	◆	
Small skipper												◆	◆	◆	◆
Essex skipper													◆	◆	◆
Lulworth skipper											♦	♦	♦	◆	◆

Source: BUTT (1986).

Notes: [a] Turf must be very sparse. [b] Shrubs are also needed.

when conditions are more extreme (Thomas, 1999).

The degree to which the species composition of the vegetation influences grassland invertebrates is variable; some groups have close associations with particular species, others are less specialised feeders (Curry, 1994). For instance, some invertebrates are restricted to a single plant species for food: the larval food plant of the narrow-bordered bee hawk moth is devil's-bit scabious, though the adult is more generalist and visits various flowers for nectar (UK Biodiversity Group, 1999b). It is not uncommon for different invertebrate life-cycle stages to have completely different requirements (BUTT, 1986)

Species Box 2.14: Large blue butterfly *Maculinea arion*

The range of the large blue butterfly is declining rapidly in Europe and it is listed as a 'Globally Threatened' species by IUCN/WCMC. It once occurred in about ninety colonies in Great Britain, but became extinct in 1979. It has since been reintroduced using Swedish stock. In 1999 populations occurred at eight sites, although some colonies were stronger than others. Further populations will be re-established as part of the recovery programme planned for the species. Female large blue butterflies lay eggs on the flower buds of wild thyme. Caterpillars first feed on flower heads of thyme, but in August, when a caterpillar has reached its fourth and final instar, it abandons the food plant and is adopted by a worker of the host ant species, *Myrmica sabuleti*. The ant milks the caterpillar and, perhaps mistaking it for an ant larva, takes it back to the nest. The butterfly caterpillar lives in the ants' nest for ten to eleven months, where it feeds on the ant larvae, then pupates in the nest and emerges as an adult butterfly in synchrony with the flowering of wild thyme. The best way to ensure established populations of the large blue is to manage habitats to optimise the density of *Myrmica sabuleti*, and the distribution of wild thyme.

Sources: UK Biodiversity Steering Group (1995); Skinner and Allen (1996); Thomas *et al.* (1998); Thomas (1999); Thomas and Elmes (2001)

(Species box 2.15). However, the association is usually broader than that. Plant species composition may be relatively unimportant in comparison with plant family (e.g. species with long corolla flower types, belonging to the pea family and deadnettle family, are preferred by bumblebees), flower colour, the time of nectar availability, plant structure and position with respect to microclimate (temperature, sunlight), condition and water and nitrogen content. Whatever the association, enough host plants in the right condition and situation should be present to maximise the opportunities for egg laying and for feeding of larval and adult stages (Crofts and Jefferson, 1999; Morris, 2000). Some grassland plants, such as cowslip, seem to be poorly exploited by invertebrates (although it is the food plant of the Duke of Burgundy butterfly), while others such as knapweeds, thistles and common bird's-foot trefoil have over fifty species of invertebrate associated with them, not including generalist herbivores (Crofts and Jefferson, 1999; Oates, 2000).

Properties that influence grassland soil invertebrate fauna include soil moisture, organic matter content, structure and pH. Soil animals in general are not particularly sensitive to pH, although some groups such as woodlice have high calcium requirements and are most common on alkaline soils. However, soil biological activity tends to be greater in neutral or slightly alkaline soils. Under conditions where soil aeration and/or low pH do

Species Box 2.15: Common field grasshopper *Chorthippus brunneus*

Grasshoppers are almost entirely vegetarian, and are generally active only in sunshine. Most have a single generation a year. The majority of European species pass the winter in the egg stage, and juveniles pass through several nymphal stages before reaching maturity. There is no pupal stage. The common field grasshopper is one of several similar species. Individuals may be grey, green, brown, purple or black. There is a small bulge on the front of the forewing, and this species is very hairy below the thorax. It feeds on grasses and inhabits a broad range of grasslands but is typically found in dry sites with short swards. Grasshoppers sing or stridulate by rubbing the hind legs against the forewings. The male common field grasshopper produces three main types of song: normal, courtship and copulation. The female sings before mating. Common field grasshoppers lay their egg pods (oviposition) in superficial layers of the soil and show a preference for bare, dry, compact soil, particularly ant hills. Young nymphs are typically found very near the sparsely vegetated oviposition sites, but nymphal development is completed in areas of longer grass, better suited to feeding. When females are mature they move back to bare sites for oviposition. Most grasshopper species probably do not move far during a single stage in the life cycle, and flight and jumping seem to be used mainly as a result of disturbance. *C. brunneus* is one of the most widespread grasshopper species in Britain, and populations from around the British Isles show considerable ecotypic differentiation in response to climatic variation. For example, when reared under standard conditions, grasshoppers collected from northern sites grew faster but had shorter periods of development than individuals collected from warmer, sunnier or more southerly locations. As a result, adults from northern sites were smaller than adults from southern sites. Responses to climate change in the future may depend on the ability of each grasshopper ecotype to adapt.

Sources: Chinery (1986); Brown (1990); Telfer and Hassall (1999)

not favour rapid decomposition a surface organic layer develops. In these conditions earthworms are scarce or absent and the invertebrate fauna tends to be dominated by microarthropods, mainly mites (Acari) and springtails (Collembola). In neutral to alkaline, well aerated soils, decomposition tends to be more rapid, and organic material is mixed with the soil. In these soils earthworms tend to be abundant, and earthworm activity is important in the incorporation of surface litter into the soil. The soil invertebrate fauna can modify soil structure considerably, but their main function in most situations is to promote the decomposition and mineralisation of organic matter (Curry, 1994).

There are a number of rare and scarce invertebrate species associated with lowland grasslands. There are 160 Red Data Book (endangered, vulnerable, rare, indeterminate or insufficiently known) invertebrate species and 350 scarce (notable) species associated with lowland grasslands (including calcareous, acid, dry, neutral, wet and coastal grasslands) (Kirby, 1994). Grasslands are especially rich in scarce species of crickets and grasshoppers, heteropteran bugs, Auchenorhyncha (which include leaf, plant and frog hoppers), pyralid micro-moths, butterflies, dung beetles (Scarabaeidae), phytophagous beetles such as leaf beetles, hoverflies and tephritid flies (Kirby, 1994). Certain molluscs are also well represented (McLean, 1990). Many invertebrate species are associated with dung, cowpats in particular, including thirty-eight species of beetle and seven species of fly that are rare or scarce in Britain. Each cowpat may contain 1,000 developing insects, principally fly larvae (Cox, 1999). Rare and scarce butterflies associated with lowland semi-natural grassland include small blue, marsh fritillary, Duke of Burgundy fritillary (associated with scrub margins on calcareous grassland), silver-spotted skipper, Adonis blue, chalkhill blue, large blue (Swedish race), silver-studded blue, Lulworth skipper and pearl-bordered fritillary. The heath fritillary is associated with species-rich grassland sites and heathland, and is listed as vulnerable (UK Biodiversity Steering Group, 1995; Crofts and Jefferson, 1999). Eight of these butterflies are included in UK Biodiversity Species Action Plans (Table 2.20). Rare and scarce grasshoppers and crickets associated with lowland semi-natural grasslands include the heath grasshopper, long-winged cone-head, wart-biter, rufous grasshopper, field cricket, mole cricket, Roesel's bush cricket and the woodland grasshopper (Crofts and Jefferson, 1999). Three of these are included in UK Biodiversity Species Action Plans (Table 2.20). Areas of calcareous and neutral grassland support species of bumblebees (Plate 6), such as the rare carder bumblebee and shrill carder bee, which have declined as a result of agricultural intensification (Table 2.20) (Carvell, 2002). In addition to butterflies, grasshoppers and crickets and bees and wasps, there are numerous invetebrate species associated with grassland that are included in UK Biodiversity Species Action Plans, including ants, flies, sixteen species of moth and at least ten species of beetle. A number of factors have contributed to the decline in these species, but habitat loss, agricultural improvement and inappropriate management are particularly damaging (Table 2.20).

Amphibians and reptiles

Grassland forms part of habitat mosaics used by all widespread species of amphibians and reptiles. For example, common frogs, common toads, smooth newts, palmate newts and great crested newts all use grassland to forage for food and for shelter where they occur close to breeding ponds (Crofts and Jefferson, 1999). The great crested newt is quite widespread in Britain, and the British population is among the largest in Europe. However, in recent years the population has declined, owing to loss of suitable breeding pools and loss and fragmentation of terrestrial habitats, including grasslands (UK Biodiversity Steering Group, 1995). Adders and common lizards are associated with grassland, especially where it borders heathland. Slow worms and grass snakes also often occur in grasslands (Crofts and Jefferson, 1999). Short, open swards of acid grassland are a component of habitats that are important for three rare species, the smooth snake, the sand lizard and the natterjack toad (Sanderson, 1998). The sand lizard and natterjack toad also occur on sand dunes, for example the Sefton Coast sand dunes, Merseyside, where they are associated with early dune successional stages. The sand lizard and natterjack toad are thought to have inhabited the area for about 9,500 years. The

Table 2.20: Invertebrates associated with grassland, featured in Species Action Plans

Name	*Type of grassland and other requirements*	*Factors causing loss or decline in relation to grassland*
Moths		
Reddish buff *Acosmetia caliginosa*	Open grassy swards, rich in sawwort (*Serratula tinctoria*) [a]	Scrub encroachment
Straw belle *Aspitates gilvaria*	Chalk grassland	Habitat loss, inappropriate grazing management, burning
Marsh moth *Athetis pallustris*	Unimproved grassland on frequently waterlogged ground	Drainage, development, heavy grazing
Basil thyme case-bearer *Coleophora tricolor*	Unimproved grassland, including road verges, grazed only by rabbits	Decline in food plant due to reduction in rabbit grazing intensity, loss of unimproved Breck grassland
Striped lychnis *Cucullia lychnitis*	Roadside verges and open grassland	Inappropriately timed cutting of larval food plant, dark mullein (*Verbascum nigrum*)
Bordered gothic *Heliophobus reticulata marginosa*	Mainly open calcareous grassland	Not known
Narrow-bordered bee hawk-moth *Hemaris tityus*	Unimproved wet, acidic and chalk grassland, larval food plant is devil's-bit scabious (*Succisa pratensis*) [a]	Agricultural improvement and inappropriate management
Silky wave *Idaea dilutaria*	Open calcareous grassland	Inappropriate grassland management
Belted beauty *Lycia zonaria britannica*	Base-rich coastal grassland such as machair and dune grassland	Vegetation succession, trampling, development, low dispersal ability
Double line *Mythimna turca*	Wet grassland and coastal grassland	Agricultural improvement of wet grassland, overgrazing
Lunar yellow underwing *Noctua orbona*	Grassy areas in woodland [a]	Not known
Pale shining brown *Polia bombycina*	Scrubby grassland on light, calcareous soils	Not known
Chalk carpet *Scotopteryx bipunctaria*	Chalk and other limestone grassland	Inappropriate grazing management, loss and fragmentation of grassland habitat
Black-veined moth *Siona lineata*	Lowland calcareous grassland	Inappropriate grassland management leading to scrub encroachment, burning
Four-spotted moth *Tyta luctuosa*	Grassland on well-drained soils, larva feeds on field bindweed (*Convolvulus arvensis*)	Agricultural intensification and development, inappropriate grassland management

Table 2.20: contd

Name	*Type of grassland and other requirements*	*Factors causing loss or decline in relation to grassland*
New Forest burnet *Zygaena viciae*	Relatively long grassland	Accidental sheep grazing, isolation of single colony, collecting
Butterflies		
Northern brown argus *Aricia artaxerxes*	Limestone grassland, up to 350 m altitude	Inappropriate grazing management, afforestation
Pearl-bordered fritillary *Boloria euphrosyne*	Unimproved grassland with scrub or bracken	Cessation of grazing on unimproved grassland and abandonment of traditional gorse and bracken management
Marsh fritillary *Eurodryas aurinia*	Damp neutral or acid grassland, dry chalk and limestone grassland, food plant is the devil's bit scabious (*Succissa pratensis*)	Agricultural improvement, afforestation and development, changes in grazing stock and practice, habitat fragmentation and isolation
Silver spotted skipper *Hesperia comma*	Chalk downland grassland	Insufficient grazing by stock and rabbits, loss and fragmentation of grassland habitat
Adonis blue *Lysandra bellargus*	Unimproved calcareous grassland, especially chalk downland, larvae tended by ants	Inappropriate grazing intensity due to changes in stock and rabbit populations, loss and fragmentation of habitat
Large blue butterfly *Maculinea arion*	Steep, south-facing downland. Requires thyme (*Thymus polytrichus*) and the ant *Myrmica sabuleti*	Loss of habitat, lack of appropriate grazing management
Heath fritillary *Mellicta athalia*	Species-rich grassland [a]	Abandonment of appropriate management of grassland
Silver-studded blue *Plebejus argus*	Calcareous grassland [a]	Fragmentation and isolation of habitat, inappropriate management, quarrying
Crickets and grasshoppers		
Wart-biter *Decticus verrucivorus*	Calcareous grassland, requires a finely balanced habitat mosaic of bare ground/short turf for egg laying, grass tussocks for concealment, sward rich in flowers and invertebrates for food [a]	Inappropriate management, leading to loss of habitat quality, small population size, bird predation
Mole cricket *Gryllotalpa gryllotalpa*	Damp meadows	Drainage, lack of suitable grazing or cutting management, use of insecticide

Table 2.20: contd

Name	*Type of grassland and other requirements*	*Factors causing loss or decline in relation to grassland*
Field cricket *Gryllus campestris*	Short, warm, tussocky grassland	Inappropriate site management reducing bare ground, invasion of acidic grassland by bracken, possibly climate change
Beetles		
A ground beetle *Amara strenua*	Grassland near the coast	Loss of coastal wet grassland through reclamation
A ground beetle *Anisodactylus poeciloides*	Coastal grazing marsh	Loss of habitat through development, erosion, construction of sea walls, intensive grazing
Broad-nosed weevil *Cathormiocerus britannicus*	Usually short, herb-rich grassland at the edge of cliffs	Lack of grazing of coastal cliff slopes
Leaf beetle *Cryptocephalus primaries*	Calcareous grassland on dry, warm hillsides	Loss of calcareous grassland, inappropriate grazing regimes
Ground beetles *Harpalus cordatus* and *H. parallelus* (joint species statement)	Coastal dune grassland, inland chalk grassland	Dune stabilisation and recreational use, inappropriate management of chalk grassland
Ground beetle *Harpalus dimidiatus*	Calcareous grassland	Loss of calcareous grassland through agricultural improvement and development, inappropriate management
Ground beetle *Harpalus punctatulus*	Grassland on chalky or sandy soils	Not known
False soldier beetle *Malachius aeneus*	Grassland and grassy areas in woodland	Not known
Crucifix ground beetle *Panagaeus crux-major*	Flood meadows and dune systems	Lack of grazing on wet pasture
A weevil *Protapion ryei*	Grassland, including machair	Not known
Flea beetle *Psylliodes sophiae*	Grassland, particularly on the sandy soils of Breckland	Stabilisation of Breck dunes, herbicides
Ants		
Black-backed meadow ant *Formica pratensis*	Rough grass [a]	Inappropriate management and scrub encroachment
Bees and wasps		
Banded mining bee *Andrena gravida*	Flower-rich sandy grassland where it forages for pollen	Loss of habitat

Table 2.20: contd

Name	*Type of grassland and other requirements*	*Factors causing loss or decline in relation to grassland*
A mining bee *Andrena lathyri*	Tall swards of calcareous or mesotrophic grassland supporting large populations of vetches	Loss of habitat, scrub encroachment
Great yellow bumblebee *Bombus distinguendus*	Extensive areas of meadow, supporting a large number of species with long corolla flower types, e.g. belonging to the pea family and deadnettle family	Loss of extensive, herb-rich grassland
Large garden bumblebee *Bombus ruderatus*	Extensive areas of meadow, supporting a large number of species with long corolla flower types	Loss of herb-rich grassland through agricultural intensification, non-native forms of *Bombus* used for pollination in greenhouses may pose a threat
Carder bumblebee *Bombus humilis*	Areas of grassland supporting a large number of species with long corolla flower types	Loss of herb-rich grassland through agricultural intensification, non-native forms of *Bombus* used for pollination in greenhouses may pose a threat
Short-haired bumblebee *Bombus subterraneus*	Extensive areas of grassland, including sand dunes, supporting a large number of species with long corolla flower types	Loss of herb-rich grassland through agricultural intensification, non-native forms of *Bombus* used for pollination in greenhouses may pose a threat
Shrill carder bee *Bombus sylvarum*	Herb-rich rough grassland	Loss of herb-rich grassland through agricultural intensification
A solitary wasp *Cereris quadricincta*	Sandy grassland on south-facing slopes, it provisions its nest with weevils	Loss of open ground for nesting, and flower-rich sandy grassland for foraging
A solitary wasp *Cerceris quinquefasciata*	Flower-rich sandy grassland, it provisions its nest with weevils	Loss of open ground for nesting, and flower-rich sandy grassland for foraging
The northern colletes *Colletes floralis*	Machair grassland and marram zone of coastal dunes	Loss of herb-rich dune grassland, could be negatively affected by warming of the UK climate
A cuckoo bee *Nomanda armata*	Lays its eggs in the nest of the mining bee *Andrena hattorfiana*. The host bee is closely associated with the flowers of field scabious where it forages for pollen. *Nomanda armata* also uses the flowers for nectar	Loss of calcareous grassland due to agricultural improvement, inappropriate grazing management

Table 2.20: contd

Name	*Type of grassland and other requirements*	*Factors causing loss or decline in relation to grassland*
A cuckoo bee *Nomada errans*	Presumed to be the parasite of the mining bee *Andrena nitidiusculus* in grassland habitats on chalk and sand. Host bee collects pollen from wild carrot and hogweed	Coastal cliff stabilisation, scrub encroachment
Flies		
Hornet robber-fly *Asilus crabroniformis*	Unimproved grassland, fly larvae believed to prey on larvae of large dung beetles [a]	Loss of unimproved grassland, use of persistent stock parasite treatments (e.g. ivermectins) that kill dung beetle hosts, changes in stock management
A wasp-mimic hoverfly *Doros profuges/conopseus*	Scrub or wood edge on calcareous grassland, larvae believed to have a commensal or predatory relationship with the ant *Lasius fuliginosus*	Not known
Picture-winged fly *Dorycera graminum*	Grassland	Not known

Sources: UK Biodiversity Steering Group, (1995); UK Biodiversity Group (1999b, d).

Notes: Species may also be associated with other habitats; [a] Also associated with heathlands.

great crested newt was introduced in the 1970s. All three species are protected under the Habitats Directive and Wildlife and Countryside Act (Smith, 2000), and are the subject of Species Action Plans (UK Biodiversity Steering Group, 1995).

Mammals

Several mammal species are associated with semi-natural lowland grassland, but are not necessarily confined to this habitat. They include the bank vole, hedgehog, brown hare, badger, harvest mouse, field vole, stoat, weasel, rabbit, greater horseshoe bat, common shrew, pygmy shrew and mole (Macdonald and Barrett, 1993; Crofts and Jefferson, 1999). With the exception of the greater horseshoe bat, which declined significantly throughout northern Europe during the twentieth century and is the subject of a Species Action Plan (Table 2.21), the majority of these species are fairly common. However, the population of the brown hare appears to have undergone a substantial decline in numbers since the early 1960s, with current estimates varying between 817,500 and 1,250,000. Similar declines appear to have occurred throughout Europe, and for this reason, the brown hare is the subject of a Species Action Plan (Table 2.21). Several factors have contributed to its decline and these include agricultural intensification – for example, loss of grassland to arable, increased disturbance from higher stocking densities and a change from hay to silage production. Hares lie up in

Table 2.21: Grassland mammals featured in Species Action Plans

Name	*Type of grassland and other requirements*	*Factors causing loss or decline in relation to grassland*
Brown hare *Lepus europaeus*	Grassland as part of the farmland landscape	Conversion of grassland to arable, a move from hay to silage
Greater horseshoe bat *Rhinolophus ferrumequinum*	Old permanent pasture	Reductions in insect prey abundance, especially loss of old pasture due to high intensity agricultural systems, loss of insect-rich feeding habitats and flyways due to the conversion of permanent pasture to arable

Source: UK Biodiversity Steering Group (1995).

Note: Species may also be associated with other habitats.

silage fields during the day, where they are not disturbed by livestock. However, silage production involves more and earlier grass cuts per season, and silage cutting coincides with the hares' breeding season. Large numbers of adults and leverets (young hares) are killed by farm machinery. This can be avoided by adopting mowing patterns that allow hares and other wildlife to escape (Hutchings and Harris, 1996; Scottish Natural Heritage, 2001).

The greater horseshoe bat forages over a number of habitats, including meadow and grazed pasture (Species box 2.16). Over 75 per cent of foraging can occur over these habitats (Robinson *et al.*, 2000). It is now recognised that food resources around maternity roosts in summer are crucial in sustaining large numbers of breeding female bats, particularly during lactation. The greater horseshoe bat is selective and conservative in its diet, even in variable habitats. Its key prey items include cockchafers, dung beetles, moths (Lepidoptera) and craneflies (Tipulidae) (Ransome, 1997, 2000). In May–June bats feed mainly over meadows (44.2 per cent) and to a lesser extent over grazed pasture (25.5 per cent), when their diet probably consists largely of cockchafers, craneflies and large moths. In July bats forage mainly over grazed pasture (91.2 per cent), when dung beetles form an important part of the diet (Robinson *et al.*, 2000). The availability of grazed pasture within 3 km or 4 km of the roost site may be particularly important for maintaining large colonies (Ransome, 1997, 2000). Reductions in insect prey have contributed to the decline of this species, mainly through habitat loss (Table 2.21). However, pesticides may also play a role. Ivermectin is an antiparasitic drug from the avermectin group that is used to treat cattle for nematode and arthropod parasites. Most ivermectin is excreted with dung and initially retains its insecticidal activity. The rate of ivermectin degradation varies greatly, depending on environmental conditions, but in Denmark ivermectin may persist in dung for up to two months after deposition. Ivermectins can have lethal and sub-lethal effects on non-target coprophilous (dung-feeding) invertebrates such as dung beetles, which form a very important component of the diet of greater horseshoe bats. The consequences of ivermectin use for greater horseshoe bats are potentially very significant (Lumaret *et al.*, 1993; Cox, 1999).

Species Box 2.16: Greater horseshoe bat *Rhinolophus ferrumequinum*

The horsehoe bats can be distinguished from other British bats by the presence of a horseshoe-shaped nose leaf round the nostrils, which is related to their echolocation system. The greater horseshoe bat is one of the largest British species, and is 57–71 mm in length. Individuals hang by the claws of their feet and rest with the wings wrapped round the body. Greater horseshoe bat numbers are greatest in sites with access to steep, south-facing slopes covered with mixed deciduous woodland and permanent pasture grazed with cattle, where prey insects are abundant. Appropriate roost sites dispersed among suitable feeding areas are also needed. Horseshoe bats often roost communally, mainly in caves and disused mines during winter, and typically in buildings during summer. Winter roosts may be up to 50 km from summer breeding roosts. Females usually become sexually mature in their third year, males in their second or third year. Females may not breed every year, but both sexes can live for over twenty years and still breed. The maximum age recorded is thirty years. Mating occurs during the autumn, when females disperse to visit males occupying territorial sites, and this can contribute to gene flow between colonies. Maternity colonies gather from May to July. The young are born in mid-July and are fully weaned at seven weeks. Greater horseshoe bats emerge from their roosts within half an hour of sunset. Between May and August they usually return to their roost after about an hour, and then forage again around dawn. From late August they may remain out all night. Insects are taken in flight, or occasionally from the ground. Hunting flight is low, fairly slow and follows regular flight paths. Greater horseshoe bats hibernate in winter roosts from late September to mid-May, but they awake from hibernation at frequent intervals and may move between hibernation sites. They feed in winter during mild weather. The greater horseshoe bat has declined significantly throughout northern Europe, particularly owing to reductions in prey abundance and loss of insect-rich feeding habitats, especially old, unimproved permanent pasture. In the UK the species is restricted to south-west England and south Wales, and its distribution is possibly limited by the need to feed in winter. There are thirty-five recognised maternity and all-year roosts and 369 hibernation sites. The greater horseshoe bat is protected under the Wildlife and Countryside Act (1981), and is the subject of an English Nature Species Recovery Programme and a UK Biodiversity Species Action Plan. The aim of the plan is to increase the population by 25 per cent by 2010.

Sources: Corbet and Harris (1991); Roberts and Hutson (1993); UK Biodiversity Steering Group (1995); Rossiter *et al.* (2000)

The mole is abundant in deciduous woodland, arable fields and permanent pasture (Species box 2.17). It is viewed as a pest of farms, gardens, sports fields and nature reserves. This is mainly because it forms molehills. For example, soil from molehills can contaminate silage, but silage additives can resolve this (Macdonald and Barrett, 1993). Peak molehill production occurs in spring and autumn, and few molehills are formed at other times of year. In a study of factors influencing molehill distribution in grassland, Edwards *et al.* (1999) revealed that molehill production in grazed areas

Species Box 2.17: Mole *Talpa europaea*

Moles occur throughout mainland Britain and on some islands, but are absent from Ireland. They are found in most habitats where the soil is deep enough to allow tunnel construction, but are uncommon in coniferous forests, on moorland and in sand dunes, probably because their prey is scarce. They spend most of their lives in an extensive system of permanent deep and semi-permanent shallow tunnels. Nests are constructed within the tunnel system for sleeping and raising young. The body is cylindrical to facilitate movement through the tunnels, and the forelimbs are adapted for digging. Earthworms are the major component of the diet, particularly in winter, and worms may be immobilised and stored alive in special chambers. Males and females occupy exclusive territories and are solitary for most of the year, but males extend their tunnels in search of females at the beginning of the breeding season. Females give birth to three or four young in the spring. The young disperse from the mother's territory at the age of five to six weeks, and become sexually mature by the spring following birth. Moles typically live for up to three years, but may survive for up to six years. Predators include owls, buzzards, stoats, cats and dogs. Moles are frequently perceived as pests of agricultural and amenity land, but the damage attributed to them is often slight. For example, in a 1992 survey less than 1 per cent of respondents reported that 10 per cent or more of their silage was seriously affected by mole activity, but 49.5 per cent carried out mole control measures.

Sources: Corbet and Harris (1991); Atkinson *et al.* (1994); Edwards *et al.* (1999)

was one-third of that in hay meadows, and that grassland with characteristics that resulted in fewer earthworms, such as low pH, also had fewer molehills.

The hedgehog occurs in deciduous woodland or scrub, moist pastures and meadows and other grassland types, although it particularly favours borders between these open and wooded habitats. It feeds largely on invertebrates at ground level, such as earthworms, ground beetles and slugs. Occasionally it eats vertebrates such as frogs and bird nestlings, as well as eggs (Macdonald and Barret, 1993). Hedgehogs were introduced to the Outer Hebrides (machair landscape) in 1974 and they are now a major predator of the eggs of ground-nesting birds (JNCC, 2001).

Rabbits are found in a variety of habitats, but they prefer short grass, for example on dry heath and pasture (Species box 2.18). At low population densities, single burrows are dug, while large groups of rabbits create warrens (systems of underground tunnels) (Macdonald and Barrett, 1993). Rabbits play an important role in maintaining grassland through arresting ecological succession and preventing the dominance of rank grasses and the establishment of scrub and tree seedlings. Their role in maintaining calcareous grassland has been particularly important (Hillier *et al.*, 1990). Chalk grassland turf grazed by rabbits is characteristically short and springy. Rabbits avoid species such as thyme, clustered bellflower, rockrose, nettle, ragwort, creeping thistle and elder (Thompson, 1994). Unpalatable species

Species Box 2.18: Rabbit *Oryctolagus cuniculus*

Rabbits originate in the western Mediterranean and were introduced to Britain in the twelfth century. In the middle of the nineteenth century, numbers increased dramatically owing to a combination of large-scale planting of hedgerows following enclosure Acts, increased cereal production and predator control. In the early 1950s the British rabbit population was estimated to be between 60 million and 100 million. In 1953 the virus myxomatosis reached Britain and within two years 99 per cent of the rabbit population had died. Rabbits have become more resistant to myxomatosis, but outbreaks still occur. The population has largely recovered and rabbits are now widespread throughout Britain and Ireland, but are absent from Rum and the Isles of Scilly. Rabbit damage is estimated to be over £100 million per year, but grazing by rabbits can be very beneficial for maintaining the diversity of habitats such as chalk grassland and heathland. Rabbit management involves extermination and fencing. In 1994 rabbit viral haemorrhagic disease (VHD) was found for the first time in wild rabbits. The implications of the disease are not yet known. Rabbits, predominantly females, produce a network of tunnels and dens called a warren. The most suitable habitats for rabbits are areas of short grass, for example on dry heath or closely grazed pasture, where burrows can be located adjacent to food sources. Hedgerows and woodland edge are particularly suitable, although open warrens are maintained on areas of short grass. Social groups vary from a single pair to up to thirty individuals sharing a warren. Within large groups there is a distinct social hierarchy, where dominant males (bucks) have priority access to females (does), and dominant females occupy the best nest sites. During the breeding season (January to August) females may produce one litter of three to seven young per month. Males can mate at the age of four months, females at three and a half months. Rabbits do not usually live for more than three years, and over 90 per cent die in their first year, often in the first three months of life. Predators include badgers, buzzards, weasels, foxes, cats, stoats and polecats.

Source: Sumption and Flowerdew (1985); Corbet and Harris (1991); Macdonald and Barrett (1993); Mammal Society (2001)

may become common close to burrows, where their growth is encouraged by rabbit disturbance of the substrate and nutrient enrichment. In some cases, disturbance by large numbers of rabbits allows ragwort to become more widespread.

As a result of myxomatosis, the rabbit population declined significantly in the 1950s (Hillier *et al.*, 1990; Thompson, 1994). There were significant ecological effects of the rabbit reduction in Britain. On chalk grassland, the cover and height of herbs and grasses increased initially, and flowering and seedling establishment increased. With time, diversity declined as microhabitats dependent on rabbit activity disappeared, grasses such as upright brome, tor grass and red fescue were no longer held in check by grazing and became dominant, and seedlings of woody plants escaped from rabbit control (Sumpton and Flowerdew, 1985; Hillier *et al.*, 1990). As noted earlier, changes in the structure and compostion of the vegetation also had a marked effect on animals associated with chalk grassland, such as the large blue butterfly. However, insect populations of different species may react to changes in grazing in different ways. For example, the Adonis blue butterfly declined on calcareous grassland following myxomatosis, while the Lulworth skipper increased. The food plant of the Adonis blue is horseshoe vetch, but it can use only plants that are grazed very short, creating a warm microclimate on the soil surface, where it is tended by ants. The food plant of the Lulworth skipper is tor grass, but the larvae feed only on tall, mature tussocks. Both species were present in the turf of grazed calcareous grassland, but tor grass was kept in check. Following myxomatosis, many swards became overgrown, and although horsehoe vetch initially remained abundant, it grew too tall for Adonis blues, while tor grass was released from grazing and grew to a suitable height for Lulworth skippers. In the longer term horseshoe vetch declined in many calcareous grasslands and tor grass spread (Thomas, 1990).

Birds

Lowland dry grasslands support relatively few breeding bird species, but these include skylark (Table 2.22), meadow pipit, grey partridge, kestrel, green woodpecker and rook. A mosaic of sward heights, from bare ground to 25 cm, meets the nesting and feeding requirements of skylarks. Less common, more locally distributed species include stone curlew (Table 2.22), lapwing, barn owl, wheatear, corn bunting (largely associated with arable habitats, Donald *et al.*, 1997), buzzard, quail, curlew, woodlark and winchat. Many of the species listed above are generally more common in other open habitats. However, the highest densities of stone curlews occur on lowland dry grassland and the Breckland grass heaths, where they nest on bare or sparsely vegetated stony or sandy ground. Feeding habitats include grasslands, manured arable fields and arable headlands (Crofts and Jefferson, 1999). Curlews are characteristic of the upland fringe in Britain, but nest in tall, tussocky lightly grazed or ungrazed areas of lowland dry grassland (Robson *et al.*, 1994; Crofts and Jefferson, 1999). Nightjar and Montagu's harrier also breed at a very few dry grassland sites. Hay meadows that form part of the machair landscape in the Hebrides are an important habitat for corncrake (Table 2.22) and machair grassland is valuable for waders such as lapwing. Corncrake numbers have been in decline in Britain and Ireland for more than 100 years. The species has been seriously affected by changes in grassland management, particularly the loss of hay meadows and the earlier mowing of grass crops for silage, which increases losses of adults, nests and chicks during mowing (Green and Stowe, 1993). Choughs have a fragmented distribution in Europe. They are associated with extensive coastal pastures in the Hebrides, where they feed on soil-living, surface-active and dung-associated invertebrates (Hindmarch and Pienkowski, 2000). Where machine use is infrequent and stocking densities are low, improved grassland may retain a range of ground nesting birds such as lapwing (UK Biodiversity Steering Group, 1995).

British lowland dry grasslands also support non-breeding bird populations and can be

Table 2.22: Grassland birds featured in Species Action Plans

Name	*Type of grassland and other requirements*	*Factors causing loss or decline in relation to grassland*
Skylark *Alauda arvensis*	Widespread, but breeding population on lowland farmland declined by 54% between 1969 and 1991	Conversion of lowland grassland to arable, intensive management of grassland, early silage cutting, which destroys nests and exposes skylarks to predators
Stone curlew *Burhinus oedicnemus*	Semi-natural grassland, grass heaths, especially Breckland	Loss of semi-natural grassland to arable farming, reduced grazing by rabbits and livestock, egg collecting, collisions with utility lines and fences, shooting in European countries while on migration
Corncrake *Crex crex*	Meadows, mostly in north and west Scotland and Ireland, e.g. associated with the machair landscape of the Hebrides, although corncrakes tend to be birds of the croftland and hay meadow rather than machair grassland	Loss of traditional grassland habitat mosaics, especially tall vegetation throughout the breeding season, changes in grass management and cutting techniques, e.g. earlier cutting, predation and disturbance
Grey partridge *Perdix perdix*	Historically, low abundance in intensive grassland, and populations declined here by up to 95% between 1969 and 1990	Nest destruction caused by early mowing

Sources: UK Biodiversity Steering Group (1995); Benstead *et al.* (1999).

Note: Species may also be associated with other habitats.

important for wintering birds such as lapwing, golden plover, fieldfare, redwing, gulls and crows. In winter, hen harrier and short-eared owls hunt for small mammals over downland, while merlin feed on small birds taken in flight. Dry grasslands are staging areas for migrants such as dotterel and ring ouzel (Crofts and Jefferson, 1999). Interest in the UK's lowland dry grassland birds has increased in recent years because of habitat and species losses elsewhere in Europe. Many of the species listed are of conservation concern in the UK, because they are rare, their breeding populations are declining, or they have European conservation status under the EC Bird Directive (Chapter 3, 'Opportunities'). A number are Biodiversity Action Plan priority species and are included in Species Action Plans (Table 2.22).

A number of England's rare or threatened bird species are either wholly or partly dependent on lowland wet grasslands, such as the ruff and black-tailed godwit. Wet grasslands can support high breeding populations of other species such as lapwing, curlew, snipe and redshank, and overwintering species such as Bewick's swan, whooper swan, barncale goose and pintail. Water-filled ditches associated with lowland grassland are also important for breeding wildfowl. Many breeding and non-breeding species associated with lowland wet grasslands are of high conservation priority (Rodwell, 1992; Ausden and Treweek, 1995; UK Biodiversity Steering Group Report, 1995; Jefferson and Grice, 1998).

Fauna of heathlands

Invertebrates

Lowland heathland is a very important invertebrate habitat in Britain. It is renowned for the richness of its invertebrate fauna, and the habitat supports many species at the edge of their European range. The most important heaths for invertebrates are in southern England. They support 50 per cent or more of the British species of spiders, dragonflies and true bugs (Kirby, 1992, 2001). Kirby (1994) lists 133 Red Data Book (endangered, vulnerable, rare, indeterminate or insufficiently known) invertebrate species and 210 scarce (notable) species associated with lowland heathlands (including wet and Breckland heaths). However, these figures underestimate the number of invertebrates associated with lowland heaths, because they do not include either species that are frequent on heathland but which also occur in a wide range of other dry sunny situations, or some species associated with wet heaths that are classed as wetland species (Kirby, 1994). It is not clear whether species associated with coastal heaths are included in the assessment.

Invertebrates associated with heathland can be broadly divided into two groups: those that feed on heathland plants such as heather, heath species and gorse, and whose distribution is determined by the availability of their food plants, and those whose distribution is a consequence of their requirements for particular conditions that are characteristic of heathland, such as warm, open spaces, bare ground such as sandy soil for burrowing, or the structural diversity of dwarf shrubs (Webb, 1986; Key, 2000) (Species box 2.19). For example, the food plants of the grayling are grasses, but it requires the warm, dry, open spaces found on heathland. The species is also associated with these conditions on patches of sparse turf in chalk grassland (Webb, 1986). The structural diversity of heathland includes all stages from bare ground to scrub, and a range of heather phases, e.g. from pioneer to degenerate (Haysom and Coulson, 1998). Heathlands are associated with relatively high numbers of predatory species, probably because of the structural diversity of the vegetation and the patchiness of the habitat. Dwarf shrubs, such as heather, have a complex structure that provides a range of niches that can be exploited by invertebrates. They provide a firm structure for web-spinning spiders, good shelter for invertebrates and litter that harbours a range of species (Kirby, 1992). Invertebrates associated with bare sandy areas include nest-burrowing bees and wasps, tiger beetles, dung beetles (for example, minotaur beetles that feed on dung, including rabbit droppings, and dor beetles that feed on a wide variety of dung), ants and spiders. The larvae of a number of tiger beetles excavate vertical burrows in the sandy soils of heathland, while adult tiger beetles are hunters over the surface of bare ground (Key, 2000) (Species box 2.20). In contrast, heathland vegetation contains relatively few species of plant to support phytophagous invertebrates. For example, relatively few species are totally dependent on heather, although, as its name suggests, larval and adult stages of the heather beetle feed predominantly on this species (Gimingham, 1992). Common gorse is the food plant of a number of species, and these in turn support a range of specialist and general predators and parasites. Grassy patches, or areas with nectar-bearing flowers, are also valuable components of heathland (Webb, 1986; Kirby, 1992).

In contrast to the above-ground invertebrate fauna, the soil invertebrate fauna of heathland is rather restricted. Heathland

Species Box 2.19: Bee-killer digger wasp or beewolf *Philanthus triangulum*

The beewolf is a solitary digger wasp. Until fairly recently the beewolf was considered a rarity in Britain. Records for the last few years indicate that its range is expanding, and it has become locally common to abundant in an increasing number of sites in southern England, with a single record for north Wales. This wasp typically occurs in lowland heaths and on sand dunes, where it nests in level, sandy exposures and in vertical soil faces. As many as 15,000 burrows may occur together. The main nest burrow may be up to 1 m long, with between three and thirty-four short burrows at the end, each terminating in a brood cell. Eggs are laid in brood cells. Cells are provisioned with paralysed prey as food for the larvae, then sealed. The major prey species of the beewolf wasp is the worker honey bee (*Apis mellifera*), but other bee species may also be utilised. The beewolf paralyses the prey by stinging it through the articular membranes behind the front legs. Female offspring need to be provisioned with at least three prey items (honey bees) in the brood cell in order to develop, while males require only one. Mother beewolves lick the body surface of their prey during the period of excavation of the brood cell, and this treatment delays fungal growth on the prey. Mothers also deposit a substance in the brood cell that appears to orientate the larvae towards the main burrow, so that young adults can dig out of the cell by the most direct route. In England a single brood is produced, while in central Europe two broods per year have been recorded. The flight period is from early July to mid-September. Nectar sources include heather, common ragwort and creeping thistle.

Sources: Strohm and Linsenmair (1994, 1997, 2001); Edwards (1997); Strohm (2000)

generates plant litter that is resistant to decomposition, because it is rich in tannins and resins, and is acidic. In addition, low soil pH does not favour rapid decomposition, so a surface organic layer develops. In these conditions deep-burrowing earthworms are scarce or absent. However, mites (Acari) and springtails (Collembola) are abundant (Gimingham, 1972; Webb, 1986; Gimingham, 1992; Curry, 1994).

Uncommon heathland invertebrates include the heath grasshopper that lives among short or sparse vegetation and lays its eggs in bare ground, the heath tiger beetle and the ladybird spider that need bare patches in mature heath (Kirby, 1992). The ladybird spider is probably the rarest spider in Britain, as only about 600 are known to exist, at a single site in Dorset. They live in burrows and catch their prey using silk trap wires (English Nature, 2001a). A study on the Dorset heathlands revealed that 117 Red Data Book and Nationally Scarce invertebrate species were associated with bare sand, in comparison with eleven associated with heather, thirty-seven with dry grassland, forty-nine with wet heath, ninety with other wetland habitats and twenty-seven with scrub (Key, 2000). Many bare-ground species such as the heath tiger beetle (Table 2.23) are declining through

Species Box 2.20: Green tiger beetle *Cicindela campestris*

Adult tiger beetles (*Cicindela* species) are slender, long-legged, fast-running predatory beetles with huge eyes and jaws. They can also combine flight and running in a form of 'hopping flight'. They hunt ants and other prey on the ground by sight, and usually live in open habitats, where they employ 'stealth and spurt tactics, like a cheetah' to attack other species. The larvae construct burrows from which they ambush prey, using their enormous jaws to drag prey into the burrow to be eaten. The green tiger beetle is found throughout the UK, mainly on heathland, sand dunes, and in other sandy places, including bare ground in grassland. It is metallic green on the underside of the abdomen, and the legs are coppery or purplish bronze. Patterns on the elytra are varied. The green tiger beetle digests its prey externally and ingests it in fluid form. The mouth parts are modified for this diet and bear extensive coverings of bristles that help distribute digestive enzymes onto the prey and facilitate the intake of fluid or semi-fluid food.

Sources: Chinery (1986); Forsythe (1987); Key (2000); Kirby (2001)

neglect of their habitat and scrub invasion (Key, 2000).

Wet heaths are different in character from dry heaths and support a distinct assemblage of invertebrates. For example, eight species of dragonfly and damselfly are almost confined to heathlands. Areas of boggy ground and pools can develop on heathland where drainage is impeded by the development of an iron pan beneath the surface (Plate 2.11, p. 81). The most valuable parts of pools on wet heathland for invertebrates are the margins and *Sphagnum* fringes. Heathland pools can be important for breeding dragonflies and damselflies, for example the white-faced darter, the keeled skimmer and the small red damselfly. The larvae of the southern damselfly live in shallow runnels, typically on heathlands in Hampshire and Dorset (Gimingham, 1992). Populations can be damaged either by deepening of the streams or by lack of management of surrounding vegetation, leading to shading and choking of runnels (Table 2.23). Quite small waterbodies can support large numbers of dragonfly and damselfly larvae, but the number of adults is restricted by territorial behaviour, for example a small pool is unlikely to support more than one large hawker male (Gimingham, 1992). As on dry heaths, structural diversity of the vegetation, from low vegetation with plants such as *Sphagnum* and sundews to taller grass communities and scrub, is important in supporting rich invertebrate communities. For example, the dingy mocha moth (Table 2.23) feeds on common shrub willows. Though its food plants are widespread, it is confined to damp heathland in Dorset and west Hampshire, and is sensitive to scrub clearance or succession to woodland (Kirby, 1992).

A number of dry and wet heathland invertebrates, including the dingy mocha

Table 2.23: Invertebrates associated with heathland, featured in Species Action Plans

Name	*Type of heathland and other requirements*	*Factors causing loss or decline in relation to heathlands*
Moths		
Reddish buff *Acosmetia caliginosa*	Open, grassy heathy swards rich in sawwort (*Serratula tinctoria*), neither strongly acidic nor strongly alkaline [a]	Establishment of conifer plantations on heathland, scrub encroachment
Speckled footman moth *Coscinia cribraria*	Dorset heathland	Habitat loss due to development, plantation forestry, drainage, fires, scrub encroachment, inappropriate habitat management
Dingy mocha *Cyclophora pendularia*	Open heathy situations in Dorset and West Hampshire	Loss of heathland due to development, forestry, agricultural improvement and road building, sucession to woodland on unmanaged sites, fire, scrub clearance during heathland restoration
Narrow-bordered bee hawk-moth *Hemaris tityus*	Drier heathland [a]	Agricultural improvement of heathland, inappropriate management
Lunar yellow underwing *Noctua orbona*	Open, sandy, heathy or calcareous sites	Not known
Butterflies		
Heath fritillary *Mellicta athalia*	Heathland	Abandonment and inappropriate habitat management
Silver-studded blue *Plebejus argus*	Lowland heathland	Loss of heathland to development and agriculture, habitat fragmentation and isolation, inappropriate heathland management
Crickets and Grasshoppers		
Wart-biter *Decticus verrucivorous*	Grassy heathland [a]	Inappropriate habitat management
Damselflies and dragonflies		
Southern damselfly *Coenagrion mercuriale*	Breeds in heathland streams	Loss of habitat due to lack of appropriate heathland management including reduced grazing and deepening of shallow breeding streams
Beetles		
A ground beetle *Amara famelica*	Open, sandy or gravelly heaths	Loss of heathland, inappropriate heathland management, scrub encroachment
A ground beetle *Anisodactylus nemoravagus*	Open, sandy heathland	Loss and fragmentation of heathland, inappropriate heathland management, leading to loss of open ground and scrub encroachment

Table 2.23: contd

Name	*Type of heathland and other requirements*	*Factors causing loss or decline in relation to heathlands*
Heath tiger beetle *Cicindela sylvatica*	Dry and sandy soils with heather	Loss of heathland, inappropriate management, especially neglect and scrub invasion
A ground beetle *Pterostichus kugelanni*	Heathland with sandy or gravely soils	Inappropriate heathland management, loss of habitat
Ants		
Dark guest ant *Anergates atratulus*	Dry lowland sandy heath, an obligate social parasite of another ant (*Tetramorium caespitum*) and survives only where host populations are high	Loss of suitable heathland habitat, inappropriate heathland management
Narrow-headed ant *Formica exsecta*	Lowland heathland in southern Britain	Loss of suitable heathland due to destruction or inappropriate management, e.g. scrub encroachment leading to shading of nests and encouragement of competitive ant species, inappropriate grazing by ponies in the New Forest and production of single-age heather stands
Black-backed meadow ant *Formica pratensis*	Lowland heath	Urban development of heath, inappropriate management and scrub encroachment, possibly leading to invasion of competitive southern wood ants
Red barbed ant *Formica rufibarbis*	Lowland or maritime heath overlying loose or sandy soils	Loss of suitable habitat, inappropriate management, excessive disturbance of nests, e.g. through trampling and mechanised scrub or heather clearance, frequent or intensive fires
Bees and wasps		
Ruby-tailed wasp *Chrysis fulgida*	Heathland sites, parasitoid on a wasp and bee	Not known
Spider-hunting wasp *Evagetes pectinipes*	Sandy heaths and open dunes, presumed to steal spiders from other spider-hunting wasps	Not known
Spider-hunting wasp *Homonotus sanguinolentus*	Lowland heathland in Dorset, Hampshire and Surrey, predator of spiders, wasp lays an egg on spider's abdomen and wasp larvae feed on its body fluids then consume its remains	Loss of southern heathland, especially grass heath

Table 2.23: contd

Name	*Type of heathland and other requirements*	*Factors causing loss or decline in relation to heathlands*
Purbeck mason wasp *Pseudepipona herrichii*	Provisions its nest with the caterpillars of a moth which feeds on heathers, especially bell heather, in early to mid-successional heathland, bell heather also a nectar source for adult wasps	Succession on heathland
Flies		
Hornet robber-fly *Asilus crabroniformis*	Unimproved heath in southern England and Wales [a]	Loss of unimproved heathland, use of persistent parasite treatments on stock (e.g. ivermectins) which kill dung beetle hosts, changes in stock management
Heath bee-fly *Bombylius minor*	Open heathland, where it is a parasitoid of solitary bees	Loss and fragmentation of heathland habitat, owing to development and scrub encroachment, inappropriate heathland management, loss of vertical sand banks which are important for the host bees
A wasp-mimic hoverfly *Chrysotoxum octomaculatum*	Dry heaths of Dorset, the New Forest and western Weald	Heathland destruction and lack of appropriate heathland management
Mottled bee-fly *Thyridanthrax fenestratus*	Heather-dominated heathland, possibly a parasitoid of either the sand wasp *Ammophila pubescens*, or of the caterpillars that the wasp collects to feed its larvae	Inappropriate heathland management, scrub encroachment, uncontrolled fires, damage to open areas by recreational or military use
Spiders		
Ladybird spider *Eresus cinnaberinus*	Dry, sandy heath with bare or lichen-covered patches, where it forms burrows	Encroachment and shading by rhododendron, pine and bracken, possibly competition from southern wood ants
A spider *Uloborus walckenaerius*	Lowland heathland on mature heather plants	Loss of heathland due to afforestation and development

Sources: UK Biodiversity Steering Group (1995); UK Biodiversity Group (1999b, d).

Notes: Species may also be associated with other habitats; [a] Also associated with grassland.

moth, are included in Species Action Plans (Table 2.23). Several factors have contributed to the decline in these species, but habitat loss and inappropriate management are particularly damaging (Table 2.23).

Amphibians and reptiles

Several amphibians occur on heathland, including the palmate newt, the smooth newt, the common frog, the common toad and the natterjack toad. The great crested newt rarely

Plate 2.11: Pools on wet heathland are valuable to inverterbrates

occurs on heathland. The palmate newt is the most common newt on heathlands and can tolerate pools with a pH as low as 3.9. Smooth newts are rarely found in waters below pH 6.0. The common frog and common toad are associated with a range of wet habitats, but the natterjack toad is restricted to sandy heathlands and coastal sand dunes (Table 2.24; Webb, 1986). The natterjack prefers mixed-age dry heath with sandy, open areas suitable

Table 2.24: Heathland amphibians and reptiles featured in Species Action Plans

Name	*Type of heathland and other requirements*	*Factors causing loss or decline in relation to heathland*
Natterjack toad *Bufo calamita*	Habitat includes heathland[a]	Loss of habitat due to development, agriculture and reduced grazing on heathlands, habitat fragmentation, leading to genetic isolation of populations, acidification and loss of breeding pools
Sand lizard *Lacerta agilis*	Mostly confined to heathland such as the dry heaths of south Dorset. Sand lizards have been reintroduced to sites in the Inner Hebrides, the New Forest, the Weald and Wales[a]	Loss, deterioration and fragmentation of heathland habitat, scrub encroachment, uncontrolled fires, shortage of suitable breeding sand on heathland sites

Source: UK Biodiversity Steering Group (1995).

Note: Species may also be associated with other habitats. [a] Also associated with sand dunes.

for burrowing. It breeds in pools with a pH between 6 and 7, and heathland pools that are more acidic are unsuitable (Beebee, 1987). Acidification of breeding pools is one of the factors that has contributed to its decline (Table 2.24). Most colonies breed in temporary pools that do not support many predators. Other factors that have caused the decline of natterjack toads on heathland include loss of habitat, in particular open, sandy ground, and habitat fragmentation.

The six native species of British reptiles all occur on lowland heaths, particularly in southern England (Species box 2.21). The sand lizard and the smooth snake are closely associated with heathland. The grass snake, common lizard, slow worm and adder all occur on heathland but are not especially associated with the habitat. The most widespread reptiles in Britain are those that are viviparous (give birth to live young). The sand lizard is an egg-laying (oviparous) species that requires particular conditions for the incubation of eggs (Webb, 1986). The decline and inappropriate management of heathland, with the associated loss of warm, open sand suitable for egg laying and incubation, is probably the main cause of population decline (Table 2.24) (Corbett and Moulton, 1995; Beebee and Rowe, 2001). Sand lizards also require the presence of mature heather for hunting and shelter (Andrews and Rebane, 1994). In addition, there is increasing evidence that disturbance associated with urbanisation,

Species Box 2.21: Adder *Vipera berus*

The adder is found throughout Britain but is absent from Ireland. Adults are usually up to 65 cm in length, occasionally almost 90 cm. Females tend to be larger than males. The adder has a bold, dark zigzag stripe down its back. Colouring varies from whitish with intense black markings in the male to shades of brown or copper with dark brown markings in the female. Entirely black individuals may also be found. The adder favours open habitats such as heathland, open woodland, field edges, dunes and sea cliffs. A study in Dorset revealed that the viper occupied high, dry ground for winter hibernation and low-lying damp river meadows and wet heaths during the active summer period. Hibernating areas are communal. Hibernation generally extends from the end of September until early March for adult males, and from the end of September until late March or early April for adult females. Males become very active during the mating period and the 'dance of the adders' is a form of combat between two males in which their heads rear up from the ground. In the south of England adult females

usually breed in alternate years. Adders remain close to the hibernating site until mating is over, and then disperse to local feeding sites. Adders feed mostly on small mammals, but also eat lizards and nestling birds, striking their prey to inject venom. The adder gives birth to live young in August or September, close to the hibernation area. Legislation in Great Britain protects adders from being killed, injured or traded.

Sources: Prestt (1971); Arnold and Burton (1978); Gasc *et al.* (1997); Inns (1999a)

such as predation by domestic cats, is a threat to the species (Haskins, 2000).

Mammals

A number of small mammals associated with open habitats occur on heathland, including the pygmy shrew, common shrew, wood mouse and field vole. Both the stoat and the weasel may also occur on heathland (Species box 2.22). Rabbits have had an important influence on heathlands, and in some areas cessation of rabbit grazing following the outbreak of myxomatosis led to scrub encroachment (Webb, 1986). Digging, scraping and grazing by rabbits are beneficial to invertebrates, since their grazing produces a mosaic of vegetation structures, nutrient-enriched and disturbed soil around their burrows supports nectar plants, and their dung and carrion support rich invertebrate communities (Kirby, 1992). Rabbits eat a wide range of plants, including heather, but avoid eating cross-leaved heath and bracken (Webb, 1986). Roe deer can be found on heathland in southern England where there is scrub encroachment. Feral populations of sika deer occur on heathland in Dorset and the New Forest. They often inhabit patches of dense undergrowth on heathlands, but may also graze in open areas such as wet heaths and mires (Webb, 1986; Macdonald and Barrett, 1993).

Species Box 2.22: Weasel *Mustela nivalis*

Weasels are widespread throughout Britain but are absent from Ireland and most offshore islands. They are the smallest, and probably most numerous, of British carnivores. Stoats are larger than weasels, and have a longer tail, which ends in a black tip. Male weasels are larger than females. Weasels occur in a range of habitats including lowland grassland, woodland and urban areas. Weasels are specialist predators of small rodents, particularly field voles, and their numbers depend on the abundance of their prey. The long, thin shape of the weasel is well suited to hunting small rodents. They do not hibernate and hunt at any time of the day or year. If rodents are scarce, additional prey such as rabbits, birds and eggs may be taken. Weasel home ranges vary according to the distribution of prey, although male ranges are larger than female ranges. A home range usually contains several dens and resting places that are visited at intervals. Resident animals of both sexes may defend territories when numbers are high. One or two litters, each of four to six young, are born per year. Family groups disperse at nine to twelve weeks. Weasels rarely survive to over two years old. Predators include hawks, owls, foxes and cats. Weasel populations are very variable and can suffer high mortality, particularly when rodent populations are low. Though local populations often experience extinctions, abandoned areas can be rapidly recolonised when conditions improve. However, populations appear to have become more patchy in recent years, and weasel habitat may have become more fragmented owing to agricultural intensification.

Sources: King (1989); Corbet and Harris (1991); McDonald *et al.* (2000); Mammal Society (2001)

Birds

Birds are relatively scarce on heathland. However, heathlands support four Red Data Book breeding bird species, the Dartford warbler, woodlark, stone curlew and nightjar (Species box 2.23), and two wintering Red Data Book species, the hen harrier and merlin (Michael, 1996). Limited areas of scrub and trees are important for a number of species. For example, nightjars use the interface between open heath and scrub, and Dartford warblers use pines in close proximity to heather, and are often

Species Box 2.23: Nightjar *Caprimulgus europaeus*

Nightjars are summer visitors to Europe. They spend the UK winter in sub-Saharan Africa, arrive in early May and are gone by October. The nightjar has a widespread distribution throughout England and Wales, but within its range it occurs locally only where suitable habitat exists. It is rare in Scotland and probably extinct as a breeding species in Northern Ireland. Nightjars require low, sparse vegetation in which to nest. More specifically, they require bare patches within heather-dominated vegetation at the base of small trees. Lowland heathland and young forestry plantations are now the most important nightjar habitats, but they also occur in woodland clearings of over 1.5 ha. Nightjars are nocturnal and crepuscular birds, and hide in low vegetation during the day, camouflaged by their grey, brown and russet plumage. The churring song of nightjar males can be heard from up to 2 km away, and it is possible to recognise individual birds on the basis of song characteristics. The species feeds on flying insects such as moths, beetles and flies, mainly at dawn and dusk, across a wide range of habitats. Nightjars are highly mobile and on average travel 3 km (but up to 6 km) from the nest site at night to feed. The range of the nightjar declined rapidly for much of the last century, and the population reached a low point of 2,100 males in 1981. The decline was probably due to a loss of heathland, and to disturbance and inappropriate management of surviving sites. For example, much planting of commercial forestry occurred on heathland, and once the trees matured, the cover became too dense to provide suitable nesting sites for the nightjar. Marked decreases have also occurred in other European countries.

There has subsequently been a partial recovery in the population, with an estimate of 3,400 males in 1992. The recovery has been assisted by increases in clear-felled and young forestry plantations as trees have reached harvestable age and sites have been replanted. In addition, specific management on some nature reserves has benefited the species. The nightjar is protected under the EC Birds Directive and is the subject of a Biodiversity Action Plan. In the short term the aim is to increase the numbers and range of the nightjar to 4,000 churring males in at least 280 10 km squares by 2003, and in the longer term to restore the nightjar to parts of its former range, for example Northern Ireland. This will best be achieved through the conservation and restoration of heathland, and the sympathetic management of forests and feeding habitat.

Sources: Cramp (1985); Morris *et al.* (1994); UK Biodiversity Group (1998a); Devon County Council (2000); Rebbeck *et al.* (2001)

Plate 1: Cressbrook Dale (Derbyshire Dales National Nature Reserve), showing improved pasture in the distance, unimproved calcareous grassland on the steep dale sides and a hay meadow community in the foreground

Plate 2: Heathland dominated by dwarf shrubs, with scattered trees and scrub

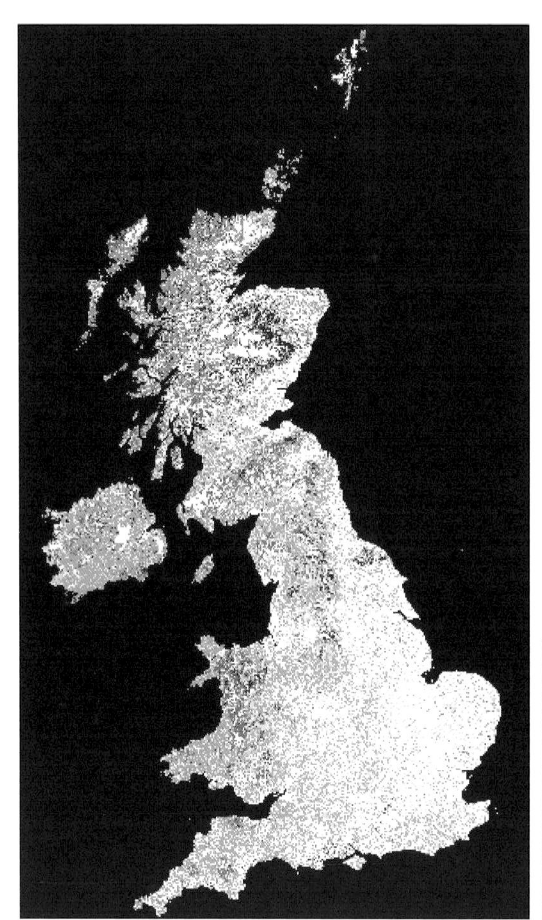

Plate 3: The extent of grassland and heathland classes (upland and lowland) in the UK from Land Cover Map 2000, CEH, Monks Wood. Bright green: improved grassland. Yellow: neutral and calcareous grassland. Olive green: acid grassland. Purple: dwarf shrub heath. Pink: open dwarf shrub heath (*Copyright © NERC*)

Plate 4: Cross-leaved heath (*left*), heather (*right*) and tormentil (*foot*)

Table 2.25: Heathland birds featured in Species Action Plans

Name	*Type of heathland and other requirements*	*Factors causing loss or decline in relation to heathlands*
Nightjar *Caprimulgus europaeus*	Lowland heathland [a]	Loss of heathland, inappropriate management so that heathland becomes unsuitable as nesting habitat, lack of feeding area close to nesting sites, possibly decline in large insects caused by changes in agricultural practice or climate change, maturation of young forestry stands
Linnet *Carduelis cannabina*	Gorse thickets, heathland and scrub, particularly near the coast	Changes in pastoral management, removal of gorse thickets
Red-backed shrike *Lanius collurio*	Lowland heathland and scrub	Habitat loss, decline in invertebrate food due to habitat loss and agricultural intensification, reduced mating opportunities as population density is reduced, egg collection and disturbance
Woodlark *Lullula arborea*	Varied habitat includes heathland in southern England and Suffolk coast [a]	Loss of lowland heathland feeding and nesting habitat, lack of appropriate management, decline in rabbit grazing, severe winter weather

Source: UK Biodiversity Group (1998a).

Notes: Species may also be associated with other habitats. [a] Also associated with dry grassland.

dependent on dense stands of common gorse (Dolman and Land, 1995). The nightjar is the subject of a Species Action Plan (Table 2.25).

Fauna summary

Lowland semi-natural grasslands and heathlands have a varied and interesting fauna. In particular, many are of considerable importance for invertebrates. Grasslands form part of habitat mosaics used by all widespread species of amphibians and reptiles, and several amphibian and reptile species occur on heathland, most notably the natterjack toad, the smooth snake and the sand lizard. Several mammals are often associated with grasslands and heathlands, but are not necessarily confined to them. Wet and dry grasslands and heathlands can support populations of feeding and breeding birds, and a number of rare or threatened birds are either wholly or partly dependent on lowland grasslands or heathlands.

3

MANAGEMENT AND CONSERVATION OF GRASSLANDS AND HEATHLANDS: PROBLEMS AND OPPORTUNITIES

•

The extent of habitat loss and decline and a number of causes of habitat and species decline were described in Chapter 2 in relation to particular community types. This chapter has a more thematic approach. It examines the key factors that have contributed to the decline of lowland grassland and heathland habitats in more detail, and evaluates the management options for habitat maintenance, enhancement, translocation, restoration and creation.

THE HISTORY OF GRASSLANDS AND HEATHLANDS

Where climate and soils are suitable, grasslands and heaths may develop if trees are removed and their regeneration is prevented (Gimingham, 1975; Ausden and Treweek, 1995). Appropriate management, that prevents succession to scrub or woodland, is fundamental to the existence of most grasslands and heathlands in the UK. Grasslands and heathlands have a long management history, from Neolithic times (about 5,000 years ago) onwards. The history of these habitats provides baseline information relevant to the management of grasslands and heathlands today.

Evidence of vegetation change comes from a variety of environmental sources, including pollen and other plant remains, land snails, insects (mostly beetles), bones (domestic animals and small vertebrates), charcoal, as well as preserved soils and land surfaces (Dimbleby, 1984; Roberts, 1989; Bell and Walker, 1992; Lowe and Walker, 1997). Pollen analysis (or pollen stratigraphy) of peats and lake deposits is the most widely used source of evidence about the changes that have taken place in the vegetation of Britain since the end of the last Ice Age. Pollen grains and spores are extremely small, but many can be identified to genus level under the microscope. Fossil pollen and spores extracted from stratified samples of sediment will provide a record of vegetation change through time (Lowe and Walker, 1997). Based on the ecological requirements of the plant groups that are recorded, it is possible to determine the sequence of changes brought about by climatic change, and by man, e.g. changes from forested to open habitats. This approach has produced widespread and consistent evidence of the role of man (Dimbleby, 1984; Roberts, 1989; Lowe and Walker, 1997). From Norman times (e.g. Domesday Book, the survey of 1086) knowledge of vegetation change depends increasingly on the methods of historical ecology, using landscape archaeology, maps and documentary records (Ratcliffe, 1984;

Rackham, 1986; Webb 1986). More recently, vegetation and habitat changes have been recorded through field survey work and satellite imagery, for example the Countryside Surveys of 1978, 1990 and 2000 and the Northern Ireland Countryside Survey 2000 (Bunce *et al.*, 1999; Firbank *et al.*, 2000; Haines-Young *et al.*, 2000). The Countryside Surveys give information on the extent, location and condition of habitats and can reveal how these have changed over time. Complementary information on trends in livestock numbers and agricultural practices is also available from organisations such as the Department for Environment, Food and Rural Affairs (DEFRA) and the Food and Agriculture Organisation of the United Nations (FAO).

Grasslands and heathlands in prehistoric times

During the last Ice Age, the ice sheet reached as far south as East Anglia in the British Isles. Conditions farther south would have been similar to open, semi-arctic grassland or tundra (Godwin, 1975; Rackham, 1986; Read and Frater, 1999). As the climate of Britain became warmer and the ice retreated (about 10,000 years ago, Bell and Walker, 1992), trees spread from the south. During the Atlantic period of 7,000 to 5,000 years ago (Fig. 3.1), most of lowland Britain was probably covered in woodland (Duffey *et al.*, 1974; Read and Frater, 1999). It is generally thought that grassland and heathland species were restricted to open sites where environmental conditions prevented tree growth or to forest clearings maintained by the grazing and trampling of wild herbivores (Duffey *et al.*, 1974; Kampf, 2000; Tubbs, 2001). In Neolithic times, livestock, including sheep, was introduced to Britain, and evidence from pollen analysis and faunal evidence from beetle and snail remains suggest that, from that time onwards, people began to clear substantial areas of forest for livestock grazing, particularly from the chalk soils, and provided opportunities for species characteristic of grassland, heathland or open ground communities to spread (Dimbleby, 1984). Woodland was replaced by short turf communities in Wessex, but by heath on the Breckland of East Anglia and Hampstead Heath in London (Bell and Walker, 1992). Woodland clearance continued throughout the Bronze Age and Iron Age (Fig. 3.1), but there were regional differences in the pattern and rate of forest clearance (Bell and Walker, 1992). Archaeological and faunal evidence dating from the Bronze Age in the Stonehenge area indicate that dry short grassland was established and was maintained by grazing animals, sheep in particular. For example, most of the insect evidence consists of species associated with dung or with open grassland, pollen from grasses is abundant, and remains of sheep bones and wool have been recorded (Ashbee *et al.*, 1989; Bell and Walker, 1992). From the Iron Age (about 2,500 years ago, Fig. 3.1), the clearance of woodland became more widespread and permanent (Duffey *et al.*, 1974; Read and Frater, 1999). In the Bronze and Iron Ages heathland formation became extensive, and from then on, these areas were used for free grazing (Webb, 1986; Bell and Walker, 1992).

Grasslands in historical times

Medieval times to 1914

Clearance of the woodland continued in historical times. Trees were felled for timber and fuel and woodland was cleared to grow crops and to support livestock that stopped woodland regeneration through grazing. Much of

GRASSLAND CHANGES	Inferred climate (climatostatographic unit/ Blytt-Sternander stage) [g, i]	Thousands of years before present (KaBP) [h]	Conventional dates [h]	Archaelogical periods [h]	HEATHLAND CHANGES
• Area & plant diversity of semi-natural grasslands cointinued to fall; road verge eutrophication [j, k, l] • BAP targets include grassland maintenance, improvement & re-establishment [l]	Sub-Atlantic (2.8/2.5 KaBP–present) (cool/wet)		2000 1990	Modern	•Evidence of increasing nutrient levels, effects of urbanisation, lack of appropriate management [l] • BAP targets include maintenance, enhancement & re-establishment of lowland heathland [l]
• 1978–1990 agriculture became increasingly intensive [k]. 1940s–1980s area of grassland declined. Many remnants deteriorated & scrub developed [d] • 1954 spead of myxomatosis – rabbit decline [d]			1980 1950		• 1984 area of lowland heath declined by 78% since 1830 [d]. Losses primarily due to agriculture, forestry, mineral extraction & development
• 1917–1918, 1939–1945 & the Agricultural Acts of 1947 & 1957 – ploughing up of long-established grasslands [b, d]			1900		• 1919 Forestry Commission established. Some planting on heathland [d]
• 1870s reversion of some arable grassland – cheap wheat imports, high milk production [a] • From 1750, pressure to extend arable areas [a]. 1780–1820 parliamentary Enclosures enclosed common & marginal land [d]			1800		• From 1750, pressure to extend arable areas [a]. 1780–1820 parliamentary Enclosures enclosed common & marginal land [d]. Heathlands were reclaimed for agriculture [e]
• Semi-natural meadows (cut for hay & grazed) & pastures (grazed & manure used to fertilise arable fields) were created by centuries of low-input, low-output management [a, d]			1700 1600		• The extent of the Dorset heaths changed little from Roman times until the mid-18th century [e] • The nutrient deficiency of heathland soils discouraged cultivation & heathland was used for rough grazing & fuel gathering [d, e]
• Black Death in 1340 led to a shift from arable cultivation to pasture [f] • Introduction of the rabbit in Norman times [b] • Medieval period, growth of the wool trade in England [b]		1	1500 1000 AD	Medieval Roman	• Royal forests set aside for hunting, e.g. the New Forest included grassland & heathland. Several of these heathlands dated from the Bronze Age [f] • Introduction of the rabbit in Norman times [b]

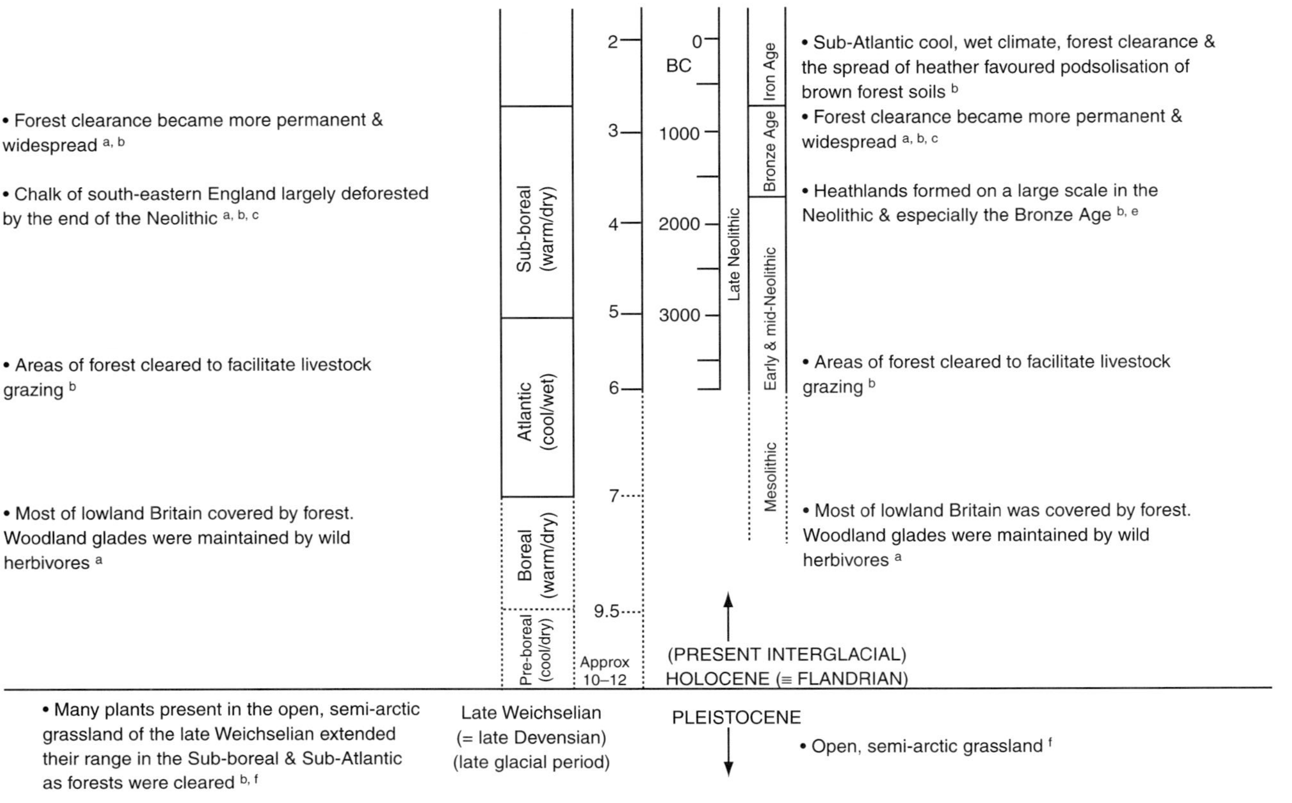

Figure 3.1: The history of grassland and heathland. *Sources*: [a] Duffey *et al.* (1974). [b] Godwin (1975). [c] Dimbleby (1984). [d] Ratcliffe (1984). [e] Webb (1986). [f] Rackham (1986). [g] Roberts (1989). [h] Bell and Walker (1992). [i] Lowe and Walker (1997). [j] Bunce *et al.* (1999). [k] Firbank *et al.* (2000). [l] Haines-Young *et al.* (2000)

the woodland was cleared to produce more grain, but grass crops were also important (Duffey *et al.*, 1974). By Anglo-Saxon times (AD 410–1066), place names suggest that meadow (grassland cut for hay and then grazed) occurred in many areas. The best grassland, that yielded a good crop from a small area, was reserved for hay (Rackham, 1986). On farms with little good land, the same fields are likely to have been managed as hay meadows for many years. For example, in Edale, in Derbyshire, the valley floor was reserved for meadow (Plate 3.1), and the valley slopes were used for grazing. On farms with a higher proportion of good land, the location of hay meadows would not have been so restricted, and hay meadows on these farms may not have had a continuous hay-meadow management history (Ausden and Treweek, 1995). Pasture (grazed grassland) was probably too common to be noted in place names (Rackham, 1986). Grassland was also important because manure was needed to maintain the fertility of arable land. Animals spent the day on downland, grass heath or valley pastures, and the night in the fold to manure the arable land (Duffey *et al.*, 1974; Smith, 1980). Rabbits were introduced at the beginning of the twelfth century, soon after the Norman Conquest, and were kept in warrens for their meat and fur. Many stands of lowland acid grassland in Breckland are associated with warrens established in late medieval times, and subsequently the rabbit became an important grazer of other grassland types (Rackham, 1986; Rodwell, 1992; Thompson, 1994).

Pastures and especially meadows were very valuable land. The importance of meadows during this period is evident from the laborious ways in which two types of meadow were managed: hay meads, such as the hay meadows (meadow foxtail–great burnet grassland, MG4) on seasonally flooded land at North Meadow at Cricklade in Wiltshire (Plate 3.2), and water meadows (crested dog's-tail–marsh marigold grassland, MG8) in the chalkland valleys of southern England (Rackham, 1986; Rodwell, 1992; Chapter 2, 'Mesotrophic grasslands'). Traditionally, hay

Plate 3.1: One of the best surviving hay meadows in the Peak District National Park, Edale

Plate 3.2: North Meadow, Cricklade, Wiltshire, National Nature Reserve, SSSI. Predominantly MG4 grassland (*Peter Price*)

meads such as North Meadow were grazed lightly in winter, the livestock were removed and the meadows were cut for hay in July. Common meads were divided into strips, the lord and each commoner having the hay on one or more strips. After the hay was cut, the meadow was usually grazed as common pasture (generally by cattle) for the rest of the season. Turning out of the stock took place around Lammas Day, in August, and these meadows are also known as lammas lands. The fertility of the meadows was maintained by the manure of grazing animals, and by the deposition of silt, organic matter and inorganic nutrients in winter floods. These hay meadows are now rare, and some of the best examples have escaped agricultural improvement thanks, often, to the maintenance of traditional management under common rights (Rackham, 1986; Rodwell, 1992). In the chalkland valleys of southern England, the beneficial effects of flood water were harnessed through active management. In the sixteenth and seventeenth centuries, water meadow management was developed to supplement spring grazing. Floodwaters were regulated through the construction of drains and sluices across streams. The water meadows were generally irrigated in February and grazed by sheep in March and April. Floodwaters were warmer than the air in early spring, and rich in nutrients. As a result, the grass in flood meadows grew up to three weeks earlier in spring than elsewhere, which helped to sustain the sheep at lambing time. This was followed by further irrigation to enhance the production of a hay crop that was harvested in July. The meadows were then set aside for grazing. Overall, the practice gave earlier grass and more hay, which allowed larger flocks to survive the winter. This was important because the larger the flock that could be used for folding on arable land, the greater the crop (Kerridge, 1953). This practice has declined since its peak between 1700 and 1850, as meadows have been neglected or drained and agriculturally improved. Water meadows managed in the traditional fashion are now very rare (Duffey *et al.*, 1974; Rackham, 1986; Rodwell, 1992).

From historical times to the present day, many changes in land management have been driven by economic factors (Figure 3.1). In the case of grassland, the ratio between area of grassland and area of arable land varied both regionally and in time in response to population flux and changes in trade patterns (Duffey *et al*., 1974). For example, there was an extension of grassland in the late fifteenth and early sixteenth centuries as the wool trade became increasingly profitable. However, from about 1750 onwards, the growth of towns encouraged landowners to increase agricultural output to supply the urban population by extending arable areas (Duffey *et al*., 1974; Rackham, 1986; Keymer and Leach, 1990). One method was to subdivide and enclose areas of common and marginal land into private plots. The parliamentary Enclosures of 1780–1820 further reduced and fragmented open common grazing areas, and by 1850 the largest expanses of open calcareous grassland were the chalk downs of southern England (Ratcliffe, 1984).

The use of clover as a fodder crop and as a means of improving the nutrient status of soils became widespread after 1780. The bacteria associated with the root nodules of clover are able to fix atmospheric nitrogen, and planting clover reduced reliance on farmyard manure in sustaining the fertility of the soil. In addition to these changes, farmers used increasing quantities of animal feedstuffs and fertilisers to improve the productivity of areas that remained under grass. Drainage allowed farmers to change the water regime of grassland, which either resulted in a change of character or permitted conversion to arable agriculture (Duffey *et al*., 1974).

1914 to the present day

As a result of the First and Second World Wars, there was an understandable drive for self-sufficiency in food production through increased agricultural intensification and output. The drive for self-sufficiency was encouraged by government support and incentives (Hindmarch and Pienkowski, 2000). For example, in 1917, and again between 1939 and 1945, the government supported the conversion of grassland, including old, established grasslands, to arable to increase food production per unit area of land. Many botanically rich grasslands were destroyed, which threatened the rare and characteristic plant species associated with them. The character of many surviving grasslands changed as their management became more intensive. Yields rose as the application of fertilisers and herbicides increased and drainage was improved. The destruction of old grassland continued, encouraged by the Agricultural Acts of 1947 and 1957, which enabled the government to provide price guarantees for farm products, and had the effect of encouraging farmers to plough up additional areas of permanent grassland for cereal production (Duffey *et al*., 1974; Ratcliffe, 1984). The post-war policy of converting suitable grassland to arable meant that many of the remaining unploughed areas of calcareous grasslands tended to be on slopes too steep to plough. These areas were often too small for sheep grazing to be economic. Overall this had the effect of reducing sheep numbers in the lowlands and removing grazers from many surviving grasslands. The effect was masked in some areas because of the simultaneous rise in the number of rabbits until myxomatosis spread in the 1950s. As a result of abandonment and a decline in rabbit grazing, many calcareous grasslands were invaded by scrub (Duffey, 1974; Ratcliffe, 1984; Keymer and Leach, 1990). Between 1939 and 1984 about 80 per cent of chalk grassland was lost or had changed in species composition (Ratcliffe, 1984). Semi-natural, neutral and lowland acid grassland also declined greatly in area after 1940, although exact figures for lowland acid grassland are not available (Ratcliffe, 1984; UK Biodiversity Group, 1998b). Some neutral grasslands were converted to arable, but the majority have been agriculturally improved. This involves ploughing and reseeding with high-yielding strains of perennial rye grass and the use of chemical fertilisers and herbicides. The development of synthetic herbicides after 1945 led to a dramatic decline in the number of species-rich hay meadows (Ratcliffe, 1984): between 1932 and 1984 unimproved neutral grassland declined by 92 per cent (Fuller, 1987; UK Biodiversity Group, 1998b). Farming became increasingly specialised, which led to large areas of the country being devoted to a single type of farming, and the east (arable)/west (pastural) division in the pattern of land use (Hutchings and Harris, 1996). Because of the separation of crops and livestock, animal waste became a pollutant rather than a fertiliser, and marginal land in arable areas was not grazed but abandoned (Baldock *et al.*, 1996). In England and Wales today, most arable farmland is situated in the lowlands of the east and south, while improved grassland is more widespread in the lowlands of the west and south-west (Haines-Young *et al.*, 2000) (Plate 3).

The trend of increasing production and yield (defined as production per hectare, or per animal) of a variety of products continued into the 1990s and beyond (Hindmarch and Pienowski, 2000). The main agricultural products associated with grassland are beef, lamb and milk. Between 1961 and 2000 the area of permanent pasture in the UK declined (Fig. 3.2). The number of cattle remained relatively stable. In contrast, in the mid-1980s, there was a marked increase in the number of sheep (Fig. 3.3). There was a general trend of increasing beef and lamb production between 1961 and the late 1990s, although the slight increase in lamb production did not mirror the large increase in the number of live animals (Fig. 3.4). About two-thirds of UK sheep and beef cows are on hill farms (DEFRA, 2001a), and the main benefit to farmers of increasing sheep numbers may be higher levels of subsidy, particularly in the uplands (see also Fielding and Haworth, 1999). Milk yields are much higher than beef or lamb yields. Milk production increased from 1961 until the mid-1980s and then levelled off (Fig. 3.5). To put these changes in context, it is clear that the main increases in production were in cereals. In the

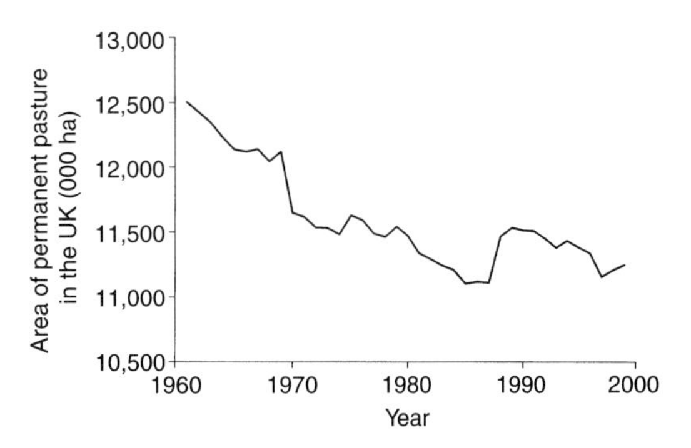

Figure 3.2: Changes in the area of permanent pasture in the UK. *Source:* FAO statistical databases: FAOSTAT agricultural data: land: landuse, FAOSTAT (2001)

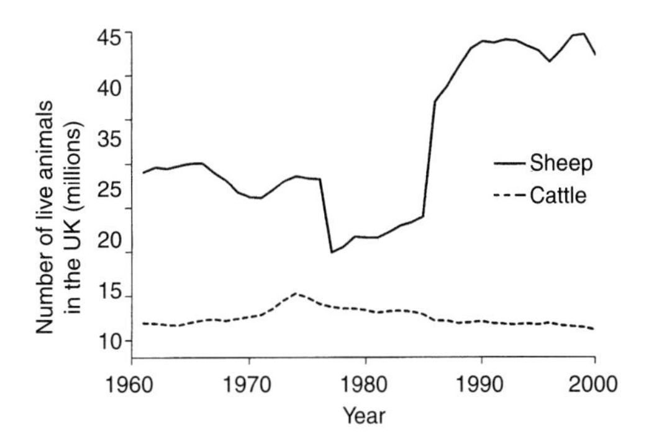

Figure 3.3: Changes in the number of sheep and cattle in the UK. *Source:* FAO statistical databases: FAOSTAT agricultural data: agricultural production: live animals, FAOSTAT (2001)

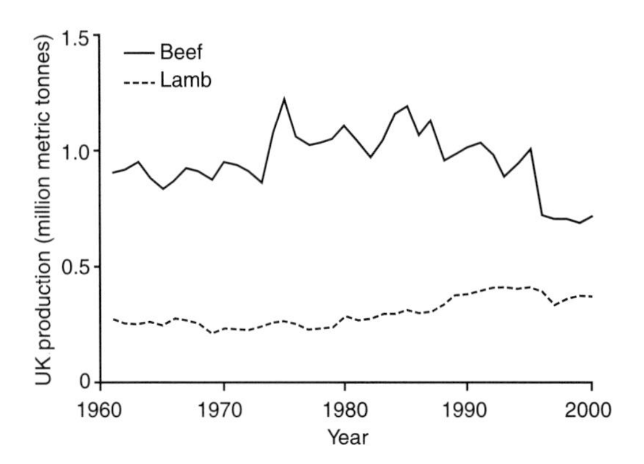

Figure 3.4: Changes in UK production of beef and lamb. *Source:* FAO statistical databases: FAOSTAT agricultural data: agricultural production: livestock primary, FAOSTAT (2001)

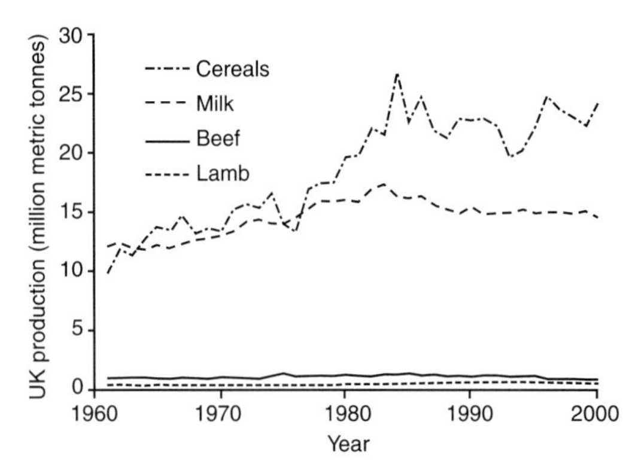

Figure 3.5: Changes in UK production of beef and lamb, compared with production of milk and cereals. *Source:* FAO statistical databases: FAOSTAT agricultural data: agricultural production: livestock primary and crops primary, FAOSTAT (2001)

UK, cereal production almost doubled between 1965 and 1986, from 14 million metric tonnes to 25 million metric tonnes (Fig. 3.5), as yield per unit area increased (Hindmarch and Pienowski, 2000). Despite these increases, the post-war strategy to become self-sufficient was only moderately successful in terms of production, in part because of the increasing population (Beebee, 2001). Between 1989 and 1991 the UK was, on average, 72.8 per cent self-sufficient in food production, and 85.1 per cent successful in indigenous food production. These figures declined somewhat in 1999 and 2000 (Table 3.1). Moreover, the increase in production relied on external inputs of energy and agri-chemicals. For example, UK fertiliser consumption almost doubled, from 1.4 million metric tonnes in 1961 to 2.7 million metric tonnes in 1986, before it declined again to 2.0 million metric tonnes in 1999 (Fig. 3.6). Between 1963 and 1994 the UK increased the proportion of imported fertilisers from 41 per cent to 66 per cent (Hindmarch and Pienkowski, 2000). Between 1990 and 1997 (the period for which figures are available), UK consumption of insecticides remained relatively stable, but consumption of herbicides increased steadily from 18,360 metric tonnes to 24,118 metric tonnes (Fig. 3.7). Although these figures do not distinguish between pastoral and arable agriculture, the overall emphasis on increasing intensity of land use is clear.

One of the most obvious effects of these land use practices has been the continuing loss of habitats and species due to both intensification and abandonment (Hindmarch and Pienkowski, 2000). For example, remaining calcareous grasslands are threatened by the aerial application of fertilisers and herbicides, nutrient enrichment from adjoining arable areas, abandonment and development. In the 1990s the area of semi-natural calcareous grassland declined by 18 per cent and the area of semi-natural acid grassland fell by 10 per cent (Keymer and Leach, 1990; Haines-Young *et al.*, 2000). Recent changes in agricultural practices on neutral (mesotrophic) grasslands include further intensification or abandonment and a shift from hay making to silage production. In the past fifty years or so, improved grasslands increased by approximately 90 per cent in area (Table 3.3). In

Table 3.1: UK self-sufficiency in food as a percentage of all food and indigenous type food

Type of food	*Average 1989–91*	*1999*	*2000*
All	72.8	66.1	66.5
indigenous type	85.1	80.4	79.0

Source: DEFRA (2001b).

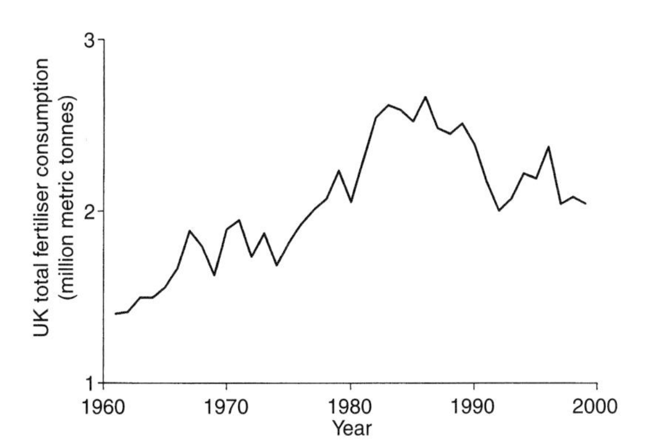

Figure 3.6: Changes in UK fertiliser consumption. *Source:* FAO statistical databases: FAOSTAT agricultural data: agricultural production: means of production: fertilisers, FAOSTAT (2001)

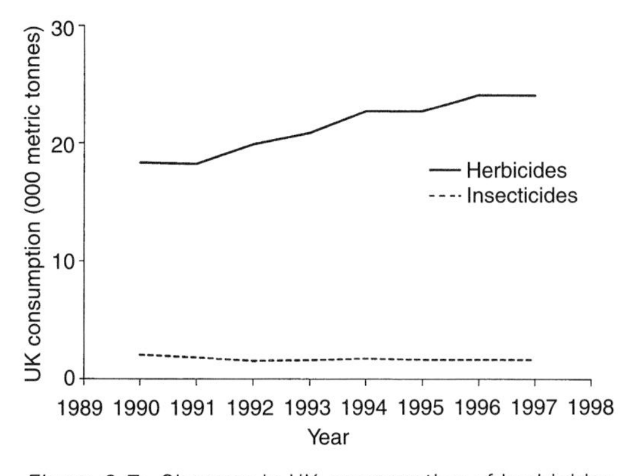

Figure 3.7: Changes in UK consumption of herbicides and insecticides. *Source:* FAO statistical databases: FAOSTAT agricultural data: agricultural production: means of production: pesticide consumption, FAOSTAT (2001)

comparison, the extent of this habitat has recently become relatively stable, although there was some increase in extent overall in the 1990s, mirrored by a decrease in the extent of neutral grassland. There were regional differences, and losses in neutral grassland tended to occur in Scotland and Northern Ireland (32 per cent decrease in Northern Ireland). The Countryside Survey recorded gains in neutral grassland in England and Wales, but the Countryside Survey neutral grassland category is a broad one that includes semi-improved as well as the more valuable unimproved grasslands (Haines-Young *et al.*, 2000; Table 2.3) and the quality of the gained areas in England is unlikely to compensate for the quality of the lost areas. In fact it is estimated that in the 1980s and 1990s the area of unimproved neutral grassland actually declined in some parts of England by between 2 per cent and 10 per cent per year (UK Biodiversity Group, 1998b), and in the 1990s plant diversity fell in some meadows by 8 per cent. Overall, the Countryside Survey 2000 shows that the area and condition of less improved grasslands with high conservation value has continued to diminish in the 1990s. In addition, road verges, which can be important refuges for grassland communities, showed evidence of increasing nutrient levels and losses in plant diversity. For example, plant diversity fell by 9 per cent in some road verges in England and Wales (Haines-Young *et al.*, 2000). Current factors affecting grasslands are described in greater detail in later sections of this chapter.

Heathlands in historical times

Medieval to 1914

Place names indicate that heathland was widespread in England in Anglo-Saxon (AD 410–1066) times (Fig. 3.1), and many heathland turf and peat-cutting rights were established in the Middle Ages (AD 1066–1536) (Rackham, 1986). The typical pattern of land use in most heathland areas of England was that cultivation and settlement occurred on areas of better soil. The heathland itself was not cultivated, except in the Breckland, where a form of shifting cultivation developed. Instead it was used to support farming on the better soils by providing rough grazing, fuel and turf. In some areas, such as Suffolk, sheep were grazed on the heath in the daytime but folded at night to manure arable land. In Dorset, cattle grazed on heath and then were fattened on better land near by (Webb, 1986). These practices prevented scrub invasion, and removed nutrients from the system. Turf and peat cutting for domestic and industrial purposes also helped to keep the nutrient content of the soil low, which maintained conditions that were unfavourable to woodland regeneration (Webb, 1986). Royal forests were established from 1066 onwards, and many of these areas contain significant mosaics of heathland, mire and grassland (Read and Frater, 1999). For example, New Forest plant communities include approximately 11,000 ha of dry, humid and wet heath and mire, and about 5,000 ha of acid, neutral and improved grassland (Tubbs, 2001). The establishment of the New Forest as a royal forest between 1066 and 1086 was crucial to the ecological history of the area because the legal status of the forest limited the expansion of settlement and resulted in pastoral land uses that persist in a modified form today (Tubbs, 2001) (Fig. 3.1). Many heathland areas had their own local breeds of hardy ponies, and the New Forest pony population has been managed since at least medieval times (Tubbs, 1986). Rabbits were introduced soon after the Norman Conquest. Warrens were often situated on heathland, so that rabbits could share the grazing with commoners' sheep and cattle, and this provided an added incentive for the maintenance of heathland (Duffey *et al.*, 1974; Rackham, 1986). Heathland warrens existed all over England and the Scottish lowlands, but the biggest concentration was in the Breckland, where acid grassland and heathland form an important part of the landscape (Rackham, 1986).

Many of the changes in economic trends that influenced grasslands also affected heathlands. For example, from about 1750 agricultural improvements began, which made it possible to cultivate poorer soils. Sheep farming had started to decline and arable areas were extended. This caused a reduction in private heathland and many open commons were reduced and fragmented following enclosure. However, the extreme nutrient deficiency of heathland podsols meant that some areas of lowland heathland were not cultivated, and where laws governing rights of common and hunting in the royal forests persisted, as in the New Forest in Hampshire, the heathland remained. Traditional heathland practices were in decline by the middle of the nineteenth century as products such as gorse, heather and rabbits became less valuable. Grazing and cutting of heaths declined below the level necessary to prevent scrub and tree invasion (Ratcliffe, 1984; Rackham, 1986; Webb, 1986). In the 1790s approximately 20 per cent of Surrey was heathland and 4 per cent was woodland. In the mid-nineteenth century much of the heathland reverted to woodland, and by the 1980s only about 3 per cent remained (Rackham, 1986). Destruction of heaths through urbanisation began in the late nineteenth century (Rackham, 1986).

1914 to the present day

In more recent times, the widespread decline in heathland has continued. As traditional land uses were abandoned, heathlands were reclaimed for agriculture or replaced by forestry (Plate 3.3). Other losses were due to mineral extraction and urban development (Ratcliffe, 1984; Rackham, 1986; Putwain and Rae, 1988; UK Biodiversity Steering Group, 1995). Large-scale afforestation of heath became an official policy when the Forestry Commission was founded in 1919, and large areas of heathland in the Breckland, Suffolk Sandlings and Dorset disappeared (Rackham, 1986). Some heathland areas were designated for Ministry of Defence use, and suffered disturbance and habitat damage (Putwain and Rae, 1988), but may have been protected from urban development. Both the Ministry of Defence and the Forestry Authority are now involved in heathland management or regeneration (UK Biodiversity Steering Group, 1995). By 1984 lowland acid heath had declined in Britain by 78 per cent in area since 1830. Characteristic heathland species declined, particularly the more local plants of wet heath. The distinctive flora of the Breckland heaths, which range from calcareous grassland to acidic heath, was also adversely affected (Ratcliffe, 1984). It is estimated that in the mid-eighteenth century about 36,000 ha of heathland existed in Dorset, but during the following 200 years dramatic losses occurred (Webb and Haskins, 1980; Haskins, 2000). Between 1920–22 and 1987–88 dry heathland on the Lleyn Peninsula in western Wales declined by 51 per cent and wet heath declined by 95 per cent (Blackstock *et al.*, 1995). Surveys of the Dorset heaths indicated that between 1978 and 1987 there were direct losses to heathland caused by conversion to farmland, forestry, urban and industrial development and, more important, succession to scrub and woodland because traditional management had ceased (Webb, 1990). Between 1987 and 1996 the total area of

Plate 3.3: Forestry plantation on former heathland. Note the heather in the foreground

Dorset heathland decreased further by 7 per cent, and the habitat became increasingly fragmented, the number of patches increasing from 142 to 151. During this period planning legislation limited direct losses of heathland to activities such as development and mineral extraction. The principal cause of the loss was once again succession to scrub and woody vegetation, which continued despite the implementation of conservation management, probably because grazing was not particularly effective at halting scrub encroachment (Rose *et al.*, 2000). Fig. 3.8 shows the extent of scrub and tree encroachment on Lindow Common, Cheshire, between 1945 and 1992. After 1992 appropriate heathland management was reinstated, scrub and trees were gradually cleared, and a significant area of heathland was restored. At present, the main threats to heathland are scrub encroachment due to lack of appropriate management and effects of urbanization such as fragmentation and disturbance (UK Biodiversity Steering Group, 1995; Haskins, 2000). In addition there is evidence of increasing nutrient levels (eutrophication) in heathland communities, suggested by an increase during the 1990s in plant species more typical of lowland grassland (Haines-Young *et al.*, 2000). Current factors affecting lowland heaths are described in greater detail in later sections of this chapter.

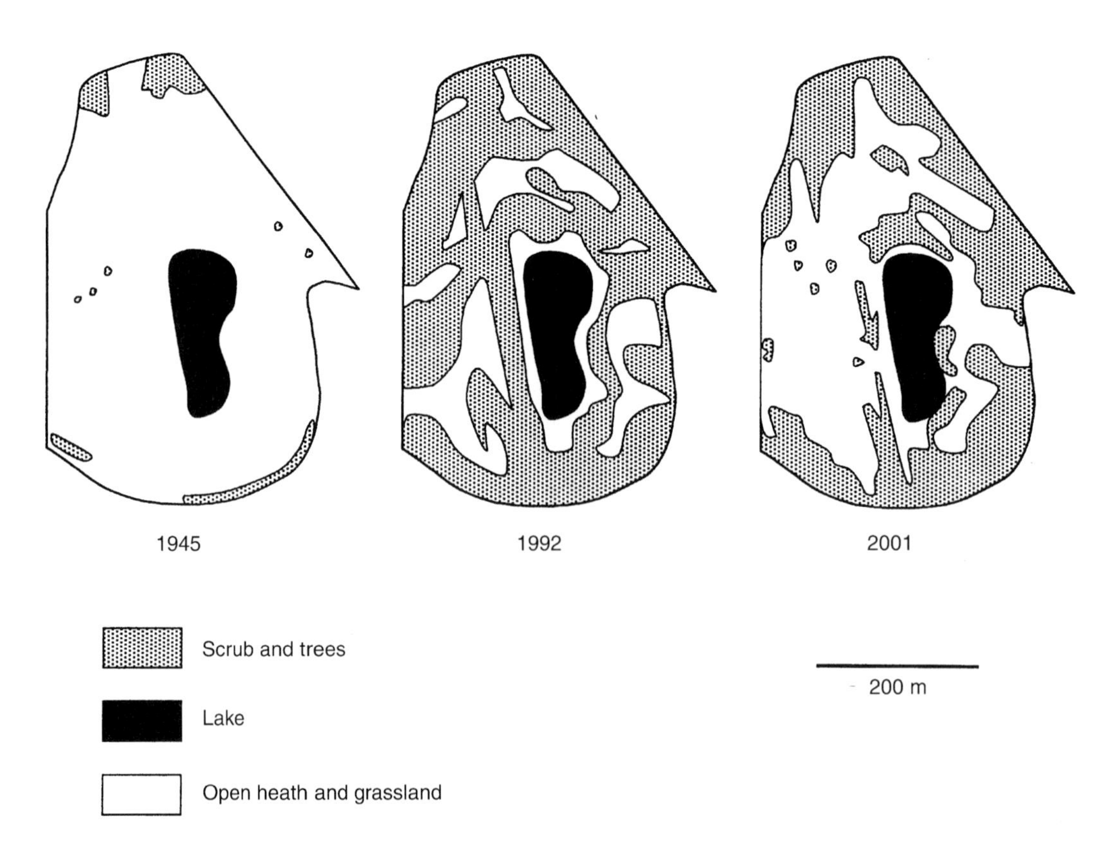

Figure 3.8: Scrub encroachment on heathland at Lindow Common, Cheshire, between 1945 and 1992, and subsequent clearance. *Source:* adapted from Lindow Common interpretation information and Multimap (2001)

Common land

Most common land is privately owned; the term derives from the fact that certain people held right of common over the land. There are 367,000 ha of common land in England (about 4 per cent of the total land area). Nearly 50 per cent of this (180,000 ha) is either wholly or partially designated as a SSSI. Over 48 per cent (178,500 ha) also lies mainly within National Parks. Registers indicated that, in the early 1990s, rights to graze cattle were registered on 20 per cent of commons, sheep on 16 per cent and horses and ponies on 13 per cent (DEFRA, 2002a). Areas of common land are often islands of semi-natural vegetation that support plants and animals that have been eliminated from surrounding areas. A survey in 1994 indicated that approximately half the area of common land in England and Wales is grassland and nearly a quarter is heathland. Inappropriate management, including agricultural improvement, overgrazing or undergrazing, has negatively affected a high proportion of commons. Many have been fragmented by new roads and reduced in size by new developments. Rubbish dumping is also a problem (Aitchinson and Medcalf, 1994). In 1998 the government produced a guide to managing common land (DEFRA, 2002a).

Summary

The development of most biologically diverse grassland and heathland habitats is a consequence of traditional, sustainable land management practices, and this has important implications for the survival of species and for the development of management programmes today (Andrews and Rebane, 1994; Hindmarch and Pienkowski, 2000).

CURRENT PROBLEMS AND OPPORTUNITIES

Problems. Semi-natural grasslands and heathlands have declined in extent in recent years, often as a result of land-use change such as conversion of grazing land to arable agriculture or as a consequence of development. Many surviving grasslands and heathlands have become fragmented and have declined in quality, often owing to inappropriate management, agricultural intensification or abandonment, although factors such as pollution, introduction of alien species, or recreation may also have an impact (Table 3.2). In addition to the effects of land use change that have already been described, predictions suggest that major causes of biodiversity change in northern temperate terrestrial ecosystems to the year 2100 will be climate change and nitrogen deposition (Sala *et al.*, 2000) ('Pollution', this chapter).

Opportunities. One of the outcomes of the Earth Summit in 1992 was that 'Biodiversity: the UK Action Plan' was launched in 1994, and established targets that address the needs of the species and habitat types (priority habitats) that are of most concern to biodiversity conservation (UK Biodiversity Steering Group, 1995; UK Biodiversity Group, 1998a, b; 1999a–d). Priority species in need of Action Plans were selected from a list of approximately 4,500 UK Species of Conservation Concern (Biodiversity Information Group, 2000). The publication of the UK Biodiversity Action Plan facilitated an integrated approach to nature conservation in the UK, and provided a framework and specific objectives for conservation management of lowland grassland and heathland habitats, which are described in Table 3.3. Several semi-natural grassland and heathland habitats, and habitats which have a grassland or heathland

Table 3.2: Factors identified in Habitat Action Plans that have caused, or are causing, loss or decline of priority lowland grassland and heathland habitats or priority habitats that contain a significant grassland or heathland component

Priority habitat	*Habitat fragmentation/loss*	*Inappropriate habitat management*	*Agricultural intensification*	*Built development*	*Introduced/ non-native/ invasive species*	*Recreation*	*Climate change*	*Pollution*
Lowland calcareous grassland	◆		◆	◆	◆	◆	◆	◆
Lowland dry acid grassland	◆	◆	◆	◆			◆	◆
Lowland meadows	◆		◆				◆	◆
Upland hay meadows	◆	◆	◆				◆	◆
Grazing marsh	◆		◆				◆	◆
Machair	◆	◆	◆		◆		◆	
Coastal sand dunes	◆					◆		◆
Maritime cliffs and slopes	◆	◆	◆	◆	◆	◆		
Lowland heathland	◆	◆	◆	◆				◆

Source: UKBAP (2001).

Notes: Lowland meadows refers to unimproved mesotrophic grassland that is grazed or cut for hay (MG4, MG5, MG8). *Upland hay meadows* refers to upland fringe meadows (MG3). *Grazing marsh* refers to coastal and floodplain grazing marsh (MG11, MG13).

component, such as maritime cliffs and slopes, have been identified as priority habitats for conservation in Habitat Action Plans. The key targets for grassland and heathland management are (1) habitat safeguard and maintenance (conservation) ('Habitat maintenance and enhancement', this chapter), (2) restoration and (3) expansion ('Habitat restoration and expansion', this chapter) (Table 3.3). The effectiveness of conservation management is often greatly influenced by international and national conservation designations and environmental schemes ('Opportunities', this chapter), and depends on appropriate monitoring of management practices and outcomes.

PROBLEMS

Habitat loss and fragmentation

Table 3.3 shows that all semi-natural grassland and heathland habitats have declined in area and become increasingly fragmented. Despite recognition of this, the process has continued in recent years. Hay meadows are one of the most threatened habitats, and in 1994 the Peak District National Park Authority set up the Hay Meadows Project to monitor and conserve as many flower-rich hay meadows as possible within the National Park. Even within the Peak District National Park (a protected landscape designation) there

Table 3.3: Summary Habitat Action Plans for lowland grasslands and heathlands

Habitat	*Status*	*Factors affecting habitat*	*Action 1997–1999*	*Action Plan objectives and costing*
Unimproved neutral (mesotrophic) grassland (UNG). Covered by Habitat Action Plans for lowland meadows (including pasture), upland meadows, coastal and floodplain grazing marsh, purple moor grass and rush pastures, e.g. NVC mesotrophic grassland communities MG3, MG4, MG5, MG8, MG11, MG13	Decrease in area and quality: 1934–1984 unimproved neutral grassland declined by 97%. 1980s and 1990s losses of 2–10% per year in some parts of England. Estimated less than 15,000 ha of species-rich neutral grassland survived in the UK in 1995. Increased fragmentation of stands. Data for upland meadows not available. Of about 300,000 ha of grazing marsh in the UK, only 10,000 ha of semi-natural habitat (inundation pasture or meadow, with ditches, e.g. MG11) remains	Losses almost entirely due to changing agricultural practice, e.g. agricultural improvement, conversion to arable, shift from hay-making to silage production, overgrazing, abandonment, reduced flooding of inundation grasslands. Atmospheric pollution	*Legal status* MG4 included in the EC Habitats Directive. The directive does not cover MG5. Several sites proposed as SACs by the UK government. England: 400 SSSIs with MG5 present, sixty-six SSSIs with MG4, seventy-five SSSIs with MG3. Several NNRs with UNG. Wales: 103 SSSIs with UNG present. Scotland: 350 SSSIs with UNG, three with MG3. Northern Ireland: a number of SSSIs and NNRs include MG5. Grazing marsh SSSI/ASSIs designated as SPAs under the EC Birds Directive and as Wetlands of International Importance under the Ramsar convention. *Management*. Management agreements maintain low-intensity farming methods on many UNG SSSIs. UNGs also included in a number of UK agri-environment schemes. Protection of remnant grassland a priority, restoration or creation more limited *Current expenditure*. Unimproved neutral grassland (lowland meadow and pasture, e.g. MG4, MG5, MG8) 1998, £1,802,000 a year; upland hay meadows (MG3) 1998, £108,000 a year; coastal and floodplain grazing marsh 1997: £4,200,000 a year.	Promote conservation and management as part of a European network. Recommend favourable measures for grassland conservation during negotiations in Europe to revise the CAP Lowland meadows and pastures and upland hay meadows: objectives cover habitat conservation, restoration and expansion: stop depletion of unimproved lowland hay meadows (ULHM) throughout the UK; within SSSIs and ASSIs, initiate rehabilitation management for ULHM in unfavourable condition by 2005 and achieve favourable status by 2010; at other localities, achieve favourable condition for 30% by 2005, and for 100% by 2015; re-establish 500 ha of lowland hay meadow and 50 ha of upland hay meadow by 2010. Develop carefully researched guidelines. Coastal and floodplain grazing marsh: maintain and enhance existing habitat extent (300,000 ha) and quality; rehabilitate 10,000 ha of grazing marsh that has become too dry or is intensively managed by 2000; begin creating 2,500 ha of grazing marsh from arable land in targeted areas *Total expenditure to 2004* (lowland meadow and pasture): £2,216,200. *Total expenditure 2004 to 2014* (lowland meadow and pasture) £6,557,700. *Total expenditure to 2004* (upland hay meadow): £326,600. *Total expenditure 2004 to 2014* (upland hay meadow) £793,600. *Expenditure 2000* (grazing marsh) £8,400,000 a year. *Expenditure 2010* (grazing marsh) £13,200,000 per year.

Table 3.3: contd

Habitat	*Status*	*Factors affecting habitat*	*Action 1997–1999*	*Action Plan objectives and costing*
Lowland dry, acid grassland below 300m (LAG). Covered by Habitat Action Plan for lowland dry acid grassland, e.g. NVC communities U1, U2, U3, U4, inland examples of SD10, SD11	Declined considerably in area and quality in the twentieth century, although exact figures are not available. Increased fragmentation of stands	Decline mostly due to agricultural intensification or neglect, although afforestation and development activities have also had an impact. Atmospheric pollution	*Legal status.* England: 271 SSSIs have LAG as principle habitat. Wales: 172 SSSIs with LAG present. Nine English and Welsh NNRs contain LAG. Scotland: forty SSSIs with LAG present. One NNR in Scotland has a considerable area of LAG A number of SPAs designated under the EC Birds Directive contain areas of LAG *Management.* Management agreements maintain low-intensity farming methods on many LAG SSSIs. LAGs also included in a number of UK agri-environment schemes. Incentives to maintain low intensity management and in some cases re-create areas of acid grassland *Current expenditure.* 1998: £190,000 year.	Promote conservation and management as part of a European network. Recommend favourable measures for grassland conservation during negotiations in Europe to revise the CAP. Objectives cover habitat conservation, restoration and expansion: stop depletion of LAG throughout the UK; within SSSIs, initiate rehabilitation management for LAG in unfavourable condition by 2005 and achieve favourable status by 2010; at other localities, achieve favourable condition for 30% by 2005, and for 100% by 2015; re-establish 500 ha of LAG by 2010. Develop carefully researched guidelines *Total expenditure to 2004*: £2,850,000 *Total expenditure 2004–2014*: £11,787,000.
Lowland calcareous grassland (LCG). Covered by Habitat Action Plan for lowland calcareous grassland, e.g. NVC communities CG1 to CG9	Declined sharply in extent and quality, e.g. 50% of chalk grassland in Dorset lost between mid-1950s and early 1990s, although comprehensive figures are not available. Increased fragmentation	Loss associated with agricultural intensification, abandonment, development activities, recreation, atmospheric pollution	*Legal status.* Identified in the EC Habitats Directive as of Community interest and European SPA for Birds include two important calcareous grassland sites. England: 616 SSSIs have LCG as principal habitat; twenty-nine NNRs contain LCG. Wales: thirty-eight SSSIs and two NNRs include LCG. 60–70% of LCG is estimated to be in SSSIs in England and Wales. Several sites are common land. Communities CG1–CG9 have not been recorded in Scotland *Management.* Management agreements maintain low-intensity farming methods on many LCG SSSIs. LCGs also included in a number of UK agri-environment schemes. The Ministry of Defence, the largest landowner of LCG, is developing management plans to include nature conservation *Current expenditure.* 1998: £2,305,900 a year.	Promote conservation and management as part of a European network. Recommend favourable measures for grassland conservation during negotiations in Europe to revise the CAP. Objectives cover habitat conservation, restoration and expansion: stop depletion of LCG throughout the UK; within SSSIs, initiate rehabilitation management for LCG in unfavourable condition by 2005 and achieve favourable status by 2010; at other localities, achieve favourable condition for 30% by 2005, and for 100% by 2015; re-establish 1,000 ha of LCG by 2010. Develop carefully researched guidelines *Total expenditure to 2004*: £6,170,500 *Total expenditure 2004–2014*: £13,956,000.

Table 3.3: Summary Habitat Action Plans for lowland grassland and heathland

Habitat	*Status*	*Factors affecting habitat*	*Action 1997–1999*	*Action Plan objectives and costing*
Maritime and coastal grasslands. Covered by Habitat Action Plans for maritime cliffs and slopes, machair, coastal sand dunes, e.g. NVC communities MC8 to MC11 (maritime cliffs), SD8 (machair), SD6–SD12 (dune grassland)	Comprehensive figures are not available, although it is estimated that machair grassland (SD8 sub-communities) is restricted to about 25,000 ha worldwide: 17,500 ha in Scotland and the remainder in western Ireland	Maritime grasslands often converted to arable, or abandoned. Decline in machair grassland quality associated with changes in agricultural practice, social changes, predation of breeding birds by introduced species, sand and shingle extraction, recreation	*Legal status*. A high proportion of hard cliff (generally vertical or steeply sloping) coast in England has been notified as SSSI. In Wales about half the total maritime cliff resource has SSSI status, but only a small proportion has been notified as ASSI in Northern Ireland. Approximately 80% of Scottish machair is notified as SSSI. Machair is listed in the EC Habitats Directive and some sites are or may become SPA under the EC Birds Directive or SACs. A few SSSIs are designated as Wetlands of International Importance under the Ramsar Convention *Management*. Agri-environment initiatives for grazing management of cliff grassland. Extensive areas of cliff coastline owned by the National Trust, RSPB and the Wildlife Trusts, or designated as Heritage Coasts (England & Wales) or National Scenic Areas (Scotland). Agri-environment schemes and SSSI management agreements for machair. A Corncrake Initiative provides financial incentives to delay mowing on some machair meadows *Current expenditure* (1998). All maritime cliff and slope habitats (including grassland): £416,800. Machair: £127,600	Maintenance of existing cliff habitat (including maritime grasslands); improve the quality of 30%; of maritime cliff and slope habitats by 2010; increase area of cliff-top semi-natural habitats by 500 ha over the next twenty years. Maintain dune grassland on the majority of dune systems. Maintain existing extent of machair. Restore agriculturally improved machair grassland to traditional mixed management. Apply appropriate management to 50% of overgrazed sites by 2005, and to 100% by 2010. Restore machair sites degraded by sand extraction *Total expenditure to 2005*. All maritime cliff and slope habitats (including grassland): £1,650,500. Machair: £1,004,000. *Total expenditure 2005–2014*: All maritime cliff and slope habitats (including grassland): £5,960,000; Machair: £3,548,000. (For expenditure on dune grasslands, *see* Dune heaths, *below*)
Rush pasture. Covered by Habitat Action Plans for purple moor grass and rush pastures, e.g. NVC rush pasture communities M23, M24	Susceptible to agricultural modification, e.g. in Devon and Cornwall (where the habitat is known as culm grassland), 8% of the area present in 1900 remains–62% of sites and 48% of total area were lost between 1984 and 1991	Agricultural improvement: drainage, cultivation and fertiliser application, inappropriate management, abandonment, development and forestry	*Legal status*. Total SSSI/ASSI status in the UK is about 3,800 ha. Listed in EC Habitats Directive *Management*. Managed under Countryside Stewardship agreements, Tir Cymen and ESA agreements *Current expenditure* (1997). Purple moor grass and rush pastures: £180,000 a year approximately	Encourage surveys in Europe to determine the international status of the habitat Maintain and enhance 13,500 ha by 2000; re-create 500 ha by 2005 *Expenditure 2000*: up to £330,000 a year. *Expenditure 2010*: up to £640,000 a year.

Table 3.3: contd

Habitat	*Status*	*Factors affecting habitat*	*Action 1997–1999*	*Action Plan objectives and costing*
Improved grassland (IG). Covered by Habitat Statement for improved grassland, e.g. NVC communities MG6 and MG7	In the past fifty years or so, improved grasslands have increased by approximately 90% in area. In recent years the area has remained relatively stable	Increase of IG at the expense of more diverse habitats is largely due to increased intensification of farming. The change from hay to silage has led to further declines in quality. A high proportion of land reclamation is to IG	*Legal status.* Areas of IG of international importance as feeding areas for wildfowl are protected within SSSIs *Management.* Some management schemes involve compensation to farmers for wildfowl damage, others cover areas of IG where they are of value for amenity *No expenditure figures available*	Enhance areas of IG that are of importance to wildlife and restore semi-natural vegetation on appropriate IG sites. Direct damaging activities such as development away from semi-natural habitats to areas of IG that have no restoration potential. Encourage environmentally sensitive farming methods *No expenditure figures available*
Lowland heathland (LH), below 300m. Covered by Habitat Action Plan for lowland heathland, e.g. NVC communities H1–H6, H8 and H9, M15, M16	A rare and threatened habitat. In England only one-sixth of heathland present in 1800 now remains Increased fragmentation of the habitat.	In the past LH was lost primarily to agriculture, forestry, mineral extraction and development. At present the main factors are scrub encroachment due to lack of appropriate management, nutrient enrichment, the effects of urbanisation such as fragmentation and disturbance, agricultural improvement	*Legal status.* A large proportion of LH has been notified as SSSI through the Wildlife and Countryside Act 1981 *Management.* The Countryside Stewardship scheme includes heathland management and recreation. Management through agri-environment and other schemes in England, Scotland and Northern Ireland. Improved protection and management in Wales in progress. English Nature's National Lowland Heathland Programme supports heathland management projects. The National Trust, MoD, County Wildlife trusts and RSPB are actively involved in LH management, and the Forestry Authority is promoting LH regeneration within woodlands *Current expenditure* (1997). Maintenance and enhancement of 58,000 ha: £300,000–£1,800,000 per year. Re-establishment of 6,000 ha: £200,000–£400,000 per year	Promote conservation and management as part of a European network. Maintain and improve by management all existing LH (58,000 ha in 1995). Encourage re-establishment of a further 6,000 ha of LH by 2005, particularly where this links existing areas of LH *Expenditure.* Maintenance and enhancement of 58,000 ha, 2000: £500,000–£3,600,000 per year; 2010: £1,800,000–£4,700,000 per year. Re-establishment of 6,000 ha, 2000: £300,000–£700,000 per year; 2010: £700,000–£1,200,000 per year.

Table 3.3: contd

Habitat	*Status*	*Factors affecting habitat*	*Action 1997–1999*	*Action Plan objectives and costing*
Maritime and dune heaths. Covered by Habitat Action Plans for maritime cliff and slopes, coastal sand dunes, e.g. NVC communities H7, H11.	Comprehensive figures are not available. The upper sections of hard cliffs on acidic rocks may support maritime heaths. On dunes that have become acidified by leaching, dune heaths develop	Information on maritime heaths is not available. Undergrazing or overgrazing of dune heath habitat can lead to community decline. Fragmentation of dune systems by golf courses. Sea defence and artificial stabilisation may also have an effect	*Legal status.* A high proportion of hard cliff coast in England has been notified as SSSI. In Wales about half the total maritime cliff resource is SSSI status, but only a small proportion has been notified as ASSI in Northern Ireland. A large proportion of the sand dune resource in the UK is designated as SSSI or ASSI in Northern Ireland. Several sand dune sites have been selected as possible SACs under the EC Habitats Directive *Management.* Extensive areas of cliff coastline owned by the National Trust, RSPB and the Wildlife Trusts, or designated as Heritage Coasts (England and Wales) or National Scenic Areas (Scotland). Many sites with fixed dune vegetation are managed by grazing with domestic livestock *Current expenditure* (1999). All maritime cliff and slope habitats (including heathland): £416,800 per year. All coastal sand dune habitats (including heathland): £19,500 per year	Promote conservation and management as part of a European network. Maintenance of existing cliff habitat (including maritime heath); improve the quality of 30% of maritime cliff and slope habitats by 2010; increase area of cliff-top semi-natural habitats by 500 ha over the next 20 years. Protect existing dune systems, restore sand dune habitat, maintain dune heath on the majority of dune systems *Total expenditure to 2005.* All maritime cliff and slope habitats (including heathland): £1,650,500. All coastal sand dune habitats (including heathland): £980,000. *Total expenditure 2005–2014.* All maritime cliff and slope habitats (including heathland): £5,960,000. All coastal sand dune habitats (including heathland): £4,100,600

Sources: UK Biodiversity Steering Group (1995), UK Biodiversity Group (1998b, 1999c).

Notes: SAC Special Area of Conservation. SSSI Site of Special Scientific Interest. *ASSI* Area of Special Scientific Interest. *NNR* National Nature Reserve. *CAP* Common Agricultural Policy. For further information refer to Table 3.7. Definitions of types of mesotrophic grassland are not always used consistently, and for practical purposes, mesotrophic grasslands have been subdivided along the following lines: improved grassland is classed separately from other types in the Countryside Survey 2000 (Haines-Young *et al.*, 2000) and in Biodiversity Action Plans (UK Biodiversity Steering Group, 1995); unimproved and semi-improved mesotrophic grasslands are classed together as neutral grassland in the BAP neutral grassland Habitat Description and the Countryside Survey 2000 (Haines-Young *et al.*, 2000); Habitat Action Plans divide well-drained permanent pastures and meadows (Rodwell, 1992) into upland and lowland types; there is a Habitat Action Plan for coastal and floodplain grazing marsh (most likely, inundation grasslands MG 11 and MG13, Rodwell, 1992) (UK Biodiversity Steering Group, 1995, 1998b). Estimates of expenditure are included for comparison. Current expenditure comes primarily from agri-environment schemes and grant schemes; 75% of additional funding is likely to come from the public sector. Public-sector expenditure may be balanced by reduced agricultural support payments.

is evidence of a major loss (50 per cent) of flower-rich meadows between the mid-1980s and 1995–1996. Most of the meadows of high conservation value (Category A) or of some conservation interest (Category B) contained an element of MG5a grassland. The quality of these meadows varied from excellent, flower-rich sites to those with an element of semi-improved grassland (MG6). Community types MG5b, MG5c, MG4, MG3 and MG8 were poorly represented, perhaps because many meadows have been affected by a degree of agricultural improvement, which tends to mask variations linked with soil or climate. Furthermore, a resurvey in 1998 of unprotected Category A and Category B meadows that were identified in 1995–1996 surveys revealed that a further 25 per cent loss or decline in habitat quality had occurred in just two or three years. Of the thirty-seven meadows that had deteriorated or been lost, seven had been ploughed and reseeded. The others had received increased quantities of inorganic or organic fertilisers, or had received heavy herbicide applications, or had been heavily grazed rather than cropped for hay. By 1997, 151 flower-rich meadows were entered into conservation agreements, but a further 259 Category A and B meadows remained unprotected. By 1999 an additional twenty-five meadows had been secured into conservation agreements, with another thirty-six under negotiation, but there appear to be significant disincentives that prevent farmers from taking up agri-environment schemes (Buckingham *et al.*, 1999).

In Dorset, between 1987 and 1996, the number of heaths continued to rise, while their size decreased, i.e. the habitat became increasingly fragmented. Eleven of sixty-two sites fell below the threshold of the 10 ha minimum size for qualification as a SSSI on the basis of habitat area, although the sites are still notified because of the presence of nationally rare or scarce species (Rose *et al.*, 2000). An important question is whether the remaining fragments are sufficient to support characteristic species in the long term. Habitat fragments may be seen as islands of suitable habitat surrounded by inhospitable areas, and it has been generally observed that there tend to be more characteristic species on larger habitat islands than smaller ones (Spellerberg, 1991; Haskins, 2000). This may be because smaller fragments have a smaller range of habitat conditions, or because larger patches should support larger populations, that are less likely to become locally extinct (see below) (Bullock and Webb, 1995).

Species such as the wartbiter cricket that requires a range of grassland conditions, including short turf, grass tussocks and a sward rich in flowering herbs, or those that are less mobile, such as the silver-studded blue butterfly, a key heathland species, are likely to be particularly vulnerable to fragmentation (UK Biodiversity Group, 1999b; Haskins 2000). A study of the distribution of the Dartford warbler revealed that territories were positively associated with larger areas of dry/humid heath and mature gorse, but negatively correlated with fragmentation (isolated and smaller heaths). Smaller or more isolated patches of heath may be less favoured for territories because they are more difficult to find, or there may be greater disturbance, or heath edges may be avoided, or food may be scarce (van den Berg *et al.*, 2001). Fragmentation and changes in land use may also disrupt important processes. For example, a 1996 survey of Dorset heaths revealed drastic changes in the composition of heathland vegetation, in particular a reduction in the area of wet heathland vegetation types, possibly due to changes in hydrological conditions. Fragmentation and land use changes such as the growth of coniferous plantations adjacent to heathland areas, or the encroachment of woodland on to heathland sites, may reduce the groundwater levels that are essential to the maintenance of wet communities (Rose *et al.*, 2000). Small habitat patches also have a greater proportion of edge than do larger habitats; a small fragment may in effect be all habitat edge (Newman, 2000). Edges may be influenced by the proximity of inhospitable habitats. For example, edge effects include herbicide drift or fertiliser run-off from adjacent agricultural land on to semi-improved grassland of roadside verges (Haines-Young *et al.*, 2000), and disturbance of breeding heathland birds such as the Dartford warbler (Species box 3.1), nightjar and woodlark, and predation of sand lizards by domestic cats from adjacent housing developments (Haskins, 2000). In addition, the majority of grasslands and heathlands depend on management for their maintenance, and options such as low-intensity grazing may be impractical on small sites.

The area of a habitat fragment is also significant because it must be large enough to support a minimum viable population of a species, i.e. the smallest population size that can be predicted to persist in the future. If the

Species Box 3.1: Dartford warbler *Sylvia undata*

In the UK the Dartford warbler is restricted to the southern counties of England, where its habitat is typically dry lowland heath. It is distinguished from other warblers by its long tail, which is often cocked when perching. Its back is dark grey or grey-brown and its underparts are dull wine-red. The species is at the extreme north of its range in England, and feeds exclusively on invertebrates, and for these reasons UK populations are particularly vulnerable to extended periods of cold weather. However, populations can recover rapidly following harsh winters. For example, national surveys estimated that there were 457 Dartford warbler territories in 1961. These were reduced to eleven territories in 1963 following two cold winters, but the population had recovered to occupy 562 territories by 1974. The number of territories in 1994 was estimated to be 1,595. Throughout its range, including in the UK, the main threat to the Dartford warbler is the loss and fragmentation of habitat. In Dorset, Dartford warbler territories are usually located in larger, less isolated areas of dry or humid heath, with mature common gorse scrub. Birds may also nest in areas of pine scrub, but the presence of common gorse is more important. This preference for gorse is probably due to the supply of invertebrate food, which is greater in gorse scrub than in dwarf shrub vegetation.

Small bare patches of ground surrounded by heathland vegetation are also important for foraging for invertebrates. Breeding territories are rarely located in fragmented areas of heathland in close proximity to woodland, urban areas or intensive agriculture. In England conservation of the species is centred on appropriate management of its lowland heath habitat.

Sources: Marchant *et al.* (1990); Cramp (1992); Svensson and Grant (1999); van den Berg *et al.* (2001)

population size falls below that number, it is at risk of extinction. As a population becomes smaller it tends to lose genetic variability through genetic drift (random changes in allele frequencies). For example, non-isolated or larger populations of sand lizards and adders in Sweden had higher genetic diversity than isolated or small populations (Madsen *et al.*, 2000). Loss of genetic variability can lead to inbreeding depression (a loss of vigour among offspring when closely related individuals mate) and a lack of evolutionary flexibility. Slight variations in reproduction and mortality rates can cause small populations to fluctuate in size and may lead to extinction. Random environmental variation or catastrophes such as fires can also kill entire small populations. The migration of individuals between populations tends to increase the amount of genetic variability in a population, and balance the effects of genetic drift (Primack, 1998; Newman, 2000). For example, a small, isolated population of adders from meadows in southern Sweden had lower genetic variation, smaller brood sizes and a higher proportion of inviable offspring than non-isolated populations. The incidence of inviable offspring was significantly reduced when males from other areas were introduced to the isolated population (Madsen *et al.*, 1996). A number of grassland and heathland butterfly species, such as the silver-studded blue and the marsh fritillary exist as metapopulations, with small populations in suitable patches of habitat linked by the occasional dispersal of adults between them. This means that the long-term survival of these species depends on the protection of several patches of habitat, because the persistence of a network of colonies may be crucial to the long-term maintenance of genetic diversity (Warren, 1994; UK Biodiversity Group, 1999b).

Landscape isolation is a term that has been used to represent both the area of a habitat patch (e.g. heathland) and its isolation (Bullock and Webb, 1995). A study of the Dorset heaths revealed that generally larger and/or less isolated heath fragments hold more heathland animal species, and small (4 ha) patches of heathland are more likely to be occupied by specialist heathland species if they are near other patches of heathland (within about 2.5–3.0 km) (Bullock and Webb, 1995). Habitat corridors, links or stepping stones provide routes for organisms to disperse between habitat fragments (Spellerberg, 1991).

Land use change

Land use change is a significant cause of grassland and heathland loss that has already been described in some detail ('The history of grasslands and heathlands', this chapter). Important aspects of land use change include agricultural improvement, inappropriate management and urbanisation.

Agricultural improvement

Processes of agricultural intensification include drainage, ploughing, reseeding, fertiliser and herbicide application. The biodiversity of Britain's lowland grasslands has declined drastically, mainly owing to processes of agricultural intensification (Crofts and Jefferson, 1999), and until the mid-1980s one of the principal causes of heathland decline was agricultural improvement, including conversion to agricultural land (UK Biodiversity Steering Group, 1995; Little, 1998) (Tables 3.2 and 3.3).

Inappropriate management

The opposite side of the coin to agricultural improvement is abandonment, and, like agricultural intensification, inappropriate management is a common cause of species decline (Chapter 2, 'Fauna of lowland grasslands and heathlands'). Areas of grassland and heathland that have not been agriculturally improved are often uneconomic to manage and become neglected. For instance, acid and calcareous grasslands tend to be less productive than mesotrophic grasslands; they are less easily agriculturally improved and are more likely to be abandoned (Crofts and Jefferson, 1999). Abandonment of grassland leads to the encroachment of coarse grasses and scrub, and a decline in species diversity (Rodwell, 1992). Currently the principal cause of heathland loss is succession to scrub and woody vegetation because traditional management practices, principally burning and grazing, have ceased (UK Biodiversity Steering Group, 1995; Rose *et al.*, 2000).

Urbanisation

Much development has been on land of high conservation value, such as semi-natural grassland and heathland, because it was not of high agricultural quality (Haskins, 2000; Beebee, 2001). For example, 50 per cent of the destruction of Dorset heathlands between 1978 and 1987 was due to development (Webb, 1990). Even where the ecological value of wildlife habitats is recognised, loss of some semi-natural habitat to development can still occur, and development close to sensitive habitats can have adverse effects, including disturbance, damage and pollution (Haskins, 2000; McClure, 2001). For instance, roads can have a significant impact on the species composition and performance of adjacent heathland vegetation. Changes in plant species composition near the road are probably caused primarily by oxides of nitrogen. The extent of the effect is related to the volume of traffic carried by the road, and near major roads, changes in the vegetation can be detected up to 200 m from the carriageway (Angold, 1997).

Pollution

Pollution may be defined as the release into the environment of harmful amounts of a substance (pollutant) as a result of human activity. The range of pollutants discharged into the environment is huge, and pollutant impacts vary in their spatial (local, regional or global) and temporal scales (Ashmore, 1997). This section looks at a number of examples. Effects of pollutants on grassland and heathland habitats include eutrophication (nutrient enrichment, for example nitrogen deposition), climate change (caused, for example, by elevated levels of atmospheric carbon dioxide) and the impact of metal pollution. The effects of climate change and nitrogen deposition are likely to become increasingly important in causing biodiversity change (Sala *et al.*, 2000).

Eutrophication

Semi-natural grasslands and heathlands are associated with low or intermediate levels of nutrient availability, particularly nitrogen, and plant species from these habitats can compete successfully only in soils with low nitrogen levels (Chapter 2, 'Geology and soils') (Bobbink and Roelofs, 1995). One of the major threats to the structure and functioning of semi-natural ecosystems is the rise in atmospheric nitrogen pollution (NH_y and NO_x) from increases in fossil fuel consumption (e.g. vehicle and power station sources) and changes in agricultural practices. In some areas atmospheric nitrogen deposition has more than doubled during the past three decades (INDITE, 1994; Pitcairn *et al.*, 1995; Power *et al.*, 1998), and there is evidence that nitrogen enrichment has led to a shift from nitrogen limitation to phosphorus limitation of plant productivity in Dutch wet heathlands (Aerts and Berendse, 1988). Atmospheric nitrogen deposition may acidify and add nutrients to ecosystems. Effects include short-term effects of nitrogen deposition on individual species, acidification and nitrogen enrichment of soils, increased susceptibility to natural stress conditions such as drought, and changes in competitive relationships between species, resulting in loss of diversity (INDITE, 1994; Bobbink and Roelofs, 1995; Lee and Caporn, 1998).

An important approach to controlling pollutant emissions is the critical loads approach. The critical load is the level of deposition of a pollutant below which it is estimated that significant harmful effects to the environment do not occur (INDITE, 1994). Critical nitrogen loads have been suggested for different ecosystems on the basis of observed changes in structure and function. For example, a transition from heather to grass can occur when levels of nitrogen deposition on lowland dry heathland exceed the critical load of 15–20 kg N $ha^{-1}yr^{-1}$ (Table 3.4) (Bobbink and Roelofs, 1995). Current levels of nitrogen deposition in many areas of Britain exceed the estimated critical load (Brown and Farmer, 1996) and there is already evidence of increasing nutrient levels and associated changes in species composition in lowland grasslands and heathlands throughout the UK (Haines-Young *et al.*, 2000). The critical load concept is a useful decision-making tool, but it may be more realistic to predict the ecological consequences of different pollutant scenarios (Shaw, 2000). In addition, critical load values are affected by management practices.

High rates of nitrogen deposition in the Netherlands have contributed to the rapid replacement of heather-dominated heathlands by grassy vegetation dominated by wavy-hair grass (on dry heaths) and purple moor grass (on wet heaths) (de Kroon and Bobbink, 1997; Bobbink *et al.*, 1998; Bakker and Berendse, 1999). Patterns in UK heaths are less consistent, but wavy-hair grass has been reported to expand rapidly within gaps created by scrub or bracken management or die-back of unmanaged heather in the Breckland heaths of East Anglia, where levels of nitrogen deposition are particularly high (30–40 kg N ha^{-1} yr^{-1}; Marrs, 1993; Pitcairn *et al.*, 1995). Lowland acid grasslands are closely related to heathlands, and a number of species have declined or become locally extinct in the Netherlands. The relative importance of eutrophication or acidification in causing these changes depends on soil characteristics (Bobbink *et al.*, 1998).

Table 3.4: Suggested nitrogen critical loads for lowland semi-natural grassland and heathland ecosystems (kg N ha^{-1} yr^{-1})

Ecosystem	*Critical load*	*Indicators of change*
Species-rich heaths/acid grassland	7–15	Decline in sensitive species
Calcareous species-rich grassland	14–25	Increase in tall grasses, decline in diversity
Lowland dry heathland	15–20	Transition from heather to grass
Lowland wet heathland	17–22	Transition from heather to grass
Neutral-acid species-rich grassland	20–30	Increase in tall grasses, decline in diversity

Source: Bobbink and Roelofs (1995)

In many wet heathlands in the Netherlands, purple moor grass has replaced cross-leaved heath as the most abundant species, and a number of rare plant species such as marsh gentian and oblong-leaved sundew have declined. These changes are probably a result of nitrogen eutrophication caused by litter accumulation following cessation of management, in combination with increased atmospheric deposition of nitrogen (Aerts and Berendse, 1988). On an annual basis, purple moor grass produces more litter and loses more nitrogen than cross-leaved heath. Under low nitrogen conditions these losses are not easily recovered by nutrient uptake, and purple moor grass remains subordinate to cross-leaved heath. However, at elevated nitrogen levels, purple moor grass has a relatively high growth rate and its above- and below-ground biomass production greatly exceeds that of cross-leaved heath. Under these conditions, purple moor grass is able to become dominant (de Kroon and Bobbink, 1997).

To study the long-term effects of enhanced nitrogen deposition on lowland heathland in the UK, experimental additions of nitrogen in the form of ammonium sulphate were applied to a nitrogen-poor dry heathland in southern England between 1989 and 1996. Background deposition at the site was estimated to be between 13 kg N ha^{-1} yr^{-1} and 18 kg N ha^{-1} yr^{-1}, a value similar to the critical load that has been suggested for the conversion of lowland heath to grassland (Table 3.4). Experimental additions of 7.7 kg N ha^{-1} yr^{-1} or 15.4 kg N ha^{-1} yr^{-1} took total deposition to above the estimated critical load. Nitrogen addition stimulated shoot growth, flowering, canopy density, biomass production and litter production. Plant, litter and soil nitrogen concentrations were also significantly increased, but the majority of added experimental nitrogen (over 75 per cent) was accumulated within the soil. Nitrogen addition also increased the rate of nutrient cycling (Power *et al.*, 1995, 1998). Although grasses are better able to respond to increased nitrogen than heather and cross-leaved heath, invasion of a dense, mature heather canopy by grasses is unlikely, even where there are high levels of nitrogen input (Ashmore, 1997). For example, grass invasion was not observed in the nitrogen addition experiment reported by Power *et al.* (1998). However, increased nitrogen deposition may also act by increasing the sensitivity of heather to factors such as drought, cold stress and attack by the heather beetle, which will lead to canopy breakdown (Ashmore, 1997). In addition, a decline in management practices that rejuvenate heather plants (delay canopy breakdown, Chapter 2, 'Cyclical change'), and maintain the low nutrient status of heathlands, may alter the balance of competition (Marrs, 1993). Increased availability of nitrogen, in combination with canopy breakdown or practices that create gaps, may give grasses the competitive edge over heather. Management practices such as cutting and burning remove nutrients from the system through the removal of above-ground plant material and a proportion of the litter, but the soil is not usually disturbed. Turf removal may be the most effective management tool to remove additional nitrogen from the system (Power *et al.*, 1995, 1998). An example of the effects of elevated nitrogen and drought on a UK heathland is given in Chapter 4.

In addition to the influence of management, critical load values are greatly affected by differences in basic nutrient limitation (i.e. whether ecosystems are nitrogen-limited or phosphorus-limited). For example, tor grass is a component of chalk grassland vegetation such as CG2, but is not abundant. In the absence of grazing, tor grass becomes dominant and the vegetation changes (e.g. to CG4) (Rodwell, 1992). In recent years, the abundance of tor grass on chalk grasslands has increased steadily, even under different management regimes, including grazing, mowing and burning. Increasing dominance of tor grass leads to a decrease in species richness of up to 50 per cent within a few years. Chalk grasslands are nitrogen-limited ecosystems that are sensitive to eutrophication, and the increase in tor grass is attributed to the effects of atmospheric nitrogen deposition (Bobbink and Willems, 1987; Bakker and Berendse, 1999). In contrast, the vascular plant species composition of U4 acid grassland and CG2 calcareous grassland in the upland fringes was slow to change in response to long-term experimental nitrogen additions (applications of 3.5–14 g N m^{-2} yr^{-1} of ammonium nitrate or 14 g N m^{-2} yr^{-1} of ammonium sulphate). No significant changes were seen in the first four years of application, but results from the last two years of nitrogen addition (1995–1996) showed dose-related losses in overall plant cover on both the acidic and calcareous plots. Individual species cover values were also affected by the treatments, with significant reductions in wild thyme and glaucous sedge on the calcareous plots, and trends towards reduced cover of tormentil and fescues on the acid plots. The effects of ammonium sulphate were greater than those of ammonium nitrate. Phosphorus limitation is probably the main factor responsible for the limited effects of nitrogen addition on species composition in this study, but grazing is also likely to play an important part in preventing a shift to more competitive grasses. The most sensitive indicators of change were bryophytes, lichens and soil processes. For example, nitrogen treatments led to significant, dose-related losses in bryophyte cover on acid grassland plots in the first year and this was maintained over a further five years of treatment (Carroll *et al.*, 1997, 2000).

Remnants of semi-natural grassland may survive on roadside verges within an agriculturally improved landscape (Plate 3.4). The majority are examples of the MG1 false oat-grass community, in which coarse-leaved grasses are generally abundant, although a

Plate 3.4: Roadside verges (*foreground*) can be a valuable refuge in an agriculturally improved landscape for semi-natural grassland communities such as the meadow cranesbill variant of false oat-grass grassland (MG1)

Plate 3.5: Meadow cranesbill

variety of species may become locally dominant, for example meadow cranesbill (Plate 3.5) in the meadow cranesbill variant of the MG1a sub-community (Rodwell, 1992). Road verges are also an important refuge for MG5 grassland, which is a valued community, typical of traditionally managed lowland unimproved hay meadows (Table 2.4). Since 1990 road verges have shown evidence of increasing nutrient levels and losses in plant diversity. Plant diversity fell by 9 per cent in some road verges in England and Wales (Haines-Young *et al.*, 2000). Eutrophication may be caused by leaching of fertilisers from adjacent fields and nitrogen deposition from exhaust emissions (Crofts and Jefferson, 1999).

Climate change

There are gases in the atmosphere that let short-wave radiation from the sun pass through, but absorb and reradiate outgoing long-wave radiation reflected from the surface of the Earth. These gases are known as greenhouse gases, and without them the temperature close to the ground would be much cooler. Greenhouse gases include water vapour, carbon dioxide, methane, nitrous oxide and ozone. The concentrations of greenhouse gases such as carbon dioxide, methane and nitrous oxide are increasing as a result of human activity, and consequently climate is changing globally. In the UK, the 1990s were about 0.5°C warmer than the 1961–1990 average (Graves and Reavey, 1996; Hulme and Jenkins, 1998; Newman, 2000). By the 2080s annual temperatures and precipitation over the UK are predicted to change further, based on scenarios for different rates of global warming (Table 3.5). Climatic fluctuations are not unusual, but the effect of current changes in climate are difficult to predict, because the rate of change is unusually rapid, and because climate change may act in combination with other factors such as changes in land use and atmospheric inputs of nitrogen (Graves and Reavy, 1996; Wolters *et al.*, 2000, Burton, 2001).

The distribution patterns of many species and communities are determined to a large extent by climatic effects, and changing patterns of climate, such as those outlined above,

Table 3.5: Possible climate future for the UK by the 2080s

Annual mean temperature	*Mean annual precipitation*
Predicted to become between 1.5°C and 3.2°C warmer than the 1961–1990 average over south-east England, and between 1.2°C and 2.6°C warmer over Scotland	Predicted to increase by 0–10% over England and Wales, and 5–20% over Scotland
Warming predicted to be more rapid in winter than in summer	Winter and autumn predicted to become wetter, spring and summer to become 10–20% dryer in the south-east and wetter in the north-west

Source: Hulme and Jenkins (1998).

are likely to change the natural distribution limits of species. For example, species may move towards higher latitudes following increases in average temperatures as a result of extinctions at the southern boundary of the range and colonisation at the northern boundary. Evidence of poleward shifts in species' ranges comes from a study of thirty-five non-migratory European butterflies, which revealed that 63 per cent have a range that has shifted or extended to the north by 35–240 km during the twentieth century, while only 3 per cent shifted to the south. Northward range extensions are likely to be the result of sequential establishment of new populations. Butterfly species found on grassland or heathland that have extended northwards include Duke of Burgundy, grayling, wall brown, meadow brown, marbled white, dingy skipper, large skipper, Essex skipper, silver-spotted skipper and brown argus (Parmesan *et al.*, 1999; Thomas *et al.*, 2001). The silver-spotted skipper has responded to climate warming by expanding into habitats in 2000 that were too cool in 1982. In 1982 the butterfly was largely restricted to south and south-west-facing chalk grassland fragments, but by 2000 it had colonised nearby east-, west- and north-facing slopes (Thomas *et al.*, 2001).

A recent study has predicted the direct impacts of climate change on certain critical species and habitats in the UK (Harrison *et al.*, 2001). A number of species were chosen from selected habitats of conservation interest. Changes in the likely availability of 'suitable climate space' for these species in relation to climate change scenarios were predicted. Species showed varying responses, making it difficult to assess the impact of climate on the habitats they represent. However, general conclusions of the study in relation to the grassland and heathland habitats that were selected are shown in Table 3.6.

In addition to the likely effects on individual species, climate change effects on the hydrology of moisture sensitive habitats were predicted. For example, water availability is likely to increase in winter throughout Britain and Ireland, which could particularly benefit wet heaths. However, wet heaths, drought-prone acid grassland and some chalk grassland could be adversely affected by lower water availability in south-east England in summer (Harrison *et al.*, 2001).

Climate is not the only factor that determines whether a location is suitable for a species, and other abiotic factors such as soil characteristics are likely to be important. Species distributions also reflect interactions with other species and patterns of dispersal and colonisation, and these will influence changes in range and abundance (Davis *et al.*, 1998).

Table 3.6: Changes in the availability of 'suitable climate space' of a number of species from selected habitats of conservation interest

Habitat	*Species responses*
Wet heath	Cross-leaved heath (*Erica tetralix*) appears to be unaffected and marsh gentian (*Gentiana pneumonanthe*) remains stable in Great Britain, but may expand into Ireland. This habitat could be robust in terms of direct climate change impacts. (However, see impacts of water availability in the main text)
Drought-prone acid grassland	Common storksbill (*Erodium cicutarium*) and Spanish catchfly (*Silene otites*) have the potential to expand their suitable climate space, but it is unclear whether these species are representative of the habitat as a whole
Lowland calcareous grassland	Dwarf thistle (*Cirsium acaule*), yellow wort (*Blackstonia perfoliata*) and common rock rose (*Helianthemum nummularium*) show a small response to climate change, with some northward expansion on their climate space
Upland hay meadows	The two species with a northern distribution that were modelled, globe flower (*Trollius europaeus*) and wood cranesbill (*Geranium sylvaticum*) lose climate space, while great burnet (*Sanguisorba officinalis*) gains it. This habitat may therefore change in species composition

Source: Harrison *et al.* (2001).

For example, direct and indirect effects of simulated climate change on St John's wort, which is at the northern edge of its range in Britain, suggest that growth and survival may actually be poorer under the projected climate conditions, owing to changes in the balance of ecological interactions such as competition and herbivory (Fox *et al.*, 1999). Crucially, habitat fragmentation can affect colonisation, and barriers such as urban and intensive agricultural land use will limit the movement of species. For instance, in response to predicted climate change scenarios, 'suitable climate space' for Spanish catchfly (Table 3.6) could be found along the Thames valley into London, but the species is unlikely to occur in such an urban situation (Harrison *et al.*, 2001). In addition, many protected areas are surrounded by inhospitable landscapes, which will limit migration of species beyond their boundaries. Availability of habitat can constrain range expansion of butterfly species, although some have successfully extended northwards across heavily cultivated landscapes (Hill *et al.*, 1999; Parmesan *et al.*, 1999). Individuals and populations that expand rapidly are most likely to be favoured. Two species of bush cricket have two forms; the long-winged cone-head has a long-winged and an extra long-winged form, while Roesel's bush cricket has a short-winged form that cannot fly and a long-winged flying form. Both species have spread northwards, and the recently established populations (established in the past twenty years) have higher frequencies of longer-winged (dispersive) individuals than do the well-established populations (Thomas *et al.*, 2001). It is not known how many species will be able to extend their range northwards across the highly fragmented landscape of northern Europe, but many sedentary and specialised species may fail to do so (Parmesan *et al.*, 1999). In the face of a changing climate, it will be necessary not only to conserve habitat patches of high value, but to link fragmented patches and create stepping stones for species movements.

The responses of species and communities are likely to be complex, and different communities may respond in different ways and at different rates. An important feature of communities is that they change with time, but anthropogenic factors, such as climate change, influence the magnitude and rate of environmental change. Some communities may be more resistant to change than others, and some may be more resilient (show a more rapid recovery) (Grime *et al.*, 2000). An example of the response of two contrasting grasslands to simulated climate change is given in Chapter 4.

Metal pollution

Metal mining spoil heaps are often polluted with concentrations of heavy metals that are toxic to most plants (Plate 3.6). However, characteristic grassland communities do occur on metal-polluted substrates (e.g. OV37) (Rodwell, 2000), often because the species are able to resist metal toxicity. Resistance to heavy metals can be achieved by mechanisms such as avoidance (e.g. exclusion) or evolution of metal tolerance (Bradshaw and Chadwick, 1980; Baker, 1987; Fitter and Hay, 1987; Walker *et al.*, 1996). Sources of metal contamination also include emissions from metal refineries or smelting works (Hunter *et al.*, 1987a, b, c; Spurgeon and Hopkin, 1999) and agricultural inputs to soils. Estimates of inputs of total zinc to UK agricultural land, such as pasture, suggest that atmospheric deposition is responsible for about 50 per cent, farm manures 35 per cent, sewage sludge 8 per cent, and other sources, such as agrochemicals, the remainder. The equivalent figures for copper are about 25 per cent, 50 per cent and 15 per cent respectively (Little, 1998). For example, the sludge left behind after sewage treatment is commonly used as a low-cost nitrogen fertiliser, particularly on pasture. A ceiling has been set on the rate of metal addition to soils, and standards for the content of zinc and cadmium have been

Plate 3.6: Spoil from Pandora mine, North Wales, contaminated with lead and zinc, and sparsely colonised by grasses

agreed (Baldock *et al.*, 1996). This is because, if sludge is applied over many years, heavy metals can reach soil concentrations that are toxic to plants, or concentrations in plants that are toxic to animals (Coghlan, 1997; Newman, 2000).

The fate of pollutants in ecosystems and their effects on organisms is a huge topic, covered in detail elsewhere, e.g. Hopkin (1989), Walker *et al.* (1996), but an example of a metal-polluted grassland ecosystem illustrates a number of important points. First, the species richness at metal-polluted sites tends to be low, owing to toxic metal concentrations; secondly, decomposer activity may be reduced, leading to low rates of nutrient cycling and a build-up of plant litter; and thirdly the fate of metals may differ, even at the same site, as a result of differences in uptake and regulation by plants and animals (Hopkin, 1989). Hunter *et al.* (1987a, b, c) made a detailed study of the amounts of copper and cadmium in soil, plants and animals from metal-contaminated grassland close to a metal refinery in Merseyside. The vegetation close to the refinery was of low species diversity and was dominated by metal-tolerant populations of creeping bent and red fescue. Highly contaminated soils were also associated with reduced earthworm and woodlice population densities and a build-up of plant litter. The concentration of copper in surface soil at the refinery site was over 700 times greater than that of cadmium, yet plant uptake of cadmium from soil was forty times greater than that of copper (Fig. 3.9). The transfer of cadmium through the invertebrate food web from living or dead vegetation, through a number of intermediate invertebrates to spiders, exceeded that of copper by between three and seven times (Hunter *et al.*, 1987a, b). Small mammals at the site, such as the carnivorous common

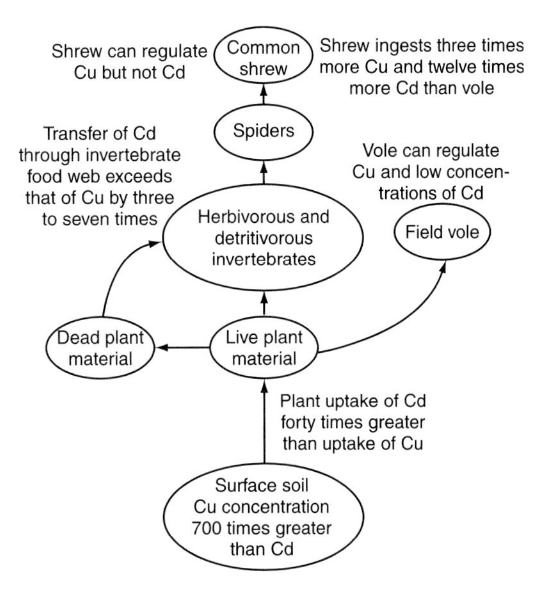

Figure 3.9: The bio-accumulation of copper and cadmium in a polluted grassland ecosystem. *Source:* adapted from Hunter *et al.* (1987a, b, c).

shrew and the herbivorous field vole, were able to control their body copper concentration, even in polluted areas, but were less able to control their body cadmium concentration (Hunter *et al.*, 1987c) (Species boxes 3.2 and 3.3). Therefore, despite high concentrations of copper in the soil, bioaccumulation of cadmium was greater because cadmium was more readily transferred to organisms and poorly regulated by them. Soil macroinvertebrates are good indicators of the impacts of pollutants. In the study by Hunter *et al.* (1987b) earthworm abundance was reduced at sites with high soil metal concentrations, and Spurgeon and Hopkin (1999) showed that reductions in earthworm abundance and diversity can be directly related to metal contamination. Earthworms may therefore be a key group for monitoring the effects of pollutants on soil biodiversity (Spurgeon and Hopkin, 1999).

Species Box 3.2: Field vole *Microtus agrestis*

Field voles are widespread and occur throughout mainland Britain, but are absent from Ireland. Their typical habitat is rough, ungrazed grassland, including young forestry plantations with a grassy ground layer, but they are also found in other habitats where grass is available. The British field vole population is estimated to be 75 million individuals, and they are thought to be the most common British mammal. However, field vole populations may be in decline because of agricultural intensification. Grass is the field vole's only food source, and grass is used to make its nests, which are about 10 cm in diameter and may be built at the base of grass tussocks or in underground burrows. The breeding season lasts from March/April to October/December, and females will normally produce five or six litters, each of four to five young, per year. The life span of a field vole is less than one year. Voles do not hibernate but produce dense fur in winter and lighter coats in spring. Predators include kestrels, owls, foxes and stoats.

Sources: Corbet and Harris (1991); McDonald *et al.* (2000); Mammal Society (2001)

Species Box 3.3: Common shrew *Sorex araneus*

The common shrew is a widespread species, found throughout mainland Britain, but absent from Ireland. It has also been introduced to many islands. It is the second most numerous of British mammals; the most recent population estimate is 41,700,000. The common shrew typically occurs in thick grass, hedgerows, scrub and deciduous woodland. It feeds on a variety of invertebrates, particularly beetles, earthworms, woodlice, spiders, slugs, snails and insect larvae. Shrews must feed every two to three hours to survive. They live in burrows and are territorial. Females give birth to one litter of five to seven young from May to September. The life span is about fifteen to eighteen months. They do not hibernate, but become less active in winter. Predators include owls, weasels, foxes, stoats and kestrels. Predators may abandon shrews because they produce an unpleasant-tasting liquid from glands on the skin. All shrews are protected under Schedule 6 of the Wildlife and Countryside Act (1981) and may not be trapped without a licence.

Sources: Corbet and Harris (1991); Mammal Society (2001)

OPPORTUNITIES

Protection and subsidies

The management of grasslands and heathlands operates within a framework of legal protection, international and national conservation designations, and agri-environment schemes. A number of conservation designations and schemes relevant to lowland grassland and heathland habitats are summarised in Table 3.7. These are not all assessed in detail here, but a number of examples are described. For example, UK agricultural policy is influenced by the Common Agricultural Policy (CAP), which was established to integrate food production among member nations and operates through a number of approaches, including mechanisms for supporting farm incomes that are broadly production-based. The CAP has been increasingly criticised because of the perceived damage it is causing to the countryside (Baldock *et al.*, 1996, Crofts and Jefferson, 1999; Hindmarch and Pienkowski, 2000).

Reform of the CAP led European Union (EU) member states to launch agri-enviromental schemes. These aimed to divert resources away from commodity price support systems, towards supporting farm incomes by providing incentives to farmers to adopt environmentally beneficial practices. As a result, the Environmentally Sensitive Areas (ESA) scheme and the Countryside Stewardship Scheme (CSS) were launched (Table 3.7). Both the ESA scheme and the CSS provide payments to landowners in return for adopting an agreed package of environmentally beneficial land management practices (Thompson *et al.*, 1999; Bignal *et al.*, 2001; Peach *et al.*, 2001). For example, farmers who enter into the South Downs and South Wessex Downs ESA schemes are paid to conserve existing areas of downland turf, and to turn arable areas into grassland (arable reversion) (Wakeham-Dawson *et al.*, 1998). Further reform (the European Commission's Agenda 2000 CAP reform) introduced a policy of providing support for rural development combined with a shift away from price support. The reform has been judged by some to provide opportunities to benefit environmentally sustainable production systems, and by others to attach too little importance to agri-environment measures (Crofts and Jefferson, 1999; Bignal *et al.*, 2001).

The level of UK agri-environment subsidies increased in the 1990s (Fig. 3.10), and unimproved lowland grassland has been a particular focus for these schemes in recognition of its high wildlife value and threatened status. Surviving heathlands largely fall outside the farming system, but heathland restoration increased in the 1980s and 1990s through the implementation of agri-environment schemes (Robertson *et al.*, 2001). Currently, funding to meet UK biodiversity Habitat Action Plan objectives comes primarily from agri-environment schemes such as CSS and ESA schemes, as well as grant schemes such as English Nature's Wildlife Enhancement Schemes that are available for SSSIs (Table 3.7) (UK Biodiversity Steering Group, 1995; Venus, 1997; UK Biodiversity Group, 1998b, 1999c). Overall, however, spending on agri-environmental policy is still a very small percentage of spending on agricultural support, uptake of agri-environment schemes in the UK has been low, and the schemes are not judged to have made fundamental changes to European agriculture (Hanley *et al.*, 1999; Crofts and Jefferson, 1999; Bignal *et al.*, 2001). Buckingham *et al.* (1999) suggest that there appear to be significant factors that prevent farmers from taking up agri-environment schemes, and that changes in the targeting and funding

Table 3.7: Examples of conservation designations, schemes and legislation relevant to lowland grassland and heathland habitats

Biosphere Reserves Areas which are internationally recognised under UNESCO's Man and the Biosphere (MAB) programme (1971). The UK's sites were designated in the 1970s. A review of MAB in the UK is in progress.
Moor House–Upper Teesdale NNR (includes northern/upland hay meadows).

Ramsar Sites Ramsar Convention (1971). Ramsar sites are wetlands of international importance, especially as waterfowl habitat.
Dorset heathlands.

EU/EC Birds Directive European Union (1979). Lists birds of special conservation concern. Linked with the designation of SPAs.
Lowland calcareous grassland, heathlands.

EU/EC Habitats Directive European Union (1992). Requires member states to identify habitats and species of interest. Linked with the designation of SACs.
Lowland calcareous grassland, heathlands.

Special Protection Areas (SPA) EC Directive on the Conservation of Wild Birds (1979). Conservation (Natural Habitats etc.) Regulations (1994). Established SPAs for birds.
Breckland, Dorset heathlands, Somerset Levels and Moors.

Special Area of Conservation (SAC) EC Habitats and Species Directive (1992). Conservation (Natural Habitats etc.) Regulations (1994). Established SACs to protect habitats and (non-bird) species, to complement SPAs in a suite of sites known as Natura 2000 sites. SAC designation should be completed by member states in June 2004.
South Uist machair, North Meadow and Clattinger Farm, New Forest, Breckland.

National Park National Parks and Access to the Countryside Act (1949). Mix of ownership and land use. Most development control is by planning consents. National Park Farm Conservation Scheme increased payments to farmers prepared to tie the land to the terms of an agreement for a period of ten years.
Peak District.

Area of Outstanding National Beauty (AONB) (England and Wales), National Scenic Areas (Scotland) National Parks and Access to the Countryside Act (1949). A factor in the outcome of planning decisions.
Dorset, Sussex downs.

Pan-European Biological and Landscape Diversity Strategy Endorsed in 1995 to maintain and enhance Europe's biological and landscape diversity. It aims to support the implementation of the Convention on Biological Diversity and increase public acceptance of conservation issues (including socio-economic sectors).

National Nature Reserve (NNR) National Parks and Access to the Countryside Act (1949). Nature Conservation and Amenity lands (Northern Ireland) Order (1985). Contain some of the most important natural and semi-natural ecosystems in Britain. Usually designated for their broad ecological value. Most owned by national conservation agency or bodies such as wildlife trusts.
Derbyshire Dales NNR (parts of six separate limestone valleys), Studland and Goldings Heath NNR, Dorset.

Site of Special Scientific Interest (SSSI) (England, Scotland and Wales), Area of Special Scientific Interest (ASSI) (Northern Ireland) National Parks and Access to the Countryside Act (1949). Protection of SSSIs strengthened by the Wildlife and Countryside Act (1981). Main site protection measure in the UK. Designated on the basis of a set of quality and rarity criteria. Protection achieved by agreements to compensate landowner for not carrying out damaging operations or through positive management agreements.
Clattinger Farm, Wiltshire (unimproved meadows and pasture).

Table 3.7: contd

Wildlife Enhancement Scheme (1981, 1985 and 1998) (run by English Nature) applies only to SSSIs Wildlife Enhancement Scheme, e.g. White Peak, Derbyshire, Staffordshire and the Peak District. Payments: hay meadow mowing and aftermath grazing, £150 per ha; haymeadow restoration, £175 per ha, grassland on lead-rich soil, £90 per ha.

Local Nature Reserve (LNR) (England, Scotland and Wales), Local Authority Nature reserves (LANR) (Northern Ireland) National Parks and Access to the Countryside Act (1949).
Lindow Common, Cheshire.

Common land The term derives from the fact that certain people held right of common over the land, e.g. to the pasture etc. 80% of common land is privately owned and subject to the interests of commoners. Nearly 50% of common land is wholly or partially designated as a SSSI.

UK Biodiversity Action Plan Rio Convention on Biological Diversity (1992).
Lowland heathland, lowland and upland fringe meadows, lowland dry acid grassland, lowland calcareous grassland, grazing marsh.

Environmentally Sensitive Area (ESA) Agriculture Act (1986). Offers incentives to farmers to adopt agricultural practices that will safeguard and enhance parts of the country of particularly high landscape, wildlife and historic value.
Somerset Levels and Moors, South Downs, Breckland, North Peak, South West Peak.

Countryside Stewardship Scheme (CSS) (England), Countryside Premium (Scotland), Tir Cymen and Tir Gofal (Wales), Northern Ireland Countryside Management Scheme Countryside Stewardship is the government's main Green grant scheme. Makes payments to farmers and other land managers to enhance landscape and wildlife. Operates outside ESAs. Operates through ten-year management agreements, e.g. managing lowland hay meadows, payment of £85 per ha per yr; restoring water meadows, payment of £225 per ha per yr.

Nitrate Sensitive Areas Closed to new entrants in 1998. Compensates farmers for changing farming practices to protect ground and surface waters from nitrate pollution, e.g. premium grass scheme: extensification of existing intensively managed grass.
North Nottinghamshire.

Organic Farming Scheme Aims to encourage expansion of organic production.

Agricultural Production Grants (DEFRA) Sheep Annual Premium Scheme (annual payment of £14 (lowlands) for each breeding ewe). Suckler Cow Premium (SCPS) (£120 per adult beef cow per year – limited by quotas). Beef Special Premium (BSPS) (£90 per male bovine). Extensification payments (top-up payments on BSPS and SCPS when overall stocking density is low).

The Bern Convention Bern Convention on the Conservation of European Wildlife and Natural Habitats (1979). Ratified by the UK in 1982. Influenced the Wildlife and Countryside Act 1981 and EC Habitats and Birds Directives.

Wildlife and Countryside Act 1981. Strengthened protection for SSSIs and certain species.

Town and Country Planning Act 1990. Development may require planning permission (local planning authority). As part of the Act, a Development Plan strengthens statutory designations such as SSSIs and NNRs. Section 106 is a planning agreement where a local authority can negotiate mitigation with a developer. A new Green Paper has been published called *Planning. Delivering a Fundamental Change.*

Table 3.7: contd

Environmental Protection Act 1990. Created English Nature, Scottish Natural Heritage, the Countryside Council for Wales and the Joint Nature Conservation Committee, with effect from April 1991.

Countryside and Rights of Way Act November 2000. Should ensure that government, local authorities, landowners and charities properly manage SSSIs; underpins government's obligations under the Convention on Biological Diversity, and gives legal backing to the Biodiversity Action Planning process. Applies to England and Wales.

Environmental Impact assessment regulations for uncultivated land and semi-natural areas February 2002. Environmental Impact (EIA) legislation extended to include EIA procedures for the use of uncultivated land or semi-natural areas for intensive agricultural purposes. Land types covered include unimproved grassland and heathland.

Sources: DOE (1995); JNCC (1998); Crofts and Jefferson (1999); anon. (2000); Rose *et al.* (2000); Phillips and Huggett (2001); DEFRA (2001d, 2002b), Planning Inspectorate (2002).

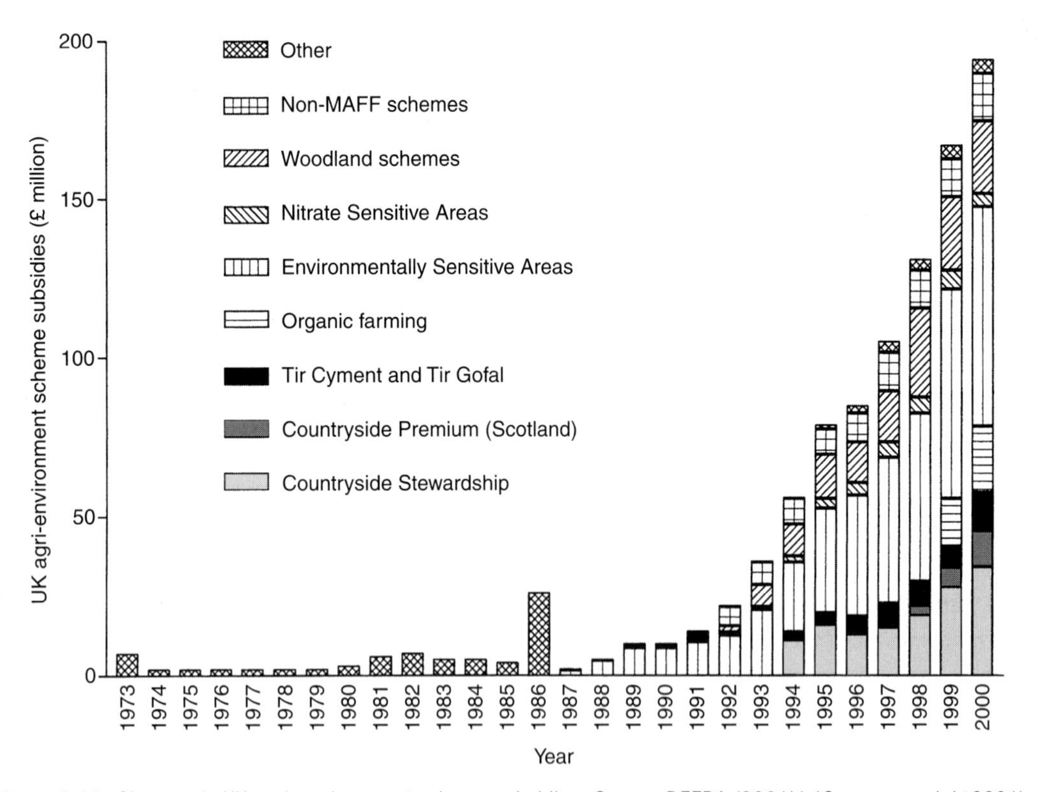

Figure 3.10: Changes in UK agri-environment scheme subsidies. *Source:* DEFRA (2001b) (*Crown copyright* 2001)

of existing schemes are needed. For example, the Countryside Stewardship Scheme is oversubscribed and has to reject about 20 per cent of applications each year owing to budget limitations (Policy Commission on the Future of Farming and Food (PCFFF), 2002). Some authors argue for closer integration of production and conservation objectives, while others favour a dual approach that supports production in some areas and biological diversity through extensification in others, because the two are rarely compatible (Buckingham *et al.*, 1999; Bignal *et al.*, 2001). The report of the PCFFF (2002) favoured broad-based environmental schemes with less demanding prescriptions, to encourage a measure of good environmental practice across a wider area than is covered by current schemes. These would supplement existing schemes, which should at least retain their current level of funding.

Whatever the case, if semi-natural habitats are to be maintained, considerable investment is needed, and financial incentives to farmers for conservation management must be attractive (Buckingham *et al.*, 1999; Hindmarch and Pienkowski, 2000; Winter *et al.*, 2000). The PCFFF (2002) report recognised that market forces cannot deliver environmental benefits, and that farmers should be rewarded from the public purse for providing them. It advocated major reform of current CAP subsidies, but also noted that even within existing rules it is possible to transfer up to 20 per cent of production payments to environmental schemes. However, such diverted payments from the EU budget must be matched by national government. In 2001–2002, only 2.5 per cent of payments were being redirected in this way. The report recommended that the bulk of any new resources for agri-environmental programmes should be spent on new, less demanding, broad-based schemes. It appeared to attach less significance to increasing support for the more exacting, oversubscribed existing schemes (PCFFF, 2002). Any increase in support for agri-environment schemes is to be welcomed, but only time will tell whether the proposed changes are appropriate to the management of semi-natural habitats. However, the recently introduced environmental impact assessment legislation for the use of uncultivated land or semi-natural areas for intensive agricultural purposes should afford increased protection to Biodiversity Action Plan priority habitats such as grasslands and heathlands (Table 3.7) (DEFRA, 2002b). For more information on UK agricultural policy and conservation management refer to Winter *et al.* (1998, 2000); Crofts and Jefferson (1999); Hanley *et al.* (1999); Hindmarch and Pienkowski (2000), Bignal *et al.* (2001) and PCFFF (2002).

The UK is also required to meet the objectives of the EC Council Directive on the Conservation of Natural Habitats of Wild Fauna and Flora, or Habitats Directive (1992). The Habitats Directive is European law, which provides for the creation of a network of protected areas across the European Union to be known as 'Natura 2000'. More specifically, Special Areas of Conservation (SACs) and Special Protection Areas (for birds) (SPAs), established under the 1979 Birds Directive, are collectively termed the Natura 2000 Series (Table 3.7). The Natura 2000 Series is intended to protect the best examples of wildlife habitat throughout Europe (Countryside Council for Wales, 1995; Thompson *et al.*, 1999). The Habitats Directive (Annex I) lists 168 natural habitat types in Europe that can be designated as SACs. A number are given 'priority' status because they are particularly at risk. Seventy-six of these habitats (including twenty-two priority habitats) exist in the UK. Of these seventy-six, a number are lowland grassland or heathland habitats (Table 3.8). As of 8 June 2001, 555 sites had been submitted by the UK as candidate SACs (cSACs). In broad terms, 217 of these UK cSACs are grassland habitats, 192 are heathland habitats and thirty-nine are maritime cliff and slope habitats (Table 3.8). It is the policy of the UK government to treat cSACs as if they were already

Table 3.8: The number of UK sites put forward as candidate SACs for grassland and heathland habitats, July 2001

Habitats Directive Annex I habitats	*UK BAP priority habitat equivalent*	*No. of sites*
Vegetated sea cliffs	Maritime cliff and slope	39
Shifting dunes with marram	Coastal sand dunes	34
Dune grassland [a]	Coastal sand dunes	32
Lime-deficient dune grassland with crowberry	Coastal sand dunes	2
Coastal dune heathland [a]	Coastal sand dunes	10
Machair	Machair	6
Inland dunes with open grey-hair grass and common bent grassland	Lowland dry acid grassland	1
Wet heathland with cross-leaved heath	Lowland heathland; Upland heathland	69
Wet heathland with Dorset heath and cross-leaved heath [a]	Lowland heathland	4
Dry heaths	Lowland heathland; Upland heathland	108
Dry coastal heaths with Cornish heath [a]	Lowland heathland	1
Grasslands on soils rich in heavy metals	No match	20
Dry grasslands and scrublands on chalk or limestone	Lowland calcareous grassland	45
Dry grasslands and scrublands on chalk or limestone, including important orchid sites [a]	Lowland calcareous grassland	14
Species-rich upland grassland with mat grass [a]	Upland calcareous grassland	22
Purple moor grass meadows	Purple moor grass and rush pastures	34
Lowland hay meadows (MG4 only)	Lowland meadows	5
Mountain hay meadows	Upland hay meadows	2

Source: DEFRA (2001c).

Notes: [a] Priority habitat. Lowland and upland heaths are grouped together, and upland calcareous grassland and upland hay meadows are included for completeness, and because some sub-communities occur on the upland fringes. Alpine/montane heaths and grasslands have been excluded from this table.

designated. Designation should be completed by June 2004 (DEFRA, 2001c).

The Habitats Directive (Annex II) also lists certain species that must be protected by law, and encourages management for wildlife outside Natura 2000 sites, for example through ESAs, CSS schemes (England) and Tir Gofal initiatives (Wales) (Countryside Council for Wales, 1995). Fifty-one species listed under the Habitats Directive have been recorded in the UK, but ten of them are extinct or visitors to the UK, and have not been used in the SAC selection process. Species listed under the Habitats Directive include the greater horseshoe bat, the southern damselfly and the marsh fritillary butterfly. For information on other policies and designations refer to Table 3.7.

Defining management objectives and monitoring outcomes

Management decision making involves defining management objectives, and these will depend on the scope of a project. An ecological impact assessment programme to evaluate the potential impacts of a particular activity on a habitat, or fulfilment of Habitat Action Plan requirements (Table 3.3) may have broader objectives than a project to assess site management techniques (e.g. Crofts and Jefferson, 1999; Treweek, 1999). Additional factors that should be taken into account when defining realistic management objectives include resources, constraints, and the role of public relations (Crofts and Jefferson, 1999).

Key management objectives for lowland grasslands and heaths incorporate the requirements of characteristic and rare species and are usually to prevent scrub encroachment and dominance of coarse grasses, to maintain structurally varied vegetation and to create areas of bare ground (Gimingham, 1992; Michael, 1993; Crofts and Jefferson, 1999). Vegetation structure requirements may vary spatially and seasonally between and within sites in order to support a variety of species. The way in which these objectives are achieved will depend on the selection of the most appropriate techniques, allowing for constraints. For example, grazing may be the most appropriate management technique, but it may be difficult to apply to urban fringe sites or in small, isolated areas of habitat. It is important to establish the availability of resources such as livestock, machinery, skills and subsidies (Table 3.7). Negotiation with the site owner or occupier is often crucial, and site interpretation information may be required to maintain public support for certain management practices such as exclusion fencing or scrub clearance (Crofts and Jefferson, 1999). Monitoring should be employed to establish whether the management regime is appropriate and whether objectives have been met.

Monitoring

The role of monitoring is to systematically record habitat or ecosystem information to determine habitat characteristics and identify how they change with time. Change can occur under normal conditions (natural variation), or following some form of perturbation, such as pollution, or a change in management. If possible, monitoring should take place before any action is taken, to produce pre-management baseline information (ideally over several years). Baseline information is particularly useful because natural year-to-year ecological variation may be considerable, and it is important to distinguish this natural variation from anthropogenic changes (Treweek, 1999). Subsequently, repeated long-term monitoring should assess the effectiveness of management practices against defined management objectives and identify any adjustments that may be required. At the planning stage, it is important to ensure that the data collected can be analysed and interpreted. If necessary, a statistician should be consulted before monitoring begins. It is essential to keep accurate records of sampling methodology, so that monitoring methods are consistent between dates and monitors, as changes in methodology may invalidate comparisons between data sets.

In the UK there are national monitoring schemes for butterflies and for birds, which make it possible to record species throughout the country, using consistent methods (Treweek, 1999), and there is an English Nature rapid assessment method for monitoring the condition of lowland grassland SSSIs according to an agreed framework of common standards (Robertson and Jefferson, 2000). There is also a national strategy and methodology for monitoring Environmentally Sensitive Areas (ESAs) (Glaves, 1998). Annual recording visits should be carried out at the same time of year to reduce seasonal variability. However, seasonal variability may be an important feature of the site or the organisms under scrutiny, in which case several recordings should be made in a year (Ausden, 1996; Crofts and Jefferson, 1999).

Monitoring of invertebrates, amphibians, reptiles, birds and mammals is an important aspect of grassland and heathland management (e.g. Drake and Denman, 1993; Ransome, 2000). Detailed information of monitoring techniques for these groups is given in Sutherland (1996). Monitoring sward composition, and vegetation structure and height (e.g. extent of vegetation cover and bare ground) can be useful for assessing the quality of grasslands and heathlands and their suitability for particular organisms, particularly invertebrates (Chapter 2, 'Fauna of lowland grasslands and heathlands'). For example, the wartbiter cricket requires bare ground or short turf for egg laying and larval development and taller tussocks where the adults can shelter and hunt (UK Biodiversity Group, 1999b).

The increase or decrease in particular species (indicator species) can indicate more widespread changes in the habitat (Crofts and Jefferson, 1999; Treweek, 1999). For instance, the spread of ragwort may be a symptom of overgrazed or badly managed pasture (Simpson, 1993; Crofts and Jefferson, 1999). Certain species have been identified as indicators of pollution, disturbance and habitat age. Furthermore, plant species can be characterised on the basis of their 'strategy' or functional type (Chapter 2, 'Ecological succession'). Competitors are associated with low-stress, low-disturbance conditions (Grime *et al.*, 1988). For instance, creeping thistle is a fast-growing competitor favoured by nutrient-rich conditions, and an increase in the proportion of such species may indicate eutrophication (Crofts and Jefferson, 1999). The three functional types (competitor, stress tolerator and ruderal) may also respond differently to environmental trends such as climate change (Chapter 4) (Hodgson *et al.*, 1995).

Detailed recordings of botanical composition can be made using visual estimates of species cover or density. The most common approach is to use quadrats to define sample areas within a study site (Chapter 5). Information on botanical composition can be used to assign a site to a particular NVC or CORINE community (Hill, 1996). This can allow the existence of a particular vegetation type to be identified, and monitored during management, or following habitat translocation, restoration or creation. A number of useful texts give detailed information on vegetation survey and monitoring techniques (Kent and Coker, 1992; Bullock, 1996; Crofts and Jefferson, 1999). Abiotic components of the ecosystem, such as details of soil type, water levels, weather data and microclimate should also be assessed. Most plant communities of nature conservation interest require a low nutrient status. Soil fertility may be modified by management practices, and microclimate can be influenced by vegetation structure. It is often also useful to obtain management information, such as grazing intensity, grazing period, rabbit control, and application rates of inputs such as farmyard manure (Crofts and Jefferson, 1999).

HABITAT MAINTENANCE AND ENHANCEMENT

The more exposed maritime grasslands and heathlands and some wet heaths are maintained by abiotic factors (Gimingham, 1992; Mitchley and Malloch, 1991; Oates, 1999), but as a rule the development of biologically diverse grassland and heathland habitats is a consequence of traditional, sustainable land-use systems ('The history of grasslands and heathlands', this chapter) (Andrews and Rebane, 1994; Webb, 1998; Crofts and Jefferson, 1999; Hindmarch and Pienkowski, 2000). In recent years the use of these management practices has declined, and many areas of lowland grassland or heathland have undergone vegetation change. In some areas, traditional practices have continued, while in others they have been reinstated in order to maintain, enhance or repair habitats. The techniques most commonly used to maintain and enhance grassland and heathland habitats are appropriate fertility management, grazing, cutting and burning. In some cases, turf stripping may be an appropriate method of nutrient removal. Another key factor at some wet grassland and heathland sites is the monitoring and management of the hydrological regime. The practicalities of grassland and heathland management are addressed in detail by Gimingham (1992), Michael (1993), Benstead *et al.* (1999) and Crofts and Jefferson (1999).

Fertilisers, lime and pesticides

Intensively managed grassland is treated to achieve consistently high yields of forage with a high nutritional quality. Dry matter yields from semi-natural grasslands can be between 40 per cent and 80 per cent of the yields of intensively managed agriculturally improved grass (Tallowin, 1997), but fertiliser and pesticide application to semi-natural habitats can cause a drastic decline in biodiversity (Crofts and Jefferson, 1999). The Park Grass Experiment was set up in 1856, originally to determine the effects of different amounts of fertiliser and organic manure on the productivity of grassland. The experiment is still in progress and is the most long-term ecological study in the world. The experiment shows that the addition of inorganic nitrogen, of phosphorus and of various combinations of nitrogen, phosphorus, potassium and calcium cause dramatic changes in the abundance of plant species. In addition, the application of nitrogen as ammonium sulphate or as sodium nitrate can have different effects because ammonium sulphate has strongly acidifying effects. Inorganic fertilisers encourage rapid growth of productive grasses and lead to the loss of species richness (Tilman *et al.*, 1994). Other studies have shown that species richness in hay meadows is reduced even by low levels (25 kg N ha^{-1}) of nitrogen fertiliser, and fertiliser inputs, particularly of phosphorus, can cause increased dominance by grasses and a reduction in many distinctive species within six years. As a result, communities such as MG4, MG5 and MG8 are replaced by more species-poor types such as MG6 (Table 2.4) (Tallowin and Smith, 1994). Similarly, the replacement of heather by grasses may be caused by artificial nutrient inputs (Gimingham, 1992).

Farmyard manure releases nutrients at a slower rate than chemical fertilisers, and if applied sparingly has less effect on plant richness. However, beyond a certain point (e.g. application rates above about 20 t ha^{-1} every three to five years, depending on site fertility) there is an increase in competitive grasses and a reduction in species richness and diversity, i.e. effects similar to those of inorganic fertilisers. Farmyard manure is a variable commodity, but as a rule it should be stored for a minimum of twelve months prior to application. Timing, rates and methods of application are assessed by Simpson and Jefferson (1996). The use of slurry can scorch or smother the sward and cause pollution problems.

Lime was applied in the past where soil acidity restricted the growth of grass through limiting nutrient availability (Chapter 2, 'Geology and soils'), whilst today it is more commonly applied to correct the acidifying effects of appling artificial nitrogen fertilisers. Liming affects the plant composition of grasslands and can eliminate species that are characteristic of acid soils, but may be required to maintain diversity in certain situations (Andrews and Rebane, 1994; Simpson and Jefferson, 1996).

A number of plant species may cause problems for farmers and conservation managers because of their capacity to dominate large areas. These include creeping thistle, spear thistle, broad-leaved dock, curled dock, common ragwort and marsh ragwort. However, all have wildlife value. For example, as many as 200 species of insect, including butterflies, bees and hoverflies, may visit ragwort (Plate 7) (Smith, 1980). The use of herbicides in semi-natural habitats should be avoided. Instead, appropriate cutting and grazing management should be used where possible to control weeds; for example, grassland should not be under- or overgrazed ('Grazing', this chapter). Ragwort can be particularly problematic because it is toxic to stock, particularly young animals, cattle and horses. If herbicide use is unavoidable, selective herbicides should be applied locally, using a weed wipe or spot spray (Simpson, 1993; Andrews and Rebane, 1994). Detailed information on grassland weed control is given in Simpson (1993).

Grazing

Grazing of grasslands and heathlands maintains existing areas of habitat, limits the ability of competitive, fast-growing species to achieve dominance by continually removing new growth, and controls encroachment by scrub, although grazing may not be sufficient to recover areas of heathland already invaded by scrub. Grazing can delay the onset of the mature and degenerate phases of heather, although it is also normally necessary to rejuvenate the vegetation periodically by cutting, burning or turf cutting. Grazing is also beneficial for grassland and heathland wildlife because it can create a mosaic of vegetation structures and areas of bare ground (Gimingham, 1992; Crofts and Jefferson, 1999; Rose *et al.*, 2000). However, management of particular species, such as the sand lizard, may require that sensitive areas are protected from grazing stock (Corbett and Moulton, 1998).

Most types of semi-natural lowland grassland and improved agricultural grassland are used for grazing, cutting (hay or silage) or a combination of the two, although acid and calcareous grasslands tend to be less productive than mesotrophic grasslands and are usually grazed rather than mown (Crofts and Jefferson, 1999). Successful meadow or pasture management that is traditional to a site should be modified only if absolutely necessary (Gibson, 1997), but unimproved grasslands are now often marginal to farming and insufficient grazing is an increasing problem (Robertson *et al.*, 2001). Grazing was once the major traditional use for lowland heaths, although the practice declined in the twentieth century in many areas, except in the New Forest, where the pastoral economy survived and common land continued to be grazed by cattle and ponies (Tubbs, 1991). Grazing management has been reintroduced to a number of lowland heath types, including dry, humid and wet heaths, and on these sites grazing reduced the cover of dwarf shrubs and scrub, and increased plant species richness, the cover of grasses, the amount of bare ground, and in some cases the cover of bryophytes and lichens (Bullock and Pakeman, 1996).

Grazing involves three important processes: defoliation (removal of above-ground plant biomass), trampling and manuring. Different types of grazing livestock and rabbits have distinct preferences and feeding characteristics, so that any change in the type of animal, the numbers of individuals present or the length of the grazing season can lead to changes in the vegetation (Watt, 1981a, b; Gimingham, 1992; Andrews and Rebane, 1994; Michael, 1994; Crofts and Jefferson, 1999).

Animals for grazing grasslands and heathlands

Grazing animals include domestic stock such as sheep, cattle, horses, goats and pigs and wild animals such as rabbits and deer (Tubbs, 1991; Crofts and Jefferson, 1999; Small *et al.*, 1999). The grazing animals best suited to a number of different NVC type grasslands are shown in Tables 2.4, 2.7, 2.8, 2.11 and 2.12. The most common grazers of heathland are cattle, sheep, horses and rabbits (Gimingham, 1992; Andrews and Rebane, 1994) (Plate 3.7). Different types of grazing animal have different grazing characteristics, and often have preferences for certain species and a dislike of others (Table 3.9), and this can influence plant species abundance in a grass sward or heathland community. Preferences can vary between sheep, cattle and ponies, between sites, between individuals, and may depend on the age, sex or the past experience of the animal (Small *et al.*, 1999). The choice of grazing animal is usually based on a number of factors. Animals are commonly selected because of their grazing behaviour (Table 3.9) or because they belong to a local farmer. This is because a large proportion of site managers rely on owners of stock who loan their animals for conservation grazing purposes (Small *et al.*, 1999). In addition, many nature reserve managers rely on commercial farmers to implement the required grazing regimes, and some farmers are responsible for managing their own SSSIs (Crofts and Jefferson, 1999).

Beef cattle are often the preferred grazing animals in conservation grazing schemes, because they are less selective than sheep, are effective at controlling coarse, vigorous grasses on grasslands and heaths and create a variable sward (Gimingham, 1992; Baxter and Farmer, 1993; Crofts and Jefferson, 1999;

Plate 3.7: New Forest pony grazing heathland

Table 3.9: Grazing characteristics of sheep, cattle, horses and rabbits in unimproved lowland grassland and lowland heathland

Sheep	*Cattle*	*Horses and ponies*	*Rabbits*
Bite			
Bite the vegetation. Graze down to ground level. Produce short swards (3 cm)	Bite, pull and tear the vegetation. Cannot graze as close to ground as sheep. Maintain longer swards (5–6 cm)	Bite the vegetation. Graze even closer to the ground than sheep. Produce very short swards (2 cm minimum)	Bite the vegetation. Graze very close to ground and can produce very short swards
Selectivity			
Selective grazers. Variation in diet between individuals. Can select a species growing in a fine mixture with other species, or from low in the grassland profile. Avoid tall plants. Flowers selected and eaten. Grass stems left. Dead material and litter left. Rough, tall sward and tussocky areas avoided and are not generally the best animals for restoration grazing	Do not graze as selectively as sheep. Less variable diet than sheep. Cannot select from fine mixtures or from low in the profile. Take tall plants. Flowers not selected, but may be eaten. Grass stems taken. Dead material and litter taken. Rough, tall sward and tussocky areas utilised. Can be used to reduce the dominance of upright brome and false oat grass, but less successful in reducing the more unpalatable tor grass. Good at opening up bracken and scrub in restoration management	Selective, patchy grazers. Variable diet. Can select a species growing in a fine mixture with other species, or from low in the grassland profile. Take tall plants. Some flowers selected and eaten, others avoided. Grass stems taken. Dead material and litter taken. Rough, tall sward and tussocky areas utilised	Highly selective, perhaps the most selective of all. Can select leaves of nutritious species from a wide range of vegetation. Flowers selected and eaten. Favour young material, but will take dead grass. Tall, rank swards are avoided
Preferences			
Preferences include silver birch, fescues, ragwort, young heather, cross-leaved heath. More resistant to ragwort and can be useful for control. Heather mainly taken in winter. Grazing reduces establishment of heather in recently restored heathland. Bracken is toxic	Preferences include silver birch, young and older heather, young purple moor grass, upright brome. Cattle browse less than horses, but consume more heather (mainly taken in winter) (20% of diet). Grazing can reduce the dominannce of purple moor grass and wavy-hair grass in favour of heather. Grazing reduces establishment of heather in recently restored heathland. Bracken and ragwort are toxic	Preferences include young purple moor grass, heather, heaths, gorse. In general however, take grass in preference to heather, but browse more than cattle. Heather mainly taken in winter (10% of diet). May depend heavily on gorse in winter. Grazing reduces establishment of heather in recently restored heathland. Ragwort is toxic	Eat a wide range of plants, but young buds, shoots and leaves are favoured. Preferences on dry heath include heather (winter), bell heather (autumn), gorse and grasses, especially purple moor grass (summer). Consume young heather and gorse seedlings. Palatable species on set aside grassland include grasses such as common couch, false oat grass and cocksfoot

Table 3.9: contd

Sheep	*Cattle*	*Horses and ponies*	*Rabbits*
Dislikes			
Species avoided include tor grass, purple moor grass, wavy-hair grass, bell heather, gorse	Species avoided include wavy-hair grass, thistles, gorse, cross-leaved heath and bell heather	Species avoided include cross-leaved heath and bell heather, ragwort	Species avoided include stinging nettles, ragwort, thyme, common rockrose, thistles, cross-leaved heath and bracken
Sward characteristics (moderate grazing levels)			
Graze preferentially in small patches; select the most palatable patches and prefer short swards to coarser vegetation which is avoided	Graze preferentially in small patches; select the most palatable patches. Cattle swards often particularly patchy and structurally diverse/tussocky (benefits invertebrates and nesting waders). Plant species richness may be greater when cattle rather than horses grazed (e.g. MG5)	Pony grazing may produce structurally varied sward that is particularly valuable for invertebrates. Consume twice as much material as cattle because food is poorly digested and so have a greater effect on heathland vegetation than cattle or deer. New Forest ponies may graze for 88% of a 24hr day in some cases. Low levels of horse grazing better for grasslands than no management	At moderate densities produce a mosaic of small patches nibbled to different heights, surrounded by taller grassland
Overgrazing			
At high stocking levels, sheep grazing produces a very short, tight grassland sward, generally poor for invertebrates, which can encourage bracken invasion in some areas. At high levels, sheep grazing can convert areas of heather into grass swards	At high densities, cattle grazing produces a shorter, more uniform sward, but not as short as that created by sheep grazing. Heavy grazing leads to a reduction in plant species richness, e.g. it may turn MG5 grassland into MG6	Overgrazing of more palatable grasses can lead to bare patches and areas of uneaten coarse grass and rank vegetation. Heavy grazing leads to a reduction in plant species richness, e.g. it may turn MG5 grassland into MG6	Heavy grazing benefits a small group of plants, invertebrates and birds such as the stone curlew, but can be damaging to other species. It reduces the cover of palatable grasses and herbs and increases the prominence of annuals. May encourage the growth of grasses at the expense of heather. Continual heavy grazing may cause destruction of the turf and allow invasion of undesirable species

Table 3.9: cont

Sheep	*Cattle*	*Horses and ponies*	*Rabbits*
Dunging			
May deposit dung in specific areas, or randomly, but do not reject adjacent vegetation. They can therefore graze vegetation to a uniform height, although local nutrient enrichment may occur	Cow pats scattered randomly, animals will not graze around pats, creating tall sward patches among shorter turf (valuable for invertebrates)	Tend to use latrine areas (creates localised patches of high nutrient levels) colonised by competitive species. On better soils, heavy grazing, dunging and urination converts heathland to neutral grassland or 'lawns' that support a number of rare species. Ericoid shrubs are susceptible to high concentrations of nitrogen and potassium in urine	Mounds of soil thrown up around burrows and latrines are enriched with urine and droppings, and support large, unpalatable herbaceous species such as nettles, thistles and mulleins
Trampling			
Agile and suitable for grazing steep slopes, although may cause trampling damage on heavy, wet soils. Wet sites associated with foot rot problems, e.g. not suitable for grazing wet heaths. Little trampling damage to old heather.	Trample ground more than sheep, which opens up the sward and allows establishment of annuals. Hoof prints can create the suitable microclimate for invertebrates, e.g. egg laying of Adonis blue. Too much trampling may damage the structure of the surface layers of the soil (poaching). Overstocking during autumn and winter can lead to infestation of grassland by docks, ragwort and thistles. Suitable for grazing wet heaths. Trampling may damage old heather.	Suitable for grazing wet heaths. Trampling may damage old heather.	Rabbit scrapes provide areas of bare ground, favoured by some invertebrates, e.g. the Adonis blue butterfly for egg laying, or where annuals such as fairy flax can germinate. Rabbits do not damage lichens by trampling as livestock do, and rabbit grazing is the preferred option for the management of lichen-rich grasslands.

Sources: Smith (1980); Sumpton and Flowerdew (1985); Webb (1986); Putwain and Rae (1988); Tubbs (1991); Gimingham (1992); Michael (1993); Simpson (1993); Andrews and Rebane (1994); Ausden and Treweek (1995); Gibson (1997); Crofts and Jefferson (1999); Small *et al.* (1999); Diaz (2000).

Oates, 1999). Browsing by cattle also suppresses birch, and trampling damages bracken. However, it may also damage heather (Gimingham, 1992; Michael, 1993). It can be difficult to persuade farmers to use their cattle for conservation grazing, because most farmers work with modern breeds that will not gain weight rapidly on low-quality forage. This is a problem, because the BSE rules state that animals must be finished (ready for sale) within thirty months (Winter *et al.*, 1998; Oates, 1999). Breeds of cattle can be broadly divided into hill, lowland and ancient or rare types. Upland cattle are well adapted to a harsh climate and low-quality grazing, and Welsh Black cattle are suited to steep unfenced coastal slopes. For lowland grasslands in a reasonable condition, most commercial beef cattle would be suitable (Crofts and Jefferson, 1999; Oates, 1999). If possible, stock treated with ivermectin should not be used to graze semi-natural habitats, as ivermectin reduces the variety of insects in dung. This may lead to a reduction in food for insectivorous birds such as curlew and chough, and bats such as the greater horseshoe bat (Chapter 2, 'Fauna of lowland grasslands and heathlands') (Andrews and Rebane, 1994).

Where grazing cannot be arranged through local farmers, sheep may be a practical option because they are cheap to purchase and easy to transport. Sheep cause less trampling damage than cattle, but are susceptible to disease on wet sites (Mitchley and Malloch, 1991; Ausden and Treweek, 1995). There are three main categories of commercial sheep breeds: hill (or mountain), upland and lowland, but primitive or rare breeds may also be useful for conservation grazing. Lowland breeds are the most productive when kept under improved conditions, but are usually unable to control coarse unpalatable grasses and are best suited to maintaining reasonable swards. Primitive, mountain and hill breeds graze less selectively than lowland breeds and will eat coarser grasses, browse scrub and can thrive with a high proportion of heather in the diet. They are also less prone to health problems (Gimingham, 1992; Michael, 1993; Crofts and Jefferson, 1999; Oates, 1999). However, Tubbs (1991) considers that ponies and cattle are more appropriate for heathland grazing than sheep, because at high densities sheep defoliate heather. Hebridean sheep tend to graze purple moor grass in preference to heather, which makes overgrazing of heather less likely (Michael, 1993). Sheep grazing was particularly important in the heaths of East Anglia (Rodwell, 1991).

Hardy, native breeds of pony are tolerant of quite poor grazing and can be useful in managing established swards and reclaiming neglected grasslands. Ponies are also useful for controlling grasses, especially purple moor grass, and promoting heather, because they tend to graze grasses in preference to it (Michael, 1993). They have a limited role in scrub control, but are effective at opening up dense bracken stands. Shetland ponies and Welsh Mountain ponies are suitable breeds for difficult cliff slopes (Oates, 1999). A study of MG5 grassland (Table 2.4) revealed that in certain circumstances there was little difference between cattle or horse-grazed swards, but that grazing intensity had a marked effect on species composition. Moderate levels of pony grazing can produce a structurally varied sward that is particularly valuable for invertebrates, but overgrazing of the more palatable grasses can lead to bare patches and areas of uneaten coarse grass and rank vegetation (Table 3.9) (Andrews and Rebane, 1994; Gibson, 1996, 1997).

Mixed grazing – for example, horses with cattle – may be beneficial, as it may be more effective in controlling unpalatable species, because different grazing animals have different feeding preferences. However, in some circumstances, mixed grazing may produce a relatively uniform sward that is poor for plants and invertebrates (Michael, 1993; Ausden and Treweek, 1995; Crofts and Jefferson, 1999).

The rabbit population is growing, and in some areas populations have returned to pre-myxomatosis levels. In such situations, overgrazing may be a problem, although rabbit populations can fluctuate greatly on an annual basis in response to disease and weather conditions. Rabbit numbers should be taken into account when planning managed grazing regimes, and it may be necessary to suspend livestock grazing if rabbit densities are high (Macdonald and Barrett, 1993; Crofts and Jefferson, 1999).

Seasonal grazing

Grasslands Grazing management has different effects at different times of year. If grazing of grassland is delayed until the end of the growing season, this will allow plants in the sward to flower and set seed, and animals to complete their life cycles. For example, the density of individuals and species richness of herbivorous heteropteran bugs on calcareous grassland was enhanced by autumn grazing and severely reduced by spring grazing (Brown *et al.*, 1990), and the eggs of ground-nesting birds may be at risk from trampling in spring and summer (Andrews and Rebane, 1994). However, winter grazing (October–April) may not remove the bulk of dead plant material, and it can be ineffective in preventing dominance by tall grasses and herbs because winter herbage is unpalatable to stock and of low nutritional value. Woody species also become tough and inedible to grazing stock from midsummer onwards and winter grazing usually allows scrub to encroach. In addition, livestock used for unsupplemented winter grazing should be able to withstand periods of nutritional deficiency (Tallowin, 1997). Alternatively, supplementary feeding should be supplied off-site, if possible, to reduce local nutrient enrichment and damage to the sward (Andrews and Rebane, 1994). Winter-only grazing may not therefore be the most appropriate technique for maintaining grasslands. The advantages and disadvantages of restricting grazing to different times of the year are summarised in Table 3.10.

Heathlands There is a need to monitor the long-term effects of different seasonal grazing regimes on lowland heaths (Dolman and Land, 1995), but studies have indicated that the effects of sheep grazing on heather differ according to time of year. Heather is particularly vulnerable to grazing damage in autumn (September and October), and is least susceptible in winter. Where grass is available, it is selected in summer, and heather is mainly grazed in the winter. Often, however, sheep are grazed on lowland heaths between April and August or March and September. Cattle may also be grazed between March and September. If animals are grazed all year round, access to grassland or supplementary feeding may be required, although this may discourage animals from tackling coarse vegetation. Supplementary feeding should be supplied off-site, if possible, to reduce nutrient enrichment (Gimingham, 1992; Michael, 1993).

Grazing pressure

Grasslands Grazing pressure is a measure of the amount of vegetation that a number of grazing animals will remove during the time that they are grazing. Grazing pressure will vary, depending on the stocking level. In agricultural systems, where the emphasis is on

Table 3.10: The effects of seasonal grassland grazing

Advantages	*Disadvantages*
Spring (April/May)	
Sward productivity high A good time to control some dominant and unpalatable grasses such as tor grass, and scrub seedlings	Can cause local extinction of early flowering plants such as fritillary. Can damage or eliminate a range of invertebrate populations. Can be associated with trampling of wader and wildfowl eggs and chicks in wet grassland
Summer (May/September)	
Sward productivity high A good time to control tall herb species and check new growth of woody scrub. Less nutrient build-up from dung as decomposition rates are high. Less poaching as soil moisture levels low	Heavy grazing pressure can remove flowers (affects invertebrate flower feeders) or prevent them from setting seed (can affect annual and biennial species). Can damage or eliminate a significant range of invertebrates
Autumn (September/October)	
Least damaging time for most sensitive invertebrates Little effect on most plant species. May help seed dispersal and establishment through trampling	Sward palatability and quality decline and vegetation is rejected, allowing competitive species and woody species to encroach. Heavy grazing may have a negative impact on some plant-feeding invertebrates
Winter (October/April)	
Most plants not affected. Less damaging to invertebrates, which usually overwinter in the base of tussocks. Moderate trampling creates bare ground for germination of annuals in spring	Low sward palatability and nutritional value. Less likely to control competitive grasses. Removes fewer nutrients from system. Stock may lose condition and supplementary feeding may be needed. Heavy trampling may lead to poaching. Hard winter grazing can destroy habitat for overwintering invertebrates

Sources: Lane (1992); Treweek *et al.* (1997); Crofts and Jefferson (1999).

production, stocking levels are usually high. Where conservation is the objective, stocking levels are usually lower. For example, the overall stocking level on unimproved mesotrophic grassland for conservation management may be a quarter of that on improved grassland for production purposes (Crofts and Jefferson, 1999). Lower stocking levels mean that some biomass is removed by stock, but a proportion remains to support invertebrate herbivore populations, or to enhance the structural diversity of the habitat. For instance, extensive grazing by cattle allows a succession of suitable forage plants and nesting sites for bumblebees (Carvell, 2002). When grazing pressure is too low, competitive grasses will not be kept in check because the amount of plant material removed is too low. This can result in tall, rank swards and a loss of biological diversity. When grazing pressure is too high, it normally results in damage to the ecological character of the sward and a decline in productivity because the amount of vegetation removed is too great (Table 3.9). For example, the species richness and abundance of leaf miners on calcareous grassland were severely reduced in heavily grazed areas, and were directly related to grazing intensity (Brown *et al.*, 1990). Heavy grazing by cattle and horses may eventually turn MG5 grassland into MG6 pasture (Gibson, 1996, 1997).

Stocking levels may be expressed in livestock units (LU or lsu), a system that allows for the different quantities of plant material removed by different ages and types of stock. However, there are a number of different schemes in use, and they are not always equivalent (Ausden and Treweek, 1995; Crofts and Jefferson, 1999). Some examples of stocking levels that are effective for grassland conservation purposes are shown in Table 3.11, but requirements vary from site to site and adjustments are usually made in response to the observed impact of grazing on a day-to-day basis. The appropriate grazing season (period) should be defined before stocking levels are considered (Lane, 1992; Andrews and Rebane, 1994; Crofts and Jefferson, 1999). As a general rule, it is usually advisable to start with low stocking rates, and increase the grazing pressure if the desired effect is not achieved, rather than start with levels that are too high, as sporadic heavy grazing can be particularly damaging to invertebrates (Ausden and Treweek, 1995).

Heathlands Heather-dominated dwarf-shrub communities are tolerant of moderate defoliation, and regular removal of heather shoots helps to maintain the plant in a juvenile state in which growth is active. Heather responds favourably to grazing only within a rather narrow range of intensities. Undergrazing fails to hold it in the building phase, while overgrazing causes damage to heather. If more than 40 per cent of the annual growth of heather is grazed, the canopy may break down (Gimingham, 1992). Grazing animals usually select more nutritious species such as grasses in preference to heather, but dwarf shrubs are more sensitive to overgrazing and trampling than grasses. Overgrazing converts heather stands to acid grasslands, dominated by wavy-hair grass or purple moor grass. On deeper soils, overgrazing may also encourage bracken invasion (Mitchley and Malloch, 1991; Gimingham, 1992; Dolman and Land, 1995). In the New Forest heathlands, areas of low-cropped heather may occur where grazing

Table 3.11: A guide to stocking levels for lowland grassland (to maintain sward for conservation) (number of animals per hectare)

No. of grazing weeks per year	*Calcareous*		*Neutral*		*Acidic*		*Wet/marshy*	
	Sheep	*Cattle*	*Sheep*	*Cattle*	*Sheep*	*Cattle*	*Sheep*	*Cattle*
2	60	15	100	25	50	12	50	12
16	7.5	2	12.5	3	6	1.5	6	1.5
36	3.5	1	5.5	1.5	3	0.5	3	–
Annual stocking rate LU/ha/yr	0.25		0.5		0.2		0.2	
Equivalent No. of sheep (ha/yr)	2		4		1.6		1.6	

Sources: Andrews and Rebane (1994); Crofts and Jefferson (1999).

Notes: Sheep, 60 kg live weight = 0.125 livestock units, Cattle, 250 kg live weight = 0.5 livestock units, i.e. in this instance, four sheep are equivalent to one beef animal. The annual stocking rate is the livestock units per hectare that can theoretically graze throughout the fifty-two weeks of the year.

suppresses growth, and on better soils concentrated grazing, trampling, dunging and urination may convert the heath into areas of neutral species-rich grassland. These grazed, poached and dunged grasslands or 'lawns' support rare plant species, such as pennyroyal, that have declined elsewhere now that greens and commons are no longer heavily grazed (Tubbs, 1991; Ausden and Treweek, 1995; UK Biodiversity Group, 1998a).

Precise stocking rates for heathland are difficult to define, as they depend on factors such as environmental conditions, heather phase and productivity, and proportions of grass heath and heather; however a stocking rate equivalent to one or two sheep per hectare per year is often used (Gimingham, 1992; Dolman and Land, 1995). In general, a change from degenerate dwarf shrub and scrub vegetation to short dwarf shrub vegetation, with a high diversity of species, can be achieved through a wide range of grazing treatments using different grazing animals, different stocking rates and with or without associated burning or cutting, although different grazing regimes may result in more subtle vegetation changes such as the suppression of some species in favour of others (Bullock and Pakeman, 1996).

Grazing systems on grasslands and heathlands

Grazing systems involve continuous grazing or sequences for moving grazing stock over an area of pasture. Set stocking or continuous grazing is the most common system on marginal semi-natural vegetation, and involves stock being present on the whole area throughout the appropriate grazing season. At lower stocking rates it produces a mosaic of tall and short vegetation and allows plants in undergrazed parts of the sward to flower and set seed, but high stocking rates may create an overgrazed turf. Rotational grazing involves dividing up the grazing area into compartments and moving the stock to fresh grazing compartments at appropriate intervals, and provides scope for generating different sward heights and vegetation structures by varying regimes between compartments (Crofts and Jefferson, 1999). The habitat requirements of the majority of invertebrates are poorly understood, and so rotational grazing compartments should be set out so that the entire population of a species, or insect life-cycle stage, is not confined to any one compartment (Ausden and Treweek, 1995). Confining animals to successive small portions of a site in winter may also be effective for grazing rank swards down to the desired level, although supplementary feeding or subsequent access to better grazing may be required (Crofts and Jefferson, 1999).

Disruption of grazing management on grasslands and heathlands

The significance of grazing management at a number of sites was illustrated by the outbreak of foot-and-mouth disease in 2001. Restrictions on the movement of livestock meant that some sites were grazed more intensively than usual, while other sites were not grazed. The possible effects of this disruption are shown in Table 3.12. These habitats may also be affected by changes in livestock farming post-foot-and-mouth disease, such as a move away from extensive livestock grazing systems (Robertson *et al.*, 2001).

Mowing and cutting

Grassland

Mowing or cutting is a form of management that is used in a variety of situations, including

Table 3.12: Possible effects of disruption of grazing management at lowland grassland and heathland sites following the foot-and-mouth disease outbreak in 2001

Grassland	*Heathland*
Lack of grazing	
Short-term. Inappropriate conditions for invertebrates and birds, e.g. loss of stock on culm grassland (series of grassland and mire habitats) may affect the marsh fritillary butterfly (*Euphydryas aurinia*), whose populations fluctuate wildly, making it prone to local extinction; e.g. withdrawal of sheep from the Malvern hills may affect the high brown fritillary butterfly (*Argynnis adippe*), which is rare and whose numbers are declining; e.g. loss of stock from permanent pasture around a greater horseshoe bat roost in Gloucestershire may reduce survival of young	*Short-term.* Inappropriate conditions for reptiles, invertebrates and birds, e.g. culling of cattle used to graze the east Devon pebbled heaths may affect the southern damselfly (*Coenagrion mercuriale*), which has benefited from grazing management
Longer-term. Spread of competitive grasses and woody species	*Long-term.* Difficult to find graziers willing to manage heathland; a move away from extensive livestock grazing systems, e.g. cattle were moved off the New Forest and about one-third of the 3,000 cattle that grazed the forest were culled in welfare culls
Increased grazing pressure	
Reductions in sward height: suppression of flowering. Heavily disturbed and enriched ground: invasion of weeds	Trampling damage and nutrient enrichment, e.g. around supplementary feeding areas on Cavenham Heath NNR

Source: Robertson *et al.* (2001).

agricultural grasslands such as meadows and wet grasslands, amenity grasslands and road-side verges. Where grazing is not practical, mowing may be an alternative management practice, but it is not selective, cuts the vegetation to a uniform height and therefore does not create the same patchiness of habitat conditions or structural diversity that is particularly important for invertebrates (Table 3.13). It is also a rapid method of removing vegetation, and can be damaging to invertebrate, amphibian, bird and mammal populations (Kirby, 1992; Benstead *et al.*, 1999). However, if cuttings are removed, mowing can maintain the low nutrient status of the soil, which can be beneficial for plant species diversity. Mowing or cutting is a traditional agricultural practice, and many grasslands are managed by farmers to produce a conserved forage crop (hay or silage) to feed livestock in winter (Plate 3.8). Haymaking has declined over the last twenty years and has been replaced by silage production, because the feed value of silage is higher than that of field-dried hay (Plate 3.9). Silage making preserves green forage crops, including grass, under airtight conditions in which bacteria convert soluble carbohydrates into lactic acid (fermentation). The acid conditions prevent further decomposition. For the best results in terms of nutritional value, fertilised grassland is cut for silage early in the season and then at six-week intervals (Crofts and Jefferson, 1999). The more traditional method of haymaking reduces the moisture content of the forage so that it can be stored. Most hay is made from improved lowland grassland swards in June or July when the digestibility (the proportion of nutrients extracted) of the

Table 3.13: Summary comparison between the effects of grazing, cutting and burning on vegetation

Effect	*Grazing*	*Cutting*	*Burning*
Removal of biomass?	Yes	Yes	Yes
Maintenance of low nutrient status of the soil?	Yes – but there will be some nutrient recycling through dung and urine. In the past, animals were often removed from the heath or grassland at night and this was more efficient in depleting nutrients	Yes – if cuttings and litter are removed	Yes
Creation of patchiness and structurally varied vegetation?	Yes – unless grazing is very heavy	No – unless different cutting regimes are used	Yes – but depends on size of controlled burn etc.
Selection and removal of particular species or plant parts?	Yes	No – but may benefit low-growing species	No
Creation of open ground for invertebrates and gaps for germination?	Yes – by trampling	No	Yes
Gradual removal of biomass?	Yes	No	No

Plate 3.8: Turning the hay, Hayfield, Derbyshire

Plate 3.9: Silage bales and improved grassland. Recent changes in agricultural practices on neutral grassland include further intensification and a shift from haymaking to silage production

grass is still high. Hay from semi-natural grasslands is usually cut later to ensure a reasonable yield, although digestibility may be reduced. In general, the digestibility of hay cut from semi-natural grasslands may be 20 per cent lower than that of grass cut earlier in the season from intensively managed grassland, and the energy value may be between 10 per cent and 40 per cent lower (Tallowin, 1997). It is clear that agricultural management objectives for high-quality hay and silage production conflict with maintaining the conservation interest of lowland semi-natural grassland, which involves no use of inorganic fertiliser and a single late cut for hay (Crofts and Jefferson, 1999).

As the intensity of management, such as the frequency of cutting or the level of grazing, increases there is a decline in the structural diversity of the vegetation, and in the abundance and biomass of some invertebrates (Drake, 1998; Bell *et al.* 2001). For instance, there is a decline in ground beetle (carabid) average body size as management intensity increases. This relationship reflects the requirement of larger ground beetles for a stable environment. Larger species are slow to mature and tend to be flightless, while smaller species mature rapidly and many are winged. These characteristics enable smaller species to colonise and exploit short-lived swards between cuts more readily than the larger species. Ground beetles are an important food source for many grassland birds, and the lack of larger prey on intensively managed sites is likely to have a negative impact on bird populations because it is more efficient for a predator to take fewer, but larger, prey items than many, small items (Blake and Foster, 1998).

For conservation management purposes the hay should not be cut before plants that depend on seed production have set seed, or before breeding birds have hatched. The timing of the cut will vary, depending on environmental factors such as altitude and climate (Table 2.4). Some late-flowering perennial species may be maintained by vegetative reproduction and intermittent recruitment from seed. To allow these species to occasionally set seed, cutting should be delayed in some years (for example, one year in five) to August or September. Lowland wet grassland and hay meadows can support populations of breeding waders and wildfowl. In the lowlands, hay cutting after 1 July and in the upland fringes after 15 July will usually ensure that chicks have hatched, although the situation should be monitored. If corncrakes are present, hay cutting should be delayed until after 31 July and mowing patterns should be employed that allow the chicks to escape. Mowing from the outside drives the birds into cover at the centre of the field, where they may be killed (Andrews and Rebane, 1994; Benstead *et al.*, 1999; Crofts and Jefferson, 1999). Appropriate cutting patterns can also allow hares to escape from silage fields (Scottish Natural Heritage, 2001). However, rigid prescriptions may not always be appropriate. For example, before a prescription to delay cutting until early August to restore corncrake numbers in the Argyll Islands ESA, the area supported a patchwork of fields cut between June and August. Following the prescription, the area became more homogeneous, consisting of uniform fields of improved grassland until early August, and uniform aftermath afterwards. The majority of these fields did not support corncrakes, but the change in management had a negative effect on choughs that formerly fed on fields that had been cut in early summer (Bignal *et al.*, 2001).

Grasslands that have been managed as hay meadow are usually less valuable for invertebrates than pasture because their vegetation structure is less varied and cutting management is less sympathetic to invertebrates (Kirby, 1992; Morris, 2000). Invertebrate diversity may be improved by a change in management, but the rich botanical community may be lost. For example, rotational cutting treatments or the conversion of meadow to continuously grazed pasture will reduce or eliminate early flowering species such as yellow rattle and snake's head fritillary. Hay meadows are traditionally used for grazing during the late summer and autumn after the hay cut. The cessation of aftermath grazing results in a decline in species richness, perhaps because livestock control competitive grasses and create gaps for regeneration by seed (Crofts and Jefferson, 1999). Where there is no history of traditional management on a conservation site, a cutting regime may be used to introduce and maintain differences in vegetation structure, through variations in the timing, frequency and patchiness of cutting, although the layout of management compartments should be planned with care (refer to 'Grazing', in this chapter) (Kirby, 1992; Benstead *et al.* 1999).

Road verges have considerable potential for encouraging wildlife, including mammals, birds, reptiles, amphibians and invertebrates, but inappropriate road verge management threatens semi-natural grassland, including surviving remnants of MG5 grassland, which may be useful sources of propagules for local restoration projects (Firbank *et al.*, 2000). Road verges are now often closely cut and herbicides are applied. Cuttings are rarely removed, and these can smother plants and increase the nutrient status of the soil. Late summer cutting is most appropriate for the conservation of herb-rich communities, and cuttings should be removed. However, safety factors are the main concern of roadside verge management, and the 1 m strip adjacent to the road is often mown up to six times a year by highway authorities. This zone is usually of limited conservation interest, owing to physical disturbance from traffic and the application of de-icing salt in winter, but the strips farther away from the road should be mown once or twice a year to maintain short meadow and tall herb zones (Crofts and Jefferson, 1999).

Heathland

Cutting of scrub, bracken, gorse and heather on lowland heaths reduces scrub encroachment, removes old growth and stimulates shoot production (Bullock and Webb, 1995). It is a more reliable method than burning and makes it easier to create small vegetation patches of different age and structure suitable for a variety of invertebrates (Species box 3.4). However, it does not remove accumulated litter or create bare ground (Kirby, 1992; Andrews and Rebane, 1994). Provided the heather is not too old, regeneration will take place from undamaged buds below the cut. Where regeneration from cut stems is poor because of the age of the heather, the plants should be cut close to the ground and the litter layer should be removed to encourage seedling regeneration. In all cases, cuttings should be removed from the site, as nutrient depletion is important for the maintenance of heathland, and to discourage the invasion of bracken and grass (Michael, 1993). Cutting should be carried out between mid-October and mid-February, when heather is not growing, ground-nesting birds are absent (e.g. woodlarks nest in March, nightjars nest in late August) and reptiles and invertebrates are hibernating or overwintering in the basal parts of plants (Andrews and Rebane, 1994). If heather seeds are required for heathland restoration, the heather should be cut between mid-October and late November, when capsules have a high seed content (Gimingham, 1992; Michael, 1993). Heathland is usually managed on a rotation of fifteen to twenty years to give a good range of vegetation ages and structures. In general, the preference for invertebrate management is to use small plots of 1 ha or less; however, larger plots may be better able to support a viable population of a species than can a smaller area. As with grasslands, plots should be arranged so that the whole of a vegetation type does not fall into a single management area. On larger sites, several management blocks can each be subdivided into small plots, so that the full range of vegetation structures is replicated a number of times across the whole heath (Kirby, 1992; Andrews and Rebane, 1994).

Scrub and bracken removal on grasslands and heathlands

Scrub encroachment is a common threat to grassland and heathland habitats, although a certain amount of scrub around or within a grassland or heathland site can provide an important habitat and food source for birds and invertebrates, and some scrub types can be of national importance (Species box 3.5). A detailed review of the conservation value of scrub has been produced by Mortimer *et al.* (2000). In general, management should aim to maintain a balance between scrub and open habitats, but scrub should not encroach at the expense of existing habitats of high conservation value. Scrub and bracken removal is also an integral part of the restoration of existing degraded semi-natural vegetation (Michael, 1993; Snow and Marrs, 1997). As succession occurs, the availability of some soil nutrients increases, and heathland restoration is likely to be more successful if excess nutrients are removed, for example through the removal of successional species and litter (Mitchell *et al.*, 2000).

Large blocks of scrub should be converted

Species Box 3.4: Crab spider *Thomisus onustus*

T. onustus is widespread in northern Europe, but prefers warm climates and is a rare British species, found only in southern England. It is usually found concealed among heather flowers, where it lies in wait to seize prey such as bees, butterflies and hoverflies. The abdomen is a distinctive shape and the first two pairs of legs are longer and stouter than the others. The spider can move forwards, backwards and sideways like a crab. The male and female differ in appearance. Male spiders are 2.5–3.5 mm long, females are 6–7 mm. They are beautifully camouflaged to hide from prey, and their colour varies considerably, especially in females. *T. onustus* may be almost white, yellowish, brownish or suffused with shades of pale pink or red. Individual spiders can slowly change colour to match their surroundings. At the end of their life cycle, during the summer of their second year, mature females weave two to four cocoons. Depending on the date of cocoon formation, spiderlings either emerge from cocoons in late summer and hibernate in vegetation, or remain in the cocoon over winter and emerge in spring. Adult spiders can tolerate food deprivation over a long period, but starved spiderlings survive for only about three weeks. Under laboratory conditions survival is increased if spiderlings feed on pollen and nectar when insect prey is scarce.

Misumena vatia is a crab spider that sits in flowers and ambushes insects in the same manner as *Thomisus*. It is widespread in northern Europe and occurs in England, Wales and southern Ireland, but is commoner in the south. It is typically found on white or yellow flowers.

This species has a rounder abdomen than *Thomisus*, and *Misumena* females may be white, yellow or greenish and can reversibly change colour to match the colour of flowers they are on. Using a model for the colour perception of bees that includes ultraviolet reflectance characteristics of the spiders, it has been shown that *Misumena* spiders are well matched to white flowers, but that the colour match on yellow flowers is not perfect. However, to bees approaching from a distance, spiders are probably well camouflaged on both white and yellow flowers.

Sources: Roberts (1985, 1995); Chinery (1986); Vogelei and Greissl (1989); Chittka (2001)

to narrow belts or scattered islands, as the junction between scrub and grassland or heathland is particularly important for invertebrates, birds and a range of grazing-intolerant plants such as bloody cranesbill which is characteristic of drought-resistant fringe communities, for example on the fringes of hazel scrub on limestone (Crofts and Jefferson, 1999). Similarly, nightjars are heathland birds that are associated with woodland edges or scattered trees. Where invasion of pine and birch presents a problem, larger trees can be cut and the stumps treated with herbicide, but hand pulling of young trees is particularly effective (Gimingham, 1992). Common gorse is often found on heath

Species Box 3.5: Hawthorn *Crataegus monogyna*

Hawthorn is a small tree or shrub up to 18 m tall. It is a very common native species growing in scrub, hedgerows or woodlands in a variety of habitats, but particularly on calcareous soils. The decline in traditional grazing and cutting management on marginal land has led to an increase in scrub development in the UK. Hawthorn scrub is particularly characteristic of the English lowlands and of marginal uplands in England and Wales. Scrub often exists as a mosaic with grassland and other open habitats. Spatial patchiness is beneficial to many plants and animals, in particular to birds and invertebrates, and those scrub areas with the greatest biodiversity are generally open and patchy, rather than closed. Lowland hawthorn and mixed scrub supports distinctive assemblages of warblers in the canopy closure stage of scrub development, and of yellowhammers, linnets and common whitethroats in the earlier stages. A total of 356 insect species have been recorded feeding on hawthorn (*Crataegus* spp.); of these, the majority (198) are species of lepidoptera. A number of these species are red data book or Biodiversity Action Plan species: sixteen red data book and two Biodiversity Action Plan insect species feed on hawthorn. However, in many cases, the structure of scrub is of more importance to associated species than its species composition. Some scrub communities, such as W21d (the wayfaring tree sub-community of hawthorn–ivy scrub) are important vegetation communities containing many rare native plant species. Hawthorn is increasingly being planted as part of wildlife enhancement schemes in the British Isles. However, a large proportion of commercially available material is sourced from continental Europe and does not establish as successfully as locally provenanced material.

Sources: Humphries *et al.* (1981); Mortimer *et al.* (2000); Jones *et al.* (2001a)

margins where the soil has been disturbed, while western gorse and dwarf gorse are low-growing plants found among heather in open heaths. Gorse supports many species of invertebrates and is used for feeding and nesting by Dartford warblers. The cover of gorse should be maintained at about 5 per cent and any litter should be removed from cleared areas to reduce nutrient enrichment (Rorison, 1986; Gimingham, 1992). In contrast, rhododendron is invasive, supports few invertebrates, can shade out open heathland communities and should be cleared by cutting, followed by removal or poisoning of the stumps. Resprouts should be treated with herbicide and the site should be cleared of rhododendron litter to promote colonisation by heath species (Gimingham, 1992) (Plate 3.10). In addition, a dense cover of bracken and the associated layer of litter can cause heathland vegetation to decline (Species box 3.6). Bracken has some value for

Plate 3.10: Former heathland, cleared of invasive rhododendron

invertebrates, but should be reduced to less than 5 per cent of total vegetation cover. Bracken control methods include cutting and/or treatment with the herbicide Asulam (Kirby, 1992; Andrews and Rebane, 1994). The most effective control treatment appears to be cutting, but, in a study carried out at Cavenham Heath in Breckland, eighteen years of continuous treatment by cutting, herbicide application or a combination of the two was not sufficient to eradicate bracken, and if bracken is to be suppressed on a long-term basis a continuous control strategy is needed (Marrs *et al.*, 1998a).

Burning on grasslands and heathlands

Burning is used infrequently to manage grassland in the UK. It is a treatment that has a sudden impact, and so has similar effects on invertebrates to cutting, except that it destroys the quantity of litter in the sward (Table 3.13) (Morris, 2000). Burning has been an important lowland and maritime heathland management technique, but was not as widely used as in the uplands. Burning, in combination with grazing, turf stripping and cutting, prevents scrub and tree colonisation, rejuvenates heather, creates a mosaic of stands of heather of different ages and maintains low soil nutrient concentrations because above-ground vegetation and litter are removed (Gimingham, 1992; Bullock and Webb, 1995; Countryside Council for Wales, 1997). Controlled burns also open up areas of bare soil, which can promote seed germination and be beneficial to invertebrates (Kirby, 1992). As with grazing, the use of burning as a management tool has declined in recent years, although it has been suggested that it should be reinstated, as other methods of management have failed to check succession adequately (Rose *et al.*, 2000).

Controlled burns are carried out in winter or early spring and are small, short-lived and of low intensity. Ideally a rotation of fire management should result in a number of stands of different ages (Rose *et al.*, 2000). Most of the heathland fires that now occur are accidental or a result of arson, and tend to cover a large area and burn intensely for long periods, especially under dry conditions. For example, since

Species Box 3.6: Bracken *Pteridium aquilinum*

Bracken is a competitive and invasive fern that produces extensive systems of rhizomes (underground stems). Fronds emerge from the rhizome from late spring onwards, and persist until autumn. Each large frond may be capable of producing over 300 million spores. Spores ripen from August to September, and are shed between August and October. They are able to survive long-distance dispersal on air currents, and may form part of a spore bank in the soil. Under suitable conditions, especially following disturbance, spores germinate to generate a heart-shaped plate of cells, which produces gametes (the gametophyte). Following fertilisation, a new bracken plant (the spore-producing stage, or sporophyte) is produced. Bracken forms large clonal patches, and although the life span of each section of the rhizome is probably less than forty years, individual plants may persist for hundreds or thousands of years. Bracken contains a number of defensive chemicals, many of which are carcinogens, and although a number of insects feed on bracken, herbivores seem to have little effect on its distribution and abundance. Bracken can survive in a wide range of conditions, and is found in shaded and unshaded habitats, particularly on deep acidic soils, but is sensitive to late frosts. Bracken stands can protect a number of rare plant species following woodland removal, and the persistence of violets under bracken contributes to the survival of the rapidly declining high brown fritillary butterfly. However, a dense bracken canopy and the associated layer of litter can cause other species to decline. Once bracken has become established, it is very difficult to eradicate because its rhizomes bear dormant buds that can produce new fronds in response to damage, and this regeneration can continue until the reserves in the rhizome are exhausted. Bracken is found on every continent except Antarctica. It is estimated to cover 443,000 ha of the UK; of this 273,000 ha are in England and Wales, 166,000 ha are in Scotland, and 4,000 ha are in Northern Ireland. The area of bracken in the UK has increased by about 6 per cent over the period 1990–1998. Expansion has mainly occurred in the uplands, at the expense of acid grassland, dwarf shrub heath and bog. Expansion of bracken is generally contrary to the objectives of the UK Biodiversity Action Plan.

Sources: Grime *et al.* (1988); Pakeman and Marrs (1992); Ennos and Sheffield (2000); Haines-Young *et al.* (2000); Warren *et al.* (2001)

1990 six fires at one urban heathland site have burned or reburned 60 per cent of the heathland area (Haskins, 2000). On a large scale, burning can create even-aged stands of heather that lack the structural variation required by many species. If the temperature of the fire is too high, or heather has been allowed to get too old before burning, it may be killed and replaced by bracken or grasses such as wavy-hair grass. Similarly, overburning of wet heaths may increase the dominance of purple moor grass (Gimingham, 1992; Andrews and Rebane, 1994). Uncontrolled burns on dry heaths can result in an increase of overland water flow, modify the soil water regime and induce soil erosion (Gurnell and Gregory, 1995). A study of the effects of accidental fires on the Dorset heaths by Bullock and Webb

(1995) concluded that, on a large scale, the mosaic of lowland heathland vegetation types appears to be fairly resilient to fire, but, more locally, fire can have a devastating effect on populations that occupy small patches of suitable habitat, particularly where heathlands are small, fragmented and isolated. In addition, repeated burns encourage a permanent change to grass-dominated heath (Bullock and Webb, 1995; Haskins, 2000).

Accidental burning can be disastrous, but managed burning is a valuable management tool. Clearly, the frequency of burning management and the size of plot are important. Burning should be restricted to the legal period and usually takes place under suitable weather conditions in late autumn or early spring, when adverse effects on invertebrates, birds and reptiles are minimised. The rotation period and plot sizes appropriate for controlled burning are similar to those for cutting heathland vegetation, i.e. a rotation typically of fifteen to twenty years, but as little as six to ten years in Breckland, and up to thirty years on some sites. Firebreaks should be created around the proposed controlled burn site, and appropriate codes and conditions should be followed (Gimingham, 1992; Kirby, 1992; Michael, 1993; Andrews and Rebane, 1994).

In summary, grazing, cutting and burning can be effective in maintaining and enhancing grassland and heathland habitats, but each method has a number of different attributes (Table 3.13).

Turf cutting or stripping

Heathlands

Traditionally, turfs were cut from lowland heaths for fuel, but turf removal can be a useful conservation management technique because it removes nutrients and provides areas of bare ground. It is best practised on a small scale with scattered plots so that heather seed can be washed or blown into the cut plots to supplement the seed bank. Turf and litter stripping may also be used to restore existing degraded semi-natural vegetation or to remove dominant grass species such as wavy-hair grass or purple moor grass and open the area to the re-establishment of heath species (Michael, 1993; Snow and Marrs, 1997). Good heather germination can be achieved if the depth of stripping is kept to a minimum, as about 90 per cent of viable heather seed is contained in the top 5 cm of soil, although sufficient seed for regeneration may be present down to about 10 cm. The seed bank is shallow because the soil fauna that normally assist the redistribution of seeds down the soil profile are not abundant in heathland soils. Turf removal should be used with caution where there are important populations of reptiles or soil-dwelling invertebrates (Putwain and Rae, 1988; Gimingham, 1992; Michael, 1993).

Grasslands

Turf removal or stripping may also be used as a means to enhance or create grasslands (following section and 'Substrate characteristics', this chapter).

Management of grasslands to enhance plant species richness

There is increasing interest in enhancing grassland swards such as MG6 semi-improved grassland communities to meet the objectives of agri-environmental schemes. Initially, fertiliser inputs should be discontinued, original groundwater levels should be restored and low-intensity traditional management should be reintroduced. However, high residual fertility may persist for a number of years, and the recovery of botanical richness may be slow (Bardgett and McAlister, 1999; Crofts and Jefferson, 1999). Botanical diversity can be increased by the introduction of appropriate seed mixes. The seed mix should aim to mimic the original composition of the community or that of unimproved grasslands characteristic of the area. Ideally, seed should be of local provenance (but see 'Source of Plant Material', this chapter). Methods of introduction include slot seeding, or oversowing following sward disturbance by harrowing, partial cultivation or turf removal. Alternatively 'plug plants' may be transplanted. The degree of disturbance and level of fertility can influence which species are able to establish from seed and persist long-term (Burke and Grime, 1996), but in general, successful establishment of introduced species tends to be greatest on sites with low nutrient status. Turf removal before sowing is usually the most effective method of increasing botanical diversity, but it may be inappropriate if the site retains some botanical interest (Crofts and Jefferson, 1999; Hopkins *et al.*, 1999).

Water levels

Grasslands

The water regime (timing, duration and degree of flooding) is an important environmental factor determining the species composition of a number of semi-natural grassland communities. These include periodically flooded well-drained meadows (MG4, Table 2.4), ill-drained permanent pastures (MG8 and MG10, Table 2.4), inundation grasslands, including coastal and floodplain grazing marsh (MG11 and MG13, Table 2.4) and rush pastures or fen meadows (M22, M23 and M24, Table 2.7). The water regime will also determine the suitability of sites for breeding waders and wildfowl (Crofts and Jefferson, 1999; Ausden, 2001). The maintenance of wet grassland depends on an adequate supply of water, appropriate distribution through a site and water-level control. Regular vegetation management through cutting or grazing is also important. Breeding waders and wildfowl are particularly good indicators of appropriate management on a site because they require a combination of low-intensity agricultural management and a high water table during the breeding period (March to June) to ensure that soil invertebrate prey remain close to the surface (Benstead *et al.*, 1999) (Species box 3.7). Where ecological monitoring information suggests that key features are not being maintained by the current water regime, modifications may be necessary. These can be achieved by changes in ditch maintenance practices, and the installation of water-level control measures such as bunds or levees, dams, sluices and pumps. The water quality may also be improved by the use of buffer strips or reed beds to minimise the input of nutrients to the site (Treweek *et al.*, 1997; Benstead *et al.*, 1999). Where a high water table cannot be restored, but conditions are otherwise suitable for nesting waders, feeding sites can be provided by the creation of extensive shallow pools (Andrews and Rebane, 1994).

Heathlands

Heathland vegetation communities are finely adjusted to soil hydrological regimes, and appropriate hydrological conditions are essential to the maintenance of wet heath and valley mire communities (Gurnell and Gregory, 1995; Rose *et al.*, 2000). If possible, wet conditions on areas that have been damaged by drainage should be restored. The quality of water draining on to the heath should be monitored and modified where necessary (Andrews and Rebane, 1994).

Species Box 3.7: Lapwing *Vanellus vanellus*

The lapwing breeding population declined by about 60 per cent between 1960 and 1987, and in 1998 the population of lapwings was estimated to be 62,923 breeding pairs, representing a further 49 per cent decline in eleven years. Arable and pastoral farmland are important for lapwings, because the optimum conditions for nesting are different from those of chick rearing, and the species favours areas where spring cereals and grass are found in close proximity. Lapwings breed in a variety of habitats, including spring cereal fields, all of which have short vegetation or bare ground. After hatching, birds lead their chicks to adjacent pasture, where livestock grazing maintains a short sward, and where grassland invertebrates, such as earthworms and leatherjackets, provide a reliable source of food. A number of changes in farming practices have affected breeding lapwing populations, including the loss of traditional mixed farming and the polarization of pastoral systems in the west of Britain and arable in the east. In addition, changing arable practices, such as an increase in autumn-sown crops, has reduced nesting site availability. The improvement of pasture, including drainage, decreased use of farmyard manure and increased use of inorganic fertilisers and pesticides, has created a taller, thicker sward that is unsuitable for feeding chicks and adults, and has reduced the diversity of invertebrate food. Improvement has also permitted higher stocking rates that have led to an increase in trampling and nest desertion. The move to silage production may also threaten chicks. High numbers of lapwings are now concentrated in reserves, indicating the importance of sympathetic management. For example, the number of lapwings on RSPB reserves increased by 220 per cent during the first seven years after water levels on low-lying grassland were raised to increase food availability. The invertebrate fauna of traditionally flooded grassland is different from that of unflooded lowland wet grasslands. When previously unflooded grasslands are artificially flooded, the abundance of breeding wader prey is reduced, but remaining prey is made more accessible. The best hydrological management for lapwings is therefore to provide a mosaic of unflooded and winter-flooded grassland and permanent and temporary pools to enhance feeding opportunities throughout the breeding season. The majority of unflooded lowland wet grasslands have been agriculturally improved, and patchy surface flooding may provide additional habitat for wetland plants and invertebrates. However, surface flooding should not normally be introduced to sites with existing conservation value, as it may damage rare invertebrate or plant communities. Measures are now required to enhance farmland habitats for lapwings in the wider countryside.

Sources: Hudson *et al.* (1994); Ausden (2001); Wilson *et al.* (2001)

Management for different species on grasslands and heathlands

The objectives of conservation management of grassland or heathland are usually to provide suitable conditions to maintain or enhance existing populations of characteristic and rare species, or to reinstate them. Maintaining or reinstating traditional management practices such as those outlined above is likely to

Species Box 3.8: Small heath butterfly *Coenonympha pamphilus*

The small heath is a member of a large family of predominantly brown butterflies characterised by prominent eye spots near the wing margins. The spots lure bird attacks away from the head of the butterfly to the wing tips. The upper side of the wings of small heaths is almost entirely orange. Larvae are green and feed on grasses. The small heath is a rather sedentary species that is found throughout the British Isles in relatively dry, open sites such as grassy places and on heathland. The species is usually scarce or absent on improved grassland, but common on unimproved grassland. Adults may be seen between April and October, normally resting on grasses or the ground, rarely on flowers. The small heath has a complex life cycle in the south of Britain, with overlapping generations. It is able to overwinter in different larval stages, so there may be staggered but overlapping emergences in spring. It is difficult to separate the long flight period of the small heath into generations. In the north of Britain there appears to be one generation, and the timing of this single generation can vary considerably from site to site. When raised under standard conditions, individuals from southern populations are more likely to produce a second generation, rather than entering diapause. Monitoring studies have indicated that small heath populations tend to be high if the previous summer was cool and wet, but warm weather in the current summer also seems to be beneficial. Males of the species actively search for females, but the small heath butterfly also has a lekking mating system, where males form aggregations that females visit only to mate. Leks occur around landmarks such as trees or bushes, where males compete to use perches. Small heath females favour dominant males on leks, which have a higher mating success than males outside leks.

Species distributions are commonly plotted as presence or absence in 10 km grid squares, but this does not reflect the proportion of each grid square that is occupied by the species. The term 'flight area' is used to identify the local distribution of butterfly species and is a more sensitive measure of butterfly distribution and abundance. Many butterfly species, including the small heath, appear to have a stable distribution at the national scale, based on occupation of grid squares. However, while the small heath remains geographically widespread, its flight area within occupied grid squares in one regional study area is estimated to have decreased by about 49 per cent between 1901 and 1997, suggesting that the small heath population has rapidly declined. The boundaries of most flight areas correspond to identifiable changes in habitat, and the observed population decline of the small heath, and other widespread butterflies, is probably due to habitat change, in particular, the loss or improvement of semi-natural grassland.

Sources: Chinery (1986); Pollard and Yates (1993); Wickman and Jansson (1997); Cowley *et al.* (2000)

support a number of characteristic species (Species box 3.8). However, different sites may require different approaches. In addition, conservation of rare species may require specialised management. Examples of the requirements of different species have been given throughout the text, but additional information on management for particular species is given in Gimingham (1992), Kirby (2001), Michael (1993), the UK Biodiversity Steering Group (1995), UK Biodiversity Group (1998a, b), Crofts and Jefferson (1999) and

UK Biodiversity Group (1999a–d). In many cases, the aims of UK Biodiversity Species Action Plans are to maintain the existing range of a rare species, maintain viable populations at these sites, regenerate populations at suitable historic sites (for instance, by means of the seed bank, e.g. pennyroyal, or through re-introduction, e.g. southern damselfly) and establish *ex situ* (off-site) breeding programmes to protect genetic diversity and create reserve populations. The implementation of Species Action Plans requires information derived from survey work and from research into species ecology, habitat and management requirements and captive breeding techniques (UK Biodiversity Steering Group, 1995; UK Biodiversity Group, 1998a). An example of species conservation (the sand lizard) is given in Chapter 4.

Summary

Lowland heathlands and semi-natural grasslands are priority habitats for conservation, because the UK has international obligations to protect them, they are at risk or are rare, and they contain important species (DOE, 1996a; UK Biodiversity Group, 1998a, b). UK Biodiversity Action Plan targets for grasslands and heathlands include habitat conservation (Table 3.3), but despite this target the extent and quality of the ecologically more valuable grasslands continues to decline (e.g. Plate 5) and heathlands are still affected by factors that lead to a decline in quality (Haines-Young *et al.*, 2000). The management of grasslands and heathlands for habitat maintenance and enhancement is well understood, but changes in the extent and management of grasslands and heathlands have generally been influenced by economic factors ('The history of grasslands and heathlands', this chapter). Today the management of grasslands and heathlands operates within a framework of legal protection and agricultural policy, and, although some advances have been made, valuable habitats are still threatened by land-use change. For example, there appear to be significant factors that prevent landowners from entering into conservation agreements, and changes in the targeting and funding of agri-environment schemes are needed (Buckingham *et al.*, 1999). In addition, regional and global changes such as atmospheric deposition of nitrogen and climate change need to be addressed at the national and international levels (INDITE, 1994; Ashmore, 1997; Hulme and Jenkins, 1998).

HABITAT RESTORATION AND EXPANSION

Land-use change has a major impact on biodiversity (Sala *et al.*, 2000) and large areas of land in the UK have been modified or degraded by human activity. For example, semi-natural habitats have been modified by agricultural intensification, and areas of land have been degraded by industrial processes or used for activities such as waste disposal (Harris *et al.*, 1996; Crofts and Jefferson, 1999; Watson *et al.*, 2000). Making safe and productive reuse of degraded or derelict land is central to achieving sustainable land use. The process of reuse is typically applied to contaminated or degraded land or landfill sites, but it is increasingly applied to arable land, improved pasture (Gibson *et al.*, 1987; Harris *et al.*, 1996; Hutchings and Booth, 1996a, b; Marrs *et al.*, 1998b; Owen and Marrs, 2000) and species-poor amenity grassland (Landlife and Urban Wildlife Partnership, 2000), in part thanks to incentives to landowners through Environmentally Sensitive Area and Countryside Stewardship schemes (Dolman and Land,

1995). Restoration of grassland and heathland habitats is an important objective of the UK Biodiversity Action Plan (Table 3.3), and, to contribute to it, 'Tomorrow's Heathland Heritage' is a £25 million, ten-year programme supported by English Nature and the Heritage Lottery Fund to restore and recreate substantial areas of heathland (English Nature, 2000). Semi-natural grassland would no doubt benefit from a similar programme. However, habitat restoration or creation is never an equal substitute for the conservation of existing semi-natural habitats (Gilbert and Anderson, 1998). For instance, although valuable grasslands can be created or restored, it appears to take at least a century for ancient 'late successional' calcareous grassland communities to develop, even under appropriate grazing management. Recently established 'early successional' grasslands contain many ruderal species, while ancient 'late successional' grasslands are characterised by the dominance of stress-tolerator species (Table 2.2) (Gibson and Brown, 1992; Brown and Gibson, 1993). Similarly, conservation of habitats *in situ* is preferable to transfer through turf or soil translocation from a donor to a recipient site, but where a site is to be lost to development, transplantation may offer a means of salvaging some of the conservation interest of the donor site.

There are a number of endpoint options available for the reclamation of damaged land, and for the establishment or restoration of vegetation to enhance biodiversity. Some of these are defined in Table 3.14, although the definitions are not always used consistently. On sites where the substrate is not too inhospitable for plant growth, intervention may not always be necessary, as bare ground will normally be colonised by vegetation, although the process may be rather slow (DOE, 1996b) (the 'do nothing' approach, Table 3.14). For example, pulverised fuel ash (PFA) is a residue produced by the combustion of coal in power stations that revegetates naturally, and ecological succession on PFA is analogous to sand dune colonisation (Crofts and Jefferson, 1999). PFA is toxic initially, containing high levels of salts and boron, but loses most of its toxic characteristics three to five years after disposal, owing to weathering. The substrate then resembles infertile calcareous silt and is colonised by a variety of species. In particular, orchids can form extensive stands on PFA between eight and twenty years after dumping. Woodland normally develops thirty years after abandonment (Shaw, 1994). In some cases, degraded land can be of considerable ecological importance, for example the biological interest of at least 13 per cent of SSSIs is associated with damaged land (DOE, 1996b) and artificial or post-industrial habitats can support a diverse invertebrate fauna including a high proportion of nationally scarce and rare species (Gibson, 1998).

In many cases, some form of management intervention will be required (Table 3.14). Restoration is a term that is widely used, and may be defined as the process by which an area is returned to its original state prior to degradation of any sort (Harris *et al.*, 1996), and in its strictest sense this means the end point should be a semi-natural community, identical to that which was there before (Bradshaw, 1996). Such an end point will rarely, if ever, be achieved, but may be the ideal target. Where something approaching the original semi-natural vegetation is achieved, it may be termed rehabilitation. Restoration may also mean operations such as scrub removal to restore existing degraded semi-natural vegetation (Alonso, 2001). In practice, the term restoration is often used in a broad sense, for example 'restoration' of heathland is usually regarded as successful if the dominant species

Table 3.14: Definitions of commonly used management and restoration terms

Term	Definition
Management	Operations (such as low-intensity grazing, cutting, burning, turf stripping) that are carried out to maintain the quality of existing grassland and heathland
Enhancement	Operations (such as grazing, cutting, burning, turf stripping) that are carried out to improve the quality of existing grassland and heathland vegetation. Similar to restoration *sensu* Alonso (2001)
Do nothing	No management, site may degenerate further, e.g. due to erosion, or succession may proceed
Speeding-up natural colonisation and succession	Facilitate the process of succession
Restoration	(1) Area returned to its original state; return to an original, identical state or perfect condition – restore the semi-natural habitat that was there before (Bradshaw, 1996). (2) Produce something fairly close to the original semi-natural vegetation, i.e. rehabilitation in the strict sense. (3) Often also defined as operations such as scrub removal to restore existing degraded semi-natural vegetation that has been partly lost to invasion of other vegetation; in this sense, management, enhancement and restoration operations may grade into one another (Alonso, 2001)
Rehabilitation	Produce something fairly close to the original semi-natural vegetation
Re-creation	Create heathland or grassland where heathland or grassland is currently absent – essentially similar to restoration/rehabilitation *sensu* Bradshaw (1996)
Replacement	To produce a substitute, e.g. agricultural land
Habitat creation	Establishment of new habitats, often mimicking existing semi-natural communities
Creative conservation	Creating simplified habitats using a few native, common species, where wildlife can develop
Community/habitat translocation/transplantation	An original community/habitat is moved from a donor to a recipient site
Remediation	To remedy, make good
Reclamation	To make land fit for cultivation or alternative function
Mitigation	To appease; damage limitation – mitigation may be an outcome of restoration, or restoration may be a requirement of mitigation

Sources: Bradshaw (1996); DOE (1996b); Snow and Marrs (1997); Gilbert and Anderson (1998); Landlife and Urban Wildlife Partnership (2000); Alonso (2001).

are re-established, and the appearance of a heathland is achieved (Gimingham, 1992; Bullock, 1998), when in the strict sense, what been achieved is 'rehabilitation' (Bradshaw, 1996). For the purposes of this book, the term 'restoration' is used in the broad sense. Part of the restoration decision-making process should involve deciding which particular community to restore to a site – for example, dry or wet heathland. This may be influenced by the characteristics of the sites available (e.g. impeded drainage), or the rarity or rate of loss of different plant communities (Bullock and Webb, 1995).

Where the aim is not to restore the original ecosystem, but to create a different one, it is termed replacement (Table 3.14). For example, a high proportion of land reclamation is to improved grassland such as amenity grassland or pasture, and agricultural grassland is the most common after-use for landfill sites (UK Biodiversity Steering Group, 1995; Simmons, 1999). In many cases it may be difficult to determine the semi-natural community that was present prior to modification, and, in cases where land is to be reclaimed primarily for conservation purposes, habitat creation or possibly creative conservation may be the most appropriate options. An important point is that the establishment of 'semi-natural' vegetation (i.e. restoration, rehabilitation, re-creation or habitat creation), or simpler systems (i.e. creative conservation), provides opportunities to enhance the landscape, increase biodiversity, link increasingly fragmented habitat patches of high value and create stepping stones for species movements (Veitch *et al.*, 1995; Gilbert and Anderson, 1998; Simmons, 1999).

Site selection

Several factors may influence the selection of sites for the restoration or creation of grassland and heathland, but it is important to select sites where habitat restoration or creation is most likely to be successful and/or may achieve the greatest increase in biodiversity (Mountford *et al.*, 1999). In addition, the conservation benefit of a newly restored habitat will probably be dependent on its position in relation to other habitat patches in the landscape (Bullock and Webb, 1995). The condition of the soil in terms of nutrient status, pH and drainage is of particular significance in determining the likely success of a project ('Substrate characteristics', this chapter) (Dryden, 1997). If conditions are not right the process is likely to fail, and it is advisable to work with or slightly modify the site conditions rather than make large, costly and possibly temporary changes to the soils (Gilbert and Anderson, 1998). Establishment of sites that extend or link existing heaths or semi-natural grasslands is likely to have the greatest benefit to wildlife. Larger sites allow more flexibility and diversity of management and reduce the possibility of local extinctions of species (Dryden, 1997). Possible impacts of climate change on habitats and species should also be considered. Sites for grassland or heathland establishment most likely to benefit wildlife are summarised in Table 3.15,

Table 3.15: Where to restore or create lowland heathland and grassland vegetation

Heathland	*Grassland*
On arable fields, improved grassland or forestry plantations that have a past history of heathland, particularly those that have been farmed less intensively	On arable fields or improved grasslands that have a past history of semi-natural grassland, particularly those that have been farmed less intensively
On land adjacent or near to existing areas of semi-natural heathland	On land adjacent to or near existing areas of unimproved, semi-natural grassland
On sites that link existing isolated blocks of heathland	On sites that link existing isolated blocks of semi-natural grassland
On non-heathland sites where soil conditions and other factors permit	Adjacent to other semi-natural habitats such as woodland, hedges or ponds

Sources: Bullock and Webb (1995); Dryden, (1997); Alonso (2001).

although new grasslands composed of native species will be of value in most situations. Geographic Information Systems (GIS) and remotely sensed data can be used to select candidate sites for restoration. For example, GIS have been used to identify areas of former heathland within 100 m of existing heaths (Veitch *et al.*, 1995).

Substrate characteristics

Whatever the planned end point of vegetation establishment (e.g. restoration, habitat creation or creative conservation), the aim is usually to create a system that is self-sustaining, in which ecosystem processes such as nutrient cycling occur, and which can develop and increase in diversity over time, although clearly if heathland or grassland is the desired target then appropriate management will subsequently be required to maintain the habitat (Bradshaw, 1996; Gilbert and Anderson, 1998). Both degraded and formerly improved soils may be unsuitable for the development of the desired vegetation type, and as a result, some form of reclamation will usually be required (Table 3.14) (Bradshaw, 1996; Harris *et al.*, 1996; Owen and Marrs, 2000). The constraints placed on ecosystem establishment in terms of substrate type, drainage, pollutants and so on will differ between mineral sites, landfill sites and agricultural land. For example, a fundamental difference between landfill sites and other derelict sites is the bioreactive nature of the underlying fill. This means that processes such as settlement of fill material and installation of gas control systems have to be integrated with the restoration process (Simmons, 1999). The problems, and appropriate amendments, of land categories such as metal mine spoil, colliery spoil, PFA or landfill site are described in detail by Harris *et al.* (1996), Wheater (1999) and Watson *et al.* (2000). Problems usually relate to physical characteristics of the substrate, toxicity (e.g. low pH or metals) and nutrient availability.

There are several approaches to dealing with toxic wastes but a rapidly developing technology is phytoremediation, which uses metal-accumulating terrestrial plants for reclamation, for example through phytoextraction, where metal-accumulating plants are used to concentrate metals from the soil into harvestable parts of roots and shoots, or phytostabilization, in which metal-tolerant plants are used to reduce the mobility of heavy metals (Salt *et al.*, 1995; Raskin and Ensley, 2000; Garbisu and Alkorta, 2001).

In many cases, huge efforts are made to establish improved grassland or an amenity mixture of species by modifying the pH and increasing the nutrient status of degraded substrates through the application of fertilisers or the introduction of legumes, such as white clover, that can accumulate nitrogen at the rate of 100 kg N ha^{-1} yr^{-1} (Bradshaw, 1996). For example, at Ince Moss, in Wigan, amenity grassland was successfully established on acidic colliery spoil by incorporating lime at a rate of 100 tonnes ha^{-1}, triple superphosphate fertiliser and NPK fertiliser at a rate of 375 kg ha^{-1} (Crombie and Sloane, pers. com.) (Plate 3.11). This approach may be appropriate in some circumstances, but in other cases there may be an opportunity to establish semi-natural vegetation types which require only slight changes in pH, low or moderate levels of nutrients, and which may link or extend isolated fragments of habitat. For instance, on substrates with low pH, such as colliery spoil, the creation of acid grassland or heathland may be more successful than the creation of improved grassland because subsequent oxidation of constituents of the spoil (iron pyrites) and leaching of applied lime from the

Plate 3.11: Grassland successfully established on reclaimed colliery spoil (*foreground*), Wigan

upper soil profile may lead to a decline in pH over time and a reduction in plant growth (Harris *et al.*, 1996). Heathland species can establish on acidic waste such as colliery spoil, some types of metalliferous waste, and china clay waste, either by natural colonisation, where there is an appropriate source of seed, or by introduction (DOE, 1996b; English Nature, 2000). In such situations an initial application of fertiliser and lime may be required to ameliorate the pH of very acid substrates and to help the establishment of heather, cross-leaved heath, bell heather, and grasses such as wavy-hair grass, but repeated applications are likely to cause suppression of heather by other species (Putwain and Rae, 1988).

In contrast, reduction of soil fertility (particularly phosphorus availability) is an essential feature of the restoration or establishment of heathland or semi-natural grassland on arable or improved grassland soils, on top-soiled landfill sites, or on soils invaded by bracken and scrub. One of the reasons is that heathlands and most unimproved semi-natural grasslands occur on soils with a low nutrient status, and high nutrient levels favour the establishment of competitive species capable of rapid biomass accumulation (INDITE, 1994; Dryden, 1997; Snow and Marrs, 1997; Tallowin and Smith, 2001). For example, on heaths where succession has occurred, those sites with soil nutrients and pH most similar to heathland values are easiest to restore to heathland (Webb *et al.*, 1995; Mitchell *et al.*, 1999). Where heathland is to be restored on areas of clear-felled forestry plantation, or areas cleared of rhododendron, litter should be stripped to encourage germination and establishment (Gimingham, 1992). On improved soils nutrient depletion may best be achieved by topsoil removal (Marrs *et al.*, 1998b). An example of nutrient removal during heathland restoration is given in Chapter 4. Topsoil inversion is a technique which uses deep ploughing to expose the underlying subsoil. This reduces surface fertility and buries the weed seed bank. A new initiative will use topsoil inversion in Cornwall to create appropriate soil conditions for heathland restoration on china clay waste colonised by bracken (Landlife, 2002).

Similarly, species-rich semi-improved MG6 sub-communities offer the best opportunities for the restoration of unimproved mesotrophic grasslands. Major differences in both vegetation and soil properties between unimproved mesotrophic grasslands and MG7 grassland suggest that significant intervention would be required for restoration (Critchley *et al.*, 2002). There is evidence that the presence of parasitic or semi-parasitic plants such as yellow rattle in a grassland sward reduces productivity by between 8 per cent and 73 per cent, reduces the proportion of grasses and increases the proportion of herbs. Sowing yellow rattle may therefore be a potential management tool to manipulate the balance between grass and herb species for the restoration of species-rich grasslands (Davies *et al.*, 1997). However, as with heathlands, the most effective method to reduce productivity may be topsoil removal, as subsoil generally has low levels of fertility (Watson *et al.*, 2000; Tallowin and Smith, 2001). Deep cultivation (topsoil inversion) using a plough also causes significant reductions in soil nitrogen and phosphorus concentrations and has a beneficial effect on the establishment of sown species (Pywell *et al.*, 2002; Landlife, 2002).

Communities may also require restricted conditions in terms of soil pH (Critchley *et al.*, 2002). To establish lowland acid grassland or heathland on improved soils a reduction in pH will be necessary, for example through the application of sulphur or acidic plant litter (Owen and Marrs, 2000). This contrasts with the situation in the uplands, where the substrate may be too acid as a result of acid deposition, and lime may need to be applied (Gilbert and Anderson, 1998). Calcareous grasslands require high pH and low nutrient levels, and are unlikely to be restored if soil conditions deviate from this (Critchley *et al.*, 2002).

Establishment of appropriate hydrological conditions is an important aspect of the restoration of wet heaths and grasslands (Critchley *et al.*, 2002). Raising water levels to reinstate a more natural hydrological regime can be achieved in a number of ways including groundwater protection and abstraction limitation, restoring river channels and raising stream beds, and bunding (creating artificial embankments) to retain water, but subsequent restoration may be hampered by the quality of the groundwater or floodwater (van der Hoek and Braakhekke, 1998; Bakker and Berendse, 1999; Benstead *et al.*, 1999).

Management of mycorrhizae can be important in ecosystem restoration, but soil storage can reduce mycorrhizal fungal density in soils. Innoculation of soils with VA mycorrizal fungi may have the potential to enhance soil fungal communities (Allen, 1992). However, commercially available fungal innoculants may not contain species that are appropriate to particular ecological situations, and some of the species may not be native to the UK (Merryweather, 2001).

Source of plant material

One approach to vegetation establishment is to use harvested or commercial seed or container-grown plants. Where possible, many advocate harvesting seed from local semi-natural grassland or heathland. Seed should be harvested in a way that does not modify the species composition of the donor community – for example, yellow rattle is an annual species that has a short-lived seed bank and can disappear if seed is collected too regularly (Gilbert and Anderson, 1998). An alternative approach is to use commercial seed, and this can have important implications for biodiversity. Biodiversity in the UK, including genetic variation within species and populations, is threatened by the loss and fragmentation of habitats, and ironically, restoration or habitat creation schemes may pose an additional threat because a proportion of plant material on sale for habitat restoration and creation is not of British native origin (Flora Locale, 1999, 2000). For example, possibly every plant of kidney vetch sown on roadsides in Britain between the early 1970s and mid-1990s belongs to a subspecies that is native to alpine valleys in central Europe (*Anthyllis vulneraria* ssp. *carpatica*), while the plant native to Britain (*Anthyllis vulneraria* ssp. *vulneraria*) is now scarce over much of the lowlands (Akeroyd, 1994). Even within individual widespread species, distinct variants frequently occur in particular geographical areas, and there is concern that seed that is not native or local in origin can create a number of problems. These include competition between native and introduced variants of the same species; crossing between native and introduced plants, leading to erosion of native genetic variation; missed opportunities to restore fragmented populations of locally adapted plants; potential consequences for associated fauna (e.g. some bumblebees are unable to feed from certain introduced cultivars of red clover because the flower has a different shape from native forms); and implications for monitoring rare species distributions (Akeroyd, 1994; Flora Locale, 1999; JNCC, 2001; Jones, 2001) (Species box 3.9). If possible, restoration and creation schemes should use native species of local origin, although information on the origin of seeds and plants is not always available, and there is a need to encourage the development of large-scale production of native, local provenance seed (JNCC, 2001). However, it is also recognised that, where appropriate, translocation of native species can have a valuable role in reversing the decline of species and movement of individuals can balance the effects of genetic drift in populations that have become isolated by fragmentation (Primack, 1998; Landlife and the Urban Wildlife Partnership, 2000; JNCC, 2001). Furthermore, species distributions are dynamic and are likely to change, for example, in response to climate change, (Harrison *et al.*, 2001; Landlife, 2002). The terms 'native origin' and 'local provenance' are not used consistently, but the definitions that Flora Locale (1999) asks suppliers to use are as follows. *Native origin*: planting material collected from parent plants whose wild origin, e.g. from an ancient meadow, is known by the supplier. *Local provenance*: material can be traced back to a wild site, but the parent plants are of unknown origin.

Vegetation establishment

A number of techniques for the establishment of grassland and heathland vegetation have been developed, including promoting natural colonisation, the use of harvested or commercial seed, the use of container-grown plants, the use of green hay or heather litter/shoots, turf transplantation and soil transfer (Table 3.16).

Animals

The emphasis of habitat restoration and creation is often to enhance biodiversity and link or expand existing areas of habitat. Most work concentrates on the establishment of vascular plants, although one of the aims of the Twyford Down M3 motorway mitigation programme was to provide habitat suitable for downland butterflies, particularly the chalkhill blue, and these expanded their populations as restoration progressed (ITE, 1997). However, micro-organisms and animals are often left to

Species Box 3.9: Garden bumblebee *Bombus hortorum*

Bumblebees are large, hairy social bees that form annual colonies. They are important pollinators of many crops and wild flowers. Only the mated queens survive the winter to start new colonies in the spring, which is an adaptation to a cool, seasonal climate. The nest is a ball of grass and moss with wax cells inside it. The young are reared on pollen and nectar. The pollen is carried back to the nest in large pollen baskets of hair on the back legs. *B. hortorum* is Britain's most widespread bumblebee. It usually nests among the bases of tall grasses, roots and plant litter just above, or just below the soil surface. It is a relatively large, long-haired bumblebee with bands of yellow hair on its thorax (the collar and scutellum) and the first abdominal segment is also yellow. The tail is white. It is abundant throughout the British Isles and is especially common in gardens. *B. hortorum* has a relatively short life cycle and colonies are quite small, often producing no more than about thirty to eighty workers. Workers and males appear similar to the queen, but workers vary in size in relation to food supply. *B. hortorum* may sometimes produce a second colony within one season. *B. hortorum* has the longest tongue of any bumblebee in the UK, and as a result it has the ability to take nectar from species with particularly deep flowers (i.e. long corollas), and specialises in these. *B. hortorum* is parasitised by the cuckoo bee *Psithyrus barbutellus*. The cuckoo bee resembles *B. hortorum*, but the abdominal plates are visible through the less dense covering of hair, and females do not have pollen baskets. Cuckoo bees do not produce workers. Females lay eggs in nests of *B. hortorum*, often killing the bumblebee queen, and bumblebee workers raise young cuckoo bees.

From the 1960s onwards, a number of bumblebee species have declined in Britain and Europe. For example the large garden bumblebee (*B. ruderatus*), which has forms similar in appearance to *B. hortorum*, has undergone a drastic reduction in range and abundance. A study on Salisbury Plain indicated that numbers of both short- and long-tongued bumblebees were positively correlated with the abundance of mouse-ear hawkweed and red clover, total flower abundance and flowering plant species richness. The most likely cause of bumblebee decline in Britain is the loss of open habitats, rich in the most rewarding food plants, particularly members of the pea family and the dead-nettle family. Certain species remain widespread, and some, such as *B. terrestris*, have impressive navigational abilities and regularly forage over 200 m, and possibly several kilometres from the nest. However, many declining bumblebee species are less mobile and are now largely confined to small remnants of semi-natural vegetation within a fragmented agricultural landscape.

Sources: Chinery (1986); Prys-Jones and Corbet (1987); Williams (1989); Osborne *et al.* (1999); UK Biodiversity Group (1999b); Goulson and Stout (2001); Carvel (2002)

Table 3.16: Summary of vegetation establishment techniques

Outline of process	*Aftercare*
Promoting natural colonization	
Amelioration of hostile conditions, or topsoil stripping. Thinly sow a nurse crop or companion species such as annuals or slow-growing grasses to provide appropriate conditions for colonising species. The range of species that colonise may be restricted by the characteristics of the site and by difficulties of immigration such as distance from suitable source populations. Introduction of a number of species-rich turfs can act as a source of seed to enrich the rest of the site. Natural colonisation creates valuable habitats for invertebrates. Most fauna will colonise naturally, but earthworms may need to be introduced	Removal of unwanted species, nutrient additions if required, protection from grazing, e.g. by rabbits, erosion control, monitoring of vegetation establishment is essential. Following establishment, appropriate grassland and heathland management should be introduced
Establishment from the seed bank	
The buried seed bank can be a valuable source of propagules for the re-establishment of vegetation, and small amounts of soil may be used as a source of seed to initiate revegetation on a site. Heathland soils contain a considerable 'bank' of buried viable seeds, and a number of heathland species such as common cotton grass can regenerate from buried dormant buds. The buried seed bank can be lost following severe fire or mechanical disturbance, but see 'Turf cutting or stripping', this chapter. Heather seedlings germinate successfully under conditions of relatively high light intensity and protection from desiccation. In contrast, the contribution of the seed bank to the process of restoration of calcareous grassland following the cessation of arable cultivation may be rather limited. Many calcareous grassland species have transient seed banks that may dwindle rapidly during cultivation and a seed bank of arable weeds accumulates in their place. Conducting a seed bank test by germinating the seeds in a number of soil samples may be a useful exercise (see Hutchings and Booth, 1996a; Gilbert and Anderson, 1998)	Removal of unwanted species, protection from grazing initially, erosion control. Monitoring of vegetation establishment is essential. Following establishment, appropriate grassland and heathland management should be introduced
Use of harvested or commercial seed	
May not accurately mimic a particular semi-natural community unless particular care is taken in making up the seed mix. The type of target community should be dictated by soil characteristics, but topsoil can be stripped or partially removed from the site, or mixed with nutrient-poor material to create a low fertility substrate. The seedbed should be prepared. Native	Germination of heather may not be immediate and can be intermittent. Disturbance of heathland sites should be avoided for a period of three or four years to allow characteristic species to establish. In the first year, grassland should be cut when the vegetation reaches 100–150 mm in height, and should be cut back to 50–75 mm. In the first year some sites

Table 3.16: contd

Outline of process	*Aftercare*
seed, ideally from a local source, should be used. Select-species mix to meet objectives, e.g. for rehabiliation or habitat creation, use a species mix based on the species frequency and abundance of a target NVC community appropriate to the area; for creative conservation select a simple mix of common, attractive, easily established species, appropriate for local conditions. Care should be taken in selecting the proportion of grasses and herbs. On fertile soil it may be advisable to omit legumes from the seed mix or reduce the proportion of grass to delay the spread of sown grasses. The quantity of seed recommended for creating grassland varies from 50 kg $ha^{-1}$to 1 kg ha^{-1}, depending on the characteristics of the site and the sowing technique used. Detailed information is given in Gilbert and Anderson (1998). Seed can be sown in spring or autumn. Seeding of heathland species may be more successful if seed is sown into an established, low-density sward	may not require cutting, others may need to be cut up to seven times. Cuttings should be removed. Following establishment, appropriate grassland and heathland management should be introduced. Monitoring is essential
Use of container-grown or plug plants or cuttings	
Method of enriching established grassland turf or establishing heather in a low-density sward of companion species. Establishment will be more rapid than from seed. *Grassland*: turf surrounding the plug should be removed, and the remaining sward should be cut and the clippings removed before planting. *Heathland*: protection from grazing is essential. Planting should take place in autumn or early spring. Impractical to cover an entire area with pot-grown plants, but patches may provide foci for subsequent spread	Undesirable species should be removed. On grassland it may be necessary to mow the competing sward several times during the first year (to about 75 mm, to avoid damage to establishing plants). On heathland, protection from grazing animals will be essential for three to five years. Following establishment, appropriate grassland and heathland management should be introduced. Monitoring is essential
Use of green hay or heather litter/shoots	
Does not involve loss to the original habitat, but uses materials removed as part of normal management practice. Appropriate for low-fertility sites (grassland) and low-fertility acidic soils (heathland). The new vegetation will rarely exactly match the original community. Heather litter or woody stems should be cut in late autumn when seed capsules are mature, green hay should be cut in summer when plants have set seed, but harvesting should be limited to avoid altering the species composition of the meadow. Heathland material can be used immediately or dried for later use. Green hay should not be stored. Heathland litter should be spread fairly thinly (5 mm) at	Undesirable species should be removed. Disturbance of heathland sites should be avoided for a period of three to four years to allow characteristic species to establish. In the first year, grasslands should be cut when the vegetation reaches 100–150 mm in height, and should be cut back to 50–75 mm, and may need to be cut between one and seven times. Cuttings should be removed. Following establishment, appropriate grassland and heathland management should be introduced. Monitoring is essential

Table 3.16: contd

Outline of process	*Aftercare*
a rate of 0.6–1.8 kg m^{-2}. Woody stems create suitable microsites for germination. Hay should be spread to a depth of 2–5cm, then removed once it has dried and dropped its seed	
Turf translocation	
Requires a donor site, and conservation *in situ* is preferable to transfer. Donor and recipient sites should be closely matched in terms of soil characteristics, site hydrology, topography and aspect. A habitat survey of vegetation and invertebrates should be carried out prior to the transfer to provide baseline information. Techniques include hand turfing and various forms of machine turfing. The greater the depth of soil that can be removed with turfs the better, e.g. the entire soil profile, or sufficient depth to avoid damage to plants. The cut turfs should be as large as possible to keep disruption to a minimum. Turfs should be individually labelled, so that they can be replaced in sequence. Turf should be moved directly from the donor to the receptor site. Translocation should take place when the soil is damp, but not very wet. The receptor site should be stripped of soil to a depth of the translocated turfs, and the soil surface should be cultivated. Turfs should be laid down in the sequence in which they were removed, close together and level with the surface of surrounding ground. Any gaps should be filled with soil from the donor site or subsoil	The newly transplanted turf may need to be watered, and undesirable species should be removed from the spaces between the turfs. Once established, previous management should continue. Monitoring of vegetation development is essential
Soil transfer/spreading (bladed material)	
Requires a donor site, and conservation *in situ* is preferable to transfer. Successful for vegetation that has a dormant seed bank. Soil may be excavated or turfs may be lifted then rotovated over twice their original area on to the receptor site. Donor and recipient sites should be closely matched. A habitat survey should be carried out prior to the transfer to provide baseline information. Soil transfer should not take place when the soil is waterlogged. The recommended depth of donor site soil stripping for heathlands is 50 mm, and for grasslands, up to about 150 mm. Topsoil at the receptor site should be removed and the surface lightly cultivated. Soil should be transferred directly. Possible to treat a larger area than the original donor site if the soil is spread at shallower depths, but causes greater disruption than turf	Undesirable species should be removed. Disturbance of heathland sites should be avoided for three to four years to allow characteristic species to establish. During the first year, grasslands should be cut when the vegetation reaches 100–150mm in height, and should be cut back to 50–75 mm. Grassland sites may need to be cut several times in the first year, and cuttings should be removed. Once established, appropriate grassland and heathland management should be introduced. Monitoring of vegetation is essential

Table 3.16: contd

Outline of process	*Aftercare*
transplantation, particularly to woody species and invertebrates. A light cover of a nurse crop may be useful to stabilise the surface	

Sources: Putwain and Rae (1988); Kirby (1992); Gimingham (1992); Michael (1993); Ash *et al.* (1994); DOE (1996b); Hutchings and Booth (1996a); Pywell *et al.* (1996); Shaw (1996); Gilbert and Anderson (1998); Manchester *et al.* (1998); Bakker and Berendse (1999); Crofts and Jefferson (1999); Warren (2000); Willems (2001).

establish naturally, and there is a delay in the recruitment and stabilisation of characteristic animal populations (Wheater and Cullen, 1997; Wheater, 1999). In particular, earthworms are often slow to colonise (Bradshaw and Chadwick, 1980). Micro-organisms and soil animals such as woodlice, springtails, mites and worms play a vital role in nutrient cycling, and, where populations are low, dead organic matter may accumulate and the availability of nutrients can decline (Hunter *et al.*, 1987b; Harris *et al.*, 1996) (Species box 3.10). In restoring or creating habitats, the opportunity should be taken to create suitable conditions to benefit animals. For example, establishing a wide variety of food plants and creating a varied topography and vegetation structure to favour invertebrates (Dolman and Land, 1995; DOE, 1996b).

Community or habitat translocation

Community translocation involves the removal of an assemblage of species from a donor site and the attempt to establish it as a functioning community at a new receptor site. The process is often termed habitat translocation or transplantation, but community translocation is a more accurate term. This method of environmental mitigation has become common in Britain, and its use is thought to be increasing. A Habitat Transplant Site Register was compiled in 1988 by the Nature Conservancy Council, which indicated that most recorded translocations in Britain have been of grassland and marsh (forty-eight translocations) and heathland (twelve translocations) (Byrne, 1990; Bullock, 1998). The technique has been proposed as offering a solution when sites are threatened by development, but few translocations have been the subject of detailed, long-term post-transplant monitoring programmes and there is some debate as to whether translocations are successful (Jefferson *et al.*, 1999; JNCC, 2001). An assessment of ten British translocation projects (nine of grassland and one of heathland) showed that in all cases there were differences between plant or invertebrate communities of donor or control sites and those of the translocated areas. In some cases, the changes were small, but many showed major changes that were greater than the usual temporal changes in untranslocated communities. These large changes were linked with disturbance during translocation, environmental differences between donor and receptor sites and poor after-care and management. Invertebrate communities always showed large post-translocation changes (Bullock, 1998). On relocated grassland in County Durham plots showed an initial change in plant and invertebrate community structure, with a decline in species characteristic of control or

Species Box 3.10: Earthworm *Lumbricus terrestris*

There are a number of British earthworm species, most of which belong to the family Lumbricidae. Some occur near the surface of the soil, while others live much deeper. *L. terrestris* commonly burrows down to 1 m and sometimes to 2.5 m. Soil temperature and moisture content seem to be the most important factors determining activity, for example, *L. terrestris* is most active in English pastures between August and December and between April and May. Earthworms also show diurnal patterns of activity, and are most active between 6.00 pm and 6.00 am. Although earthworms are hermaphrodite (individuals have both male and female reproductive organs) they usually cross-fertilise. Several egg cocoons are produced, each containing a number of eggs, but it is rare for more than one egg in a cocoon to develop further. Earthworms feed mainly on decomposing plant material in the soil, but soil microorganisms such as fungi are also an important food source. Earthworms show a preference for early successional species of fungi, which may be used as cues by earthworms to detect fresh organic resources in soil. Earthworm communities maintain soil structure and recycle nutrients, and can be indicators of good soil quality. If soil organic matter content is maintained and the use of harmful chemicals avoided, higher earthworm populations can be encouraged. For example, in one study in Suffolk earthworm diversity and abundance were 2.4 times greater in an ancient permanent pasture than in an organic field, and four times greater than in mixed and intensive arable regimes.

The New Zealand flatworm (*Arthurdendyus triangulatus*) is an introduced species that is a predator of earthworms. The flatworm appears to cause a reduction in the number of *L. terrestris* and *Aporrectodea longa*, which have

low rates of reproduction and long life cycles. Earthworm species with higher reproductive rates may be less affected by flatworm predation. *L. terrestris* and *A. longa* are both deep-burrowing species, and a reduction in their number may have long-term consequences for soil drainage.

Earthworms have a positive effect on the structure and fertility of reclaimed soils, but natural colonisation and establishment occur slowly. Increasingly, earthworms are introduced to accelerate the process of soil amelioration and reclamation. Soil amelioration is most successful if the most active soil-forming species are introduced and soils are amended with organic matter to favour earthworm establishment. Metal contamination of soils can reduce earthworm abundance and species diversity. The absence of certain sensitive species from contaminated sites suggests that earthworms may be a useful group for quantifying the effects of pollutants on soil biodiversity. Because of the annual patterns in earthworm activity, sampling should be carried out in spring or autumn to accurately determine community structure.

Sources: Wood (1995); Edwards and Bohlen (1996); Spurgeon and Hopkin (1999); Blakemore (2000); Bonkowski *et al.* (2000); Jones *et al.* (2001b)

pre-translocation limestone grassland plots, e.g. blue moor grass and *Tapinocyba pygmaea* (a spider), followed by a period of increased diversity with time (Cullen and Wheater, 1993; Bullock, 1998; Jefferson *et al.*, 1999). Dry heaths or dry grasslands appear to translocate most successfully, but for wet grasslands such as flood meadows and wet heaths there may be problems caused by disruption of hydrological patterns. In

Table 3.17: An example of the relative costs of seeding, soil spreading and turf translocation in a revegetation project by Good *et al.* (1999)

Seeding (locally collected seed)	*Turf transfer and rotovation (soil spreading)*	*Whole-turf tanslocation*
£17,000 ha^{-1}	£25,000 ha^{-1}	£52,000 ha^{-1}

translocation schemes the vegetation is usually transferred as turfs, but an alternative method is soil transfer (Table 3.16) (Byrne, 1990). Soil transfer is cheaper than turf translocation (Table 3.17). However, turf transplantation retains larger areas of intact vegetation than soil transfer (Byrne, 1990; DOE, 1996b) and the survey by Bullock (1998) showed that soil spreading causes larger changes in plant communities than turfing techniques in terms of species composition and relative abundance. Community translocation is unlikely to achieve the preservation of a complete community intact (the 'preservation' aim favoured by conservationists) but, if managed carefully, this technique may be used to create a community that resembles the pre-translocated state in order to save components that would otherwise be destroyed (the 'mitigation' aim, favoured by developers) (Bullock, 1998).

Summary

Evaluation of the success of restoration, creation and translocation schemes very much depends on the definition of the objective, e.g. 'restoration' or 'preservation' in the strict sense, or alternatively 'mitigation' (Table 3.14). These techniques will never exactly mimic a semi-natural habitat and are not a substitute for them, but, with care, they may establish communities that resemble the semi-natural original. We must adopt the dual approach of habitat conservation and expansion if we are to meet the objectives of the UK Biodiversity Action Plan (Table 3.3).

4

CASE STUDIES

•

The following case studies have been selected to illustrate, in a little more detail, aspects of some of the issues that have been identified in previous chapters, i.e. climate change, nitrogen deposition, community translocation, creative conservation, habitat restoration and species conservation. Cross-reference has been made with earlier chapters so that the case studies can more easily be placed in context.

THE RESPONSE OF TWO CONTRASTING LIMESTONE GRASSLANDS TO SIMULATED CLIMATE CHANGE

This case study is an illustration of some of the responses of grassland plant communities and associated insect herbivores to simulated climate change.

Different plant communities, when exposed to changes in temperature and precipitation, such as those predicted in climate change scenarios (Hulme and Jenkins, 1998) (Chapter 3, 'Pollution'), will respond in different ways and at different rates. The ability of a community to maintain its composition and biomass in the face of climate change may be defined as resistance, and the rate of recovery as resilience (Grime *et al.*, 2000). Field manipulations of local climate over five years at Buxton in Derbyshire and Wytham in Oxfordshire were used to investigate the potential impact of climate change on two contrasting limestone grasslands (Table 4.1). The grassland at Buxton is ancient stable sheep pasture, whereas the grassland at Wytham was under arable cultivation until 1982, and is still in a dynamic, early successional state (Gibson and Brown, 1992; Sternberg *et al.*, 1999; Grime *et al.*, 2000). The

Table 4.1: Comparison of the grassland at Buxton, Derbyshire, and Wytham, Oxfordshire

Characteristic	*Buxton*	*Wytham*
Origin	Ancient sheep pasture	Arable until 1982
Annual precipitation	1,300 mm	680 mm
Mean annual temperature	8°C	10°C
Extractable P (mg P kg soil $^{-1}$)	3.6	8.0
Total number of species recorded	60	67
Perennial grasses (No. of species)	16	9
Sedges (No. of species)	5	0
Annuals (No. of species)	0	22

Source: Grime *et al.* (2000).

Buxton site is cooler, wetter and less fertile (e.g. in terms of extractable phosphorus) than the Wytham site. The species richness of the two sites is similar, but the composition is different. Wytham is characterised by a large proportion of fast-growing annuals (ruderals, Gibson and Brown, 1992), whereas slow-growing sedges (stress tolerators) (Table 2.2) are an important component of the vegetation at Buxton (Table 4.1) (Grime *et al.*, 2000).

Three simulated climate treatments were applied. These were elevation of winter temperature from November to April each year by means of heated cables at the soil surface (warmer winters); summer drought through July and August by means of automatically operated rain shelters; and increased summer rainfall from June to September, applied as an even spray (Plate 4.1). Climate treatments were applied singly or in combination (e.g. warmer winters with summer drought, or warmer winters with increased summer rainfall). Full details are given in Grime *et al.* (2000). In addition, there was a natural drought in 1995 that allowed the impact of extreme climatic events to be evaluated (Buckland *et al.*, 1997). Responses of plants to simulated climate change were monitored at both sites. Responses of invertebrates were predominantly monitored at Wytham (Masters *et al.*, 1998).

The plant community at Wytham was more responsive to the extreme drought in 1995 and to the experimental drought and warming treatments in terms of biomass and species composition than the Buxton community. For example, the plant cover and biomass of both communities declined in response to drought, particularly when combined with winter warming, but the effect was greater at Wytham than at Buxton. Plant cover and biomass were significantly increased in Wytham plots receiving additional summer rainfall, whilst the response to extra summer rain at Buxton was less pronounced (Sternberg *et al.*, 1999; Grime *et al.*, 2000). At Wytham, species composition varied in response to treatment; for example, species richness was significantly increased in plots receiving supplementary rainfall (Sternberg *et al.*, 1999). In contrast, the Buxton community varied little in terms of species composition in response to treatment,

Plate 4.1: Experimental plots at the Buxton Climate Change Impact Laboratory, University of Sheffield

although sedges declined in abundance in response to warming and drought (Grime *et al.*, 2000), and in some species, flowering occurred up to a month earlier in warmed plots than in control plots. Several years of climate manipulation appeared to have little effect on the seedbank at either site (Akinola *et al.*, 1998; Leishman *et al.*, 2000). The results suggest that fertile, early successional grasslands composed of fast-growing or short-lived species will respond rapidly to climate warming and reduced or increased rainfall (Table 4.2). More mature or less fertile grassland composed of slow-growing, stress-tolerator species will respond more slowly, i.e. they are likely to be more resistant to climate change (Gibson and Brown, 1992; Grime *et al.*, 2000), although complementary laboratory experiments suggest that they are likely to be less resilient (i.e. slower to recover) (MacGillivray *et al.*, 1995).

The direct and indirect effects of simulated climate change on invertebrates at Wytham were investigated in a relatively well understood group of insect herbivores, the Auchenorrhyncha, which includes leaf, plant and frog hoppers (Masters *et al.*, 1998). Direct effects of change may include changes in egg hatching date in response to warmer winters. Indirect effects may include responses to climate-induced changes in food plant quality or structure, for example, at Wytham, warmer winters and supplemented summer rainfall led to an increase in vegetation cover, while warmer winters and reduced summer rainfall led to a decline (Masters *et al.*, 1998; Sternberg *et al.*, 1999; Grime *et al.*, 2000). Fewer adult Auchenorrhyncha were recorded from winter-warmed plots, particularly in spring and summer, and individuals recorded from warmed plots were significantly older than individuals recorded from unwarmed plots. There was no difference in the rate of development of nymphs (young stages) in warmed and unwarmed plots; however, nymphal hibernation was broken earlier in warmed plots, and eggs in warmed plots hatched up to six days earlier than in unwarmed plots. Young adults that mature earlier may be at greater risk from events such as late frosts, leading to a reduction in adult numbers (Table 4.3). Supplemented summer rainfall led to a large increase in the number of Auchenorrhyncha, which was directly related to an increase in vegetation cover. However, numbers were maintained or slightly increased under summer drought conditions, when the cover of vegetation decreased significantly. Drought stress can lead to an increase in food quality (increased levels of soluble nitrogen and carbohydrate) in foliage, and higher food quality is associated with a higher abundance of leafhoppers. This relationship could explain the response of Auchenorrhyncha to summer drought (Table

Table 4.2: Summary of the responses of the vegetation of two contrasting limestone grasslands to simulated climate change

Summary of responses of vegetation to simulated climate change	
Buxton	*Wytham*
Biomass less responsive to summer drought	Biomass more responsive to summer drought
Biomass less responsive to enhanced summer rainfall	Biomass more responsive to enhanced summer rainfall
Species composition varied little in response to treatment	Species composition varied greatly in response to treatment
More resistant to climate change	Less resistant to climate change

Source: Grime *et al.* (2000).

Table 4.3: Summary of direct and indirect effects of climate change on a group of insect herbivores (Auchenorrhyncha)

Direct effects	*Indirect effects*
Winter warming	*Increased summer rainfall*
▼	▼
Egg hatch and end of nymphal hibernation occur earlier	Increased vegetation cover
▼	▼
Young adults that have matured early may be at greater risk from late cold spells	Increased numbers
▼	
Reduction in adult numbers	*Decreased summer rainfall*
	▼
	Likely increase in food quality
	▼
	Numbers maintained or increased

Source: Masters *et al.* (1998).

4.3) (Masters *et al.*, 1998). Because of the apparent differences in response to the direct effects of winter warming and the indirect effects of different rainfall patterns, it is difficult to predict the likely impacts of climate change on Auchenorrhyncha. However, the effects of milder winters suggest that these insects will mature earlier, which may lead to a mismatch of seasonal patterns of growth and reproduction between Auchenorrhyncha and associated species such as food plants, predators and parasitoids. There is also evidence that populations of Auchenorrhyncha may increase whether summer rainfall increases or decreases, owing to changes in different attributes of the vegetation (Masters *et al.*, 1998).

EFFECTS OF ELEVATED NITROGEN AND DROUGHT AT BUDWORTH COMMON

The effects of nitrogen deposition and climate change, including changes in precipitation, are likely to become increasingly important in causing biodiversity change (Sala *et al.*, 2000) (Chapter 3, 'Pollution'). For example, it has been proposed that elevated nitrogen levels may increase the vulnerability of heather plants to natural stresses such as drought and herbivore attack, which lead to heather canopy breakdown, while promoting the growth of grass (Ashmore, 1997). This case study is an illustration of a field experiment at Budworth Common to test this prediction.

Budworth Common is a lowland heath SSSI in west Cheshire, owned and managed by Cheshire County Council. Map evidence shows that Budworth Common was heathland in 1831, but subsequently the common became invaded by scrub and birch, until active management for conservation and restoration purposes was introduced in 1970. The plant community in the part of the common used for field experiments has been classified as H9 heather – wavy-hair grass heath (Plate 4.2). When experiments began in 1996, the heather was at the building, actively growing stage of growth (Cawley, 2000).

In an experiment to study the effects of enhanced nitrogen deposition and drought on lowland heathland, experimental additions of nitrogen in the form of ammonium nitrate

Plate 4.2: Experimental plots at Budworth Common

were applied to experimental plots from May 1996. The experiment is ongoing. Background deposition at the site between May 1996 and May 1997 was estimated to be 20.5 kg N ha^{-1} yr^{-1}, a value close to the estimated critical load that has been suggested for the conversion of lowland heath to grassland (15–20 kg N ha^{-1} yr^{-1}, Table 3.4) (Bobbink and Roelofs, 1995). Experimental additions of 20 kg N ha^{-1} yr^{-1}, 60 kg N ha^{-1} yr^{-1} or 120 kg N ha^{-1} yr^{-1} took total deposition to well above the estimated critical load. The experimental inputs are similar to the range of nitrogen deposition recorded in parts of the UK (over 30 kg N ha^{-1} yr^{-1}) and the Netherlands (over 100 kg N ha^{-1} yr^{-1}, downwind of intensive farming areas) (INDITE, 1994). In addition to the nitrogen treatment, a summer drought treatment was established between May and September 1997 in which droughted plots were covered and received no rainfall. Subsequently, in mid-summer 1998, there was a heather beetle infestation at Budworth Common, which allowed the effects of elevated nitrogen on sensitivity to attack by heather beetle to be assessed (Cawley, 2000).

The results of the experiment appear to support the prediction that plants receiving high levels of enhanced nitrogen may suffer greater damage by drought and heather beetle than control plants receiving no additional nitrogen. At Budworth, heather shoot water potential (a measure of plant water status, Bannister, 1986) and growth were reduced more by drought in plots receiving high levels of additional nitrogen (120 kg N ha^{-1} yr^{-1}) than in plots receiving little or no additional nitrogen (Cawley, 2000). An increased supply of nitrogen typically results in a stimulation of shoot growth at the expense of root growth, which would be expected to increase plant sensitivity to drought (INDITE, 1994). In the year following the drought (1998) there were reductions in the density of heather shoots, and a simultaneous increase in wavy-hair grass in the high-nitrogen droughted plots (Cawley, 2000; Caporn, pers. com.). In addition, damage from heather beetle in the summer of 1998 was significantly greater in the high-nitrogen (60 kg N ha^{-1} yr^{-1} and 120 kg N ha^{-1} yr^{-1}) treatments than in the low-nitrogen (20 kg N ha^{-1} yr^{-1}) treatment or the untreated controls (irrespective of drought treatment), and further advanced the spread of wavy-hair grass at the expense of heather (Cawley, 2000). While it appeared in 1998 that wavy-hair grass would continue to increase at the expense of heather, this has not happened. Instead, a nitrophilic (nitrogen-loving) moss, *Hypnum cupressiforme*, has become common in the nitrogen-enriched plots (Caporn, pers. com.). It is unclear what the final outcome of this field experiment will be. The mechanisms that may cause the changes observed to date are summarised in Table 4.4.

This field experiment shows that elevated nitrogen may promote breakdown of the heather canopy. However, the effects were seen in the short term (three years) only at very high levels of additional nitrogen. Longer-term experiments are needed to determine the effects of lower levels of nitrogen deposition. A number of additional factors such as lack of management and disturbance probably operate in combination with elevated nitrogen to contribute to the decline in heathland and transition to grassland in southern Britain and the Netherlands (Marrs, 1993).

MANCHESTER AIRPORT SECOND RUNWAY ENVIRONMENTAL MITIGATION PROGRAMME

This case study is an illustration of the management of existing grassland (sixteen sites covering 19.6 ha), habitat creation (new grassland) (24 ha in thirty-four blocks of mean 0.7 ha) and grassland translocation (4.5 ha) at Manchester Airport, which formed part of the mitigation package for the development of a second runway (Table 3.14).

As a result of an assessment of runway capacity, Manchester Airport plc identified the need for a second main runway, and a planning application for the development was submitted to Macclesfield Borough Council in July 1993. In January 1997 planning permission was granted, subject to a number of conditions. As part of the proposal to develop the second runway, Manchester Airport plc, Cheshire County Council and Manchester City Council entered into agreements under Section 106 of the Town and Country Planning Act (1990) (Table 3.7). In part, these agreements required the airport to provide an enhanced package of environmental

Table 4.4: Elevated nitrogen and the vulnerability of heather plants to drought and herbivore attack

Drought	*Heather beetle*
Increased nitrogen	Increased nitrogen
▼	▼
Likely increase in proportion of heather shoots to roots	Likely increase in food quality of heather
▼	▼
More vulnerable to drought	More vulnerable to heather beetle attack
▼	▼
Breakdown of the heather canopy	Breakdown of the heather canopy
▼	▼
Spread of grass/moss	Spread of grass/moss

Sources: INDITE (1994); Ashmore (1997); Cawley (2000).

mitigation proposals. The broad aims of the mitigation plan included the creation of a habitat network around the second runway to minimise the potential barrier effect that the new runway posed, and a reduction of the impact of habitat loss (Anderson, 1994; Manchester Airport, 1997). For example, a range of ecologically valuable grasslands were identified within the area affected by the construction of the second runway. Some swards supported interesting plant communities, while others were of value as foraging areas for great crested newts and badgers. Under the terms of planning permission and Section 106, Manchester Airport plc was committed to carry out ecological mitigation work that included the management of existing grassland, grassland creation, grassland translocation and long-term monitoring. Monitoring over a fifteen-year period was designed to ensure that measures and agreements were properly implemented, to detect any problems at an early stage so that management may be modified as appropriate, and to measure the degree of success of mitigation objectives (Manchester Airport, 1997). A steering group including representatives of local councils and nature conservation bodies such as English Nature was set up to oversee the work.

Management of existing grasslands. Existing ecologically valuable grassland (e.g. MG5c, *Cynosurus cristatus–Centaurea nigra, Danthonia decumbens* sub-community; crested dog's-tail–common knapweed community, heath-grass sub-community) on land owned by Manchester Airport plc is leased to local farmers and managed under the Countryside Stewardship scheme. The management objectives for existing grassland of nature conservation value are to retain current value and enhance less diverse swards where possible. Opportunities were identified where swards would benefit from having characteristic plant species added.

Habitat creation. The aim was to create MG5c, crested dog's-tail–common knapweed community, heath-grass sub-community, but, where the ground was too wet, patches of vegetation suited to damper soils were also created. MG5 is the typical grassland of grazed hay meadows treated in the traditional fashion on neutral brown soils in lowland Britain. It is becoming increasingly rare as a result of agricultural improvement (Chapter 2, 'Mesotrophic grasslands').

Areas of improved grassland were selected for MG5c grassland creation, and prepared by soil stripping and cultivation. In 1999 new grasslands were sown, using a seed mix from local existing MG5c meadows of nature conservation value (i.e. seed of native source and local provenance). Seed was collected several times over the growing season to obtain a broad range of species. The seed mix was screened and, if some species were poorly represented, additional native seed was purchased. Seed was sown at a rate of 30 kg ha^{-1} between August and October of the year of harvesting, using agricultural equipment (Plate 4.3).

Grassland translocation. The aim of the translocation was to rescue as much as possible of valuable habitats that would otherwise be lost in the development (Anderson, 1994) (i.e. mitigation, Chapter 3, 'Community or habitat translocation'). The threatened grassland swards were translocated to specially prepared sites during autumn and winter of 1997 and early 1998. Receptor sites were prepared by topsoil stripping, and were ripped to relieve compaction and prevent waterlogging. Donor sites were surveyed to provide baseline information. Where possible, grassland was transplanted as turfs. Turfs of sufficient depth to include the root zone and a protective layer

Plate 4.3: Grassland and pond creation, Manchester Airport

were lifted by machine and individually numbered. They were then relaid in the correct order, starting from the far side of the receptor site and working backwards towards the access point to avoid damage. Gaps between turfs were avoided and any gaps that formed were filled with soil from the donor site. Where grassland would not form turfs, topsoil, including roots and vegetation, was scraped off and transported directly to the receptor site without storage. Soil was respread at the same depth and extent as at the donor site.

Management of created and translocated grassland. Management measures include the introduction of an appropriate grazing regime or late hay cut and aftermath grazing once the vegetation was established. Fencing has been erected to allow controlled grazing. Invasive and undesirable species are removed, and in all cases no herbicides, pesticides or chemical fertilisers are applied.

Monitoring. Existing grasslands are being monitored to ensure that management agreements are being implemented and that the grassland quality is being maintained or improved. Monitoring of created grasslands includes an assessment of sward growth and development, assessment of earthworm populations and assessment of management requirements. Monitoring of translocated grasslands includes detailed recording of vegetation, monitoring of surface-active invertebrates using pitfall traps, monitoring of terrestrial molluscs, assessment of earthworm populations, and monitoring for use by bats, badgers and great crested newts.

Preliminary results of grassland translocation: vegetation monitoring of translocated grassland sites, 1995–2000. Translocated grasslands were monitored in 1995, 1996 and 1997 prior to translocation to obtain baseline information, and were subsequently monitored in 1998, 1999 and 2000. Monitoring will continue for fifteen years following translocation. Six translocated areas (S1–S6) and one that remains *in situ* (S7) have been monitored (Table 4.5). The pre-translocation baseline survey results indicated that grassland plots most closely resembled a number of NVC community types, including MG5c, CG6a and MG10b, or were poor matches to NVC communities. Translocated areas were monitored using two approaches: randomly placed 30 cm × 30 cm quadrats, plus a series of larger (2 m × 2 m) quadrats to sample homogeneous patches of vegetation. The frequency classes of species within each

Table 4.5: Comparison of 1995 (pre-translocation) and 2000 (post-translocation) NVC community types for translocated and *in situ* grassland sites

Site	*NVC match with 1995 data*	*NVC match to 2000 data*	*Comments*
S1 (translocated)	MG5c	30 cm by 30 cm quadrats – no match 2 m by 2 m quadrats – mosaic of MG8 and MG5b, with indications of MG6b	Current analysis of 1995 baseline data indicates a slightly closer match with MG5a. Moved as turfs. Following transplantation, slightly moister, less acidic and increasingly fertile conditions indicated
S2, S2a and S2b (translocated)	No close match, but CG6a indicated	30 cm by 30 cm quadrats – no match 2 m by 2 m quadrats – MC9 (poor match)	S2 transferred as turfs, S2a and S2b by soil tansfer. Originate from rabbit-grazed roadside (A538) slope, similar to S7. Transferred to a gently sloping area. Match with MC9 reflects an increase in red fescue and Yorkshire fog. Following translocation, slight increase in acidity and more marked increase in fertility indicated
S3 (translocated)	No close match, but MG10b indicated	30 cm by 30 cm quadrats – similar to 1995 2 m by 2 m – MG10b and some elements of SD17	Appears to be developing towards MG10b. Following translocation, a decline in fertility indicated
S4 (translocated)	No close match	30 cm by 30 cm quadrats – MC9 2 m by 2 m quadrats – MC9, MG5, MG1	S4 was translocated because of its associated newt population. The sward had originally been sown in 1990 on an old brickworks site. The site continues to support common spotted orchids following transplantation
S5 (translocated)	No close match	30 cm by 30 cm and 2 m by 2 m quadrats – poor match, but most similarity to MG11	S5 was originally a sparse sward, translocated by soil transfer. Following translocation, an increase in fertility indicated.
S6 and S6a (translocated)	No close match	S6: 30 cm by 30 cm and 2 m by 2 m quadrats – MC9. S6a: no match, but number of species is increasing and area of bare ground is decreasing	S6 transferred as turfs, S6a, soil transfer. Match to MC9 reflects an increase in red fescue and Yorkshire fog following translocation
S7 (*in situ*)	No close match, but CG6a indicated	30 cm by 30 cm quadrats – no match, slight indication of SD16 2 m by 2 m quadrats – poor match with CG6a and SD16	Heavily rabbit-grazed roadside (A538) slope. Six more species were recorded in 2000 than in 1999 from within the 30 cm by 30 cm quadrats. Fall in fertility indicated since 1995

Source: Manchester Airport (2001).

monitored plot were estimated in accordance with NVC methodology: I, occurring in between 1–20 per cent of quadrats, II, 21–40 per cent, III, 41–60 per cent, IV, 61–80 per cent, and V, 81–100 per cent (Rodwell, 1992). Species of frequency classes IV and V are referred to as community constants. No value for abundance (percentage cover) was estimated.

Overall, the vegetation monitoring results for all translocated areas suggest that after three growing seasons the swards and individual species are still adapting to their changed situations, and it is not possible to predict how closely the swards will eventually match the baseline communities. There is evidence of increased eutrophic conditions in some of the plots, but this may be modified once grazing is introduced. For example, an area of MG5 grassland (S1) has shown a number of changes since translocation (Table 4.6) (Manchester Airport, 2001). The frequency of common bent, for example, which is typically a constant in the community, has declined since translocation. In addition, some species have been gained. However, the frequency of common knapweed, which is typically a constant species, and lends its name to the community type, has increased steadily from 1998 to 2000, although it has not yet achieved the pre-translocation frequency (Table 4.6). Analysis of 30 cm × 30 cm random quadrat data does not show similarity to any NVC community, but analysis of 2 m × 2 m quadrats placed in areas of uniform vegetation shows that the sward is developing into a mosaic of MG5b, MG8 and MG6b mesotrophic grassland (Table 4.5). Slightly moister, less acidic and increasingly fertile conditions are indicated in 2000 (Manchester Airport, 2001). Species losses and gains and indications of community change are common in the first three to four years following community translocation. The changes in the translocated plots are probably largely due to disturbance and differences in environmental conditions between the donor and recipient sites, but the consequences of translocation will not be apparent until the community has settled down to a new dynamic equilibrium (Bullock, 1998).

LANDLIFE AND CREATIVE CONSERVATION AT STOCKBRIDGE MEADOW

This case study illustrates the process of creative conservation. The aim of creative conservation is to create simplified habitats where wildlife can develop (i.e. it does not seek to mimic semi-natural communities). It can also provide an opportunity to engage people in conservation by creating accessible sites that people can enjoy (Landlife and Urban Wildlife Partnership, 2000). Stockbridge Meadow is an important creative conservation site that was established in 1993.

Stockbridge Meadow was formerly an area of amenity grassland (1.7 ha) in Knowsley, Merseyside, that was largely unmanaged. The environmental charity Landlife chose to use the creative conservation technique to enhance the site by sowing a simple mix of attractive common native species. In particular, the success of the technique relies on appropriate ground preparation and the selection of species that are suited to the site conditions (Table 4.7) (Landlife, 2001). For example, many of the desired wildflower species do not establish well if sown on fertile soils, in part owing to the seedbank of competitive weeds. Topsoil stripping is a technique that serves to reduce fertility and remove the seedbank of unwanted species.

In 1993 the ground at Stockbridge Meadow was stripped of topsoil. In this instance, the

Table 4.6: Frequency of species in grassland (area S1) in 1995 (pre-translocation) and 1998 and 2000 (post-translocation)

Species recorded in quadrats – site S1	*Frequency*		
Common name	*2000*	*1998*	*1995*
Creeping bent	IV	I	—
Creeping buttercup	IV	IV	—
Yorkshire fog	IV	IV	II
Common knapweed	IV	II	V
Crested dog's-tail	IV	III	V
Ribwort plantain	IV	IV	V
White clover	III	IV	I
Glaucous sedge	III	I	IV
Sweet vernal grass	III	V	II
Perennial rye grass	II	III	V
Timothy	II	II	II
Meadow buttercup	II	III	III
Red fescue	II	I	III
Hairy sedge	II	I	I
Red clover	II	I	II
Common bent	I	V	V
Bird's-foot trefoil	I	—	I
Autumn hawkbit	I	II	IV
Quaking grass	I	I	—
Common mouse-ear	I	—	I
Marsh thistle	I	I	I
Ash	I	I	—
Jointed rush	I	—	—
Hard rush	I	I	I
Field woodrush	I	—	—
Common sorrel	I	I	I
Dandelion sp.	I	I	III
Common sedge	I	—	—
Carnation sedge	I	—	II
Creeping thistle	I	I	I
Cat's ear	I	—	II
Bristle clubrush	I	—	—
Rough meadowgrass	I	—	—
Ragwort	I	—	—
Devil's-bit scabious	I	—	II

Source: Manchester Airport (2001).

Notes: I Occurring in 1–20 % of samples. II 21–40%. III 41–60%. IV 61–80%. V 81–100%.

Table 4.7: The process of creative conservation using the soil stripping technique, with reference to Stockbridge Meadow

Acquire authorisation and planning permission

Assess the depth of topsoil and drainage/level of the water table

Initiate a tender process with topsoil contractors

Inform the public

Remove topsoil, ideally between early May and early September, when conditions are dry

Prepare the soil for sowing as soon as possible after soil removal

Create a fine seed bed by rotovating

Sow a wildflower and grass seed mix that is suited to the conditions on site at 5 g m^{-2} (50 kg ha^{-1}), or drill at 10 kg ha^{-1}. Stockbridge Meadow was sown in 1993 (Table 4.8)

Harrow and roll to maximise seed contact with the soil

Cut annually, once in September to a height of 20–50 mm

Source: Landlife, unpublished information.

sale of the topsoil covered all costs incurred, including seed, ground preparation and sowing, and the project was completed at no cost to the local authority. However, there is considerable variation in the topsoil market, and this may not always be possible. Variation in the amount of topsoil removed can create heterogeneity on site. For example, at Stockbridge Meadow an area that retained a proportion of topsoil was more fertile than the area that had been stripped. In the spring of 1993 the Stockbridge Meadow site was prepared and sown in blocks of three or four species. A number of species were also sown in 1997 and 1999 (Tables 4.7 and 4.8). In total, twenty-two species, plus a nurse crop of Italian rye grass, were deliberately introduced.

Stockbridge Meadow was surveyed in the summer of 2000, seven years after it had been sown, and the results indicate that grassland creation was successful. The more fertile patch, where a proportion of topsoil had been retained, supported coarse grasses that were beneficial to invertebrates. However, only the more competitive wildflower species are likely to coexist with coarse grasses in the long term. The majority of the site had been soil-stripped, and here a few competitive species had colonised but were growing poorly in the infertile soil. Over the site as a whole, all the twenty-two sown species were present, although some had established more successfully than others. For example, cowslip and field scabious were locally abundant, and spreading, whilst cow parsley was rare. In total, eighty-four species of higher plant plus seven species of mosses and lichens were recorded in 2000, indicating colonisation by sixty-one species of higher

Table 4.8: Seed mix sown at low quantities by precision drill at Stockbridge Meadow in 1993

Botanical name	*Common name*	*Ratio (by mass)*
Mix A		
Leontodon autumnalis	Autumn hawkbit	500 g
Lotus corniculatus	Common bird's-foot trefoil	1.5 kg
Primula veris	Cowslip	1.0 kg
Mix B		
Daucus carota	Wild carrot	500 g
Leucanthemum vulgare	Oxeye daisy	1.0 kg
Oenothera biennis	Common evening primrose	500 g
Mix C		
Leucanthemum vulgare	Oxeye daisy	1.0 kg
Ranunculus acris	Meadow buttercup	2.0 kg
Succisa pratensis	Devil's-bit scabious	10.0 g
General mix area		
Achillea millefolium	Yarrow	100 g
Centaurea nigra	Common knapweed	200 g
Leucanthemum vulgaris	Oxeye daisy	200 g
Prunella vulgaris	Selfheal	500 g
Edge mix		
Anthriscus sylvestris	Cow parsley	400 g
Centaurea scabiosa	Greater knapweed	100 g
Dipsacus fullonum	Wild teasel	200 g
Knautia arvensis	Field scabious	100 g

Source: Landlife, unpublished information.

Additional sowings: Crested dog's-tail (*Cynosurus cristatus*) and common bent (*Agrostis cappilaris*) sown together at a rate of 3g m^{-2}, with a nurse crop of Italian rye grass (*Lolium multiflorum*) (which is annual or biennial) at a rate of 2g m^{-2}. Kidney vetch (*Anthyllis vulneraria*) was introduced as harvested seed heads strewn over the site. In addition, other species were subsequently sown: *Centaurium erythraea*, common centaury (1997); *Ophyrys apifera*, bee orchid (1997); *Dactylorhiza praetermissa*, southern marsh orchid (1999); *Rhinanthus minor*, yellow rattle (date unknown).

Note: Seed mixes were sown in blocks; information on the pattern of sowing not available.

plants in a period of seven years. Some of these species, such as eyebright and lady's bedstraw, are cultivated by Landlife and are likely to have arrived as impurities in seed mixes, but others, such as silver-hair grass, seem to have colonised naturally. Bee orchids and southern marsh orchids had established. Although southern marsh orchids were sown on site in 1999, this species typically takes several years to flower from seed, so flowering individuals recorded in 2000 probably colonised naturally (Landlife, unpublished information). Areas of wildflower grassland created by Landlife elsewhere in Merseyside also support interesting species that appear to have colonised naturally. For example, Pickerings Pasture (Plate 4.4) is a site on the banks of the Mersey that supports star-of-Bethlehem in addition to sown species. At Stockbridge Meadow patches of bare ground suitable for colonisation were still present seven years after sowing, probably owing to low substrate fertility as a result of soil stripping, and disturbance. The sown wildflowers may serve as a nurse crop, providing shelter and an appropriate microclimate for colonisers to establish (DOE, 1996b).

The benefits of creative conservation include habitat creation and public involvement, and population translocations can form a valuable part of programmes aimed at reversing the decline of once common species (JNCC, 2001). The species used for creative conservation projects by Landlife are typically common, native or naturalised ones, cultivated using seed that is sourced from a variety of sites in Great Britain (Landlife, 2001). There is increasing awareness that plant populations from different parts of Britain may be genetically distinct, and there is some concern that creative conservation may promote the translocation of populations beyond their native range (Chapter 3, 'Source of plant material'). However, local seed of formerly widespread native species is not always available, and Landlife advises against introducing wildflower seed to inappropriate sites.

HEATHLAND MANAGEMENT AND RESTORATION AT ARNE, DORSET

The Arne nature reserve occupies a peninsula situated about 6 km east of Wareham, and is owned and managed by the RSPB. Part of the area is covered in deciduous and naturally regenerated Scots pine woodland, but about 340 ha is heathland. The majority of the heathland at Arne (about 80 per cent) has been classified as the H2 heather–dwarf gorse dry heath community. In moister areas, this community grades into H3 cross-leaved heath–*Sphagnum compactum* wet heath. The area represents one of the largest remaining fragments of the formerly extensive Dorset heathland, and is particularly important because the H2 and H3 community types found here have

Plate 4.4: Creative conservation: cowslips at Pickerings Pasture

restricted distributions (Haskins, 1978; Rodwell, 1991; Gimingham, 1992). The reserve supports a number of interesting species. For example, the humid heath areas support Dorset heath and marsh gentian, both of which have very local distributions in Britain (Stace, 1997). All six British reptiles, including the smooth snake and sand lizard, occur at Arne, and there is a rich invertebrate fauna, including rare dragonflies, and a number of southern species, at or near the northern edge of their ranges. The reserve also supports populations of Dartford warbler and nightjar (Gimingham, 1992; John Day, pers. com.).

The reserve is managed to maintain the heathland habitats and to conserve the diversity of the flora and fauna (Gimingham, 1992). For example, the dry heath at Arne is managed to maintain a range of age structures. Heather is managed on a thirty to forty-year cutting/burning rotation, and gorse on a fifteen-year cutting rotation. Further details of a broad range of management practices at Arne are given in Gimingham (1992). The reserve is

Table 4.9: Nutrients removed when successional species (vegetation) and their litter are cleared from successional sites at Arne (kg ha^{-1})

Successional species	*Component*	*Potassium*	*Nitrogen*	*Phosphorus*
Birch	Vegetation	106.5	317.1	15.2
	Litter	55.9	1,029.6	19.1
	Total	*162.4*	*1,346.7*	*34.3*
	% in litter	34.4	76.5	55.7
Pine	Vegetation	238.0	597.9	21.9
	Litter	62.0	1,520.6	23.3
	Total	*300.0*	*2,118.5*	*45.2*
	% in litter	20.7	71.8	51.5
Rhododendron	Vegetation	116.8	206.2	10.3
	Litter	92.7	1,476.1	14.2
	Total	*209.5*	*1,682.3*	*24.5*
	% in litter	44.2	87.8	58.0
Common gorse	Vegetation	173.1	748.2	5.6
	Litter	46.3	1,913.4	14.0
	Total	*219.4*	*2,661.6*	*19.6*
	% in litter	21.1	71.9	71.4
Bracken	Vegetation	14.2	113.6	4.1
	Litter	24.7	448.2	8.0
	Total	*38.9*	*561.8*	*12.1*
	% in litter	63.5	79.8	66.1

Source: Mitchell *et al.* (2000).

Notes: *% in litter*: The percentage of the total nutrients removed in litter.

Plate 5: Species-rich hay meadow, June 2001. By April 2002 this meadow had been ploughed up and completely destroyed

Plate 6: Bumblebee on scabious in Deep Dale, a reserve owned by Plantlife. A number of bumblebee species have declined as a result of agricultural intensification (*Matt Brierley*)

Plate 7: Ragwort, a weed of overgrazed grassland that is toxic to stock but a valuable source of nectar to invertebrates (*Matt Brierley*)

Plate 8: Heathland restoration on former forestry plantation land at Arne, Dorset, managed by the RSPB

also managed to restore areas of heathland that have been lost to scrub invasion or forestry.

One of the major threats to heathland is the invasion of scrub and succession to woodland, despite conservation management (Rose *et al.*, 2000). As succession occurs the availability of some soil nutrients increases, and heathland restoration is likely to be more successful if excess nutrients are removed during the restoration process. This may be achieved through the removal of successional species and litter. A study to estimate nutrient removal during heathland restoration at Arne revealed that large amounts of nutrients are removed in this way (Table 4.9) (Mitchell *et al.*, 2000).

Successional vegetation types differ in terms of the amount of nutrients held in vegetation and litter; for example, the greatest concentration of nitrogen is found in gorse vegetation and litter, which is not surprising, as this species can fix atmospheric nitrogen (Grime *et al.*, 1988), while the greatest concentration of phosphorus is removed in pine vegetation and litter. Litter is clearly an important reservoir of nutrients, although the proportions held in litter differ with species and nutrient (Table 4.9). There are likely to be variations in the nutrient composition of vegetation and litter between sites, but the results from Arne indicate that vegetation and litter removal is an effective way to remove nutrients. This management technique may serve to mitigate the effects of enhanced nitrogen deposition, although turf stripping may also be required (Power *et al.*, 1995, 1998; Mitchell *et al.*, 2000).

An understanding of the use of management regimes to reduce nutrient levels is important because the feasibility of restoring heathland is greatest on those sites with soil nutrients and pH most similar to heathland values (Webb *et al.*, 1995; Mitchell *et al.*, 1999) (Chapter 3, 'Substrate characteristics'). This is illustrated by the different rates of heathland regeneration on two sites at Arne. The first is a site that is reverting from arable cultivation, the second is a site regenerating after forestry clearance. Although levels of nutrients and pH have not been recorded, and the sites may differ in a number of ways, differences between the two sites in terms of soil nutrient status and pH can be inferred on the basis of past management.

Plate 4.5: Heathland reversion on formerly arable land at Arne, managed by the RSPB

Arne parish has a history of cultivation on the better soils and rough grazing on the heath (Webb, 1986), but in 1966 an area of heathland was reclaimed to arable. The land was ploughed, lime and fertilizer were applied and the plot was cropped for just two years and then abandoned. Plate 4.5 shows the former arable plot in the foreground, and typical Arne heathland in the background. It is clear that even after over thirty years since the brief period of arable cultivation ceased, the vegetation has not reverted to typical former heathland, although there has been considerable regeneration, probably from the residual seed bank (Webb *et al.*, 1995). A number of studies have shown that reduction of soil fertility (particularly phosphorus availability) and pH is an essential feature of the restoration of heathland on arable soils, as high fertility may allow grasses to persist and hinder the establishment of heathland species (Webb *et al.*, 1995; Dryden, 1997; Snow and Marrs, 1997). At

Species Box 4.1: Sand lizard *Lacerta agilis*

The UK has the sub-species *L. agilis agilis*. The sand lizard relies on areas of exposed sand for egg incubation and is mostly confined to heathland and sand dunes. It is found locally in England, and there is one long-established introduced colony in Scotland. The species measures up to 20–22 cm in length, including the tail. Sand lizard markings are extremely variable, although they are generally grey or brown in colour, and the males' flanks become vivid green during the breeding season. In the UK sand lizards from heathland habitats are typically spotted, while those from dunes tend to be striped. Sand lizards are active during the day, and spend the night in sandy burrows they have excavated themselves, or that have been dug by mammals. The same burrows are used for hibernation. Sand lizards have overlapping home ranges, each of several hundred square metres. Mating occurs in late April and May. Eggs are laid in early June in unshaded, exposed sand, and the young emerge in August and September. In warm years, females may lay two clutches of eggs. Sand lizards hibernate for six or seven months between September–October and March–April. Males and females become sexually mature at two and three years respectively, and can live for eight to ten years. Within heathland and sand dune habitats the species is particularly associated with sunny topographical features in mature dry heath, and dense marram grass stands on coastal frontal

sand dunes, where interfaces between vegetation and open sand provide adjacent areas suitable for basking, feeding, egg incubation and shelter. Habitat management for sand lizards focuses on these favourable 'hot spots'. It includes bracken and scrub clearance, fencing to reduce disturbance and exclude grazing livestock, and sand exposure by blading or turfing to provide egg-laying and incubation sites, to link populations and to create firebreaks. Sand lizards are naturally slow to colonise, and conservation management also involves translocation and captive breeding of populations.

Sources: House and Spellerberg (1983); Nicholson and Spellerberg (1989); UK Biodiversity Steering Group (1995); Corbett and Moulton (1995, 1998); Inns (1999b)

Arne there has been no intervention to restore heathland on this plot because the grassy patches support nectar-bearing flowers that are a valuable food source for invertebrates (Chapter 2, 'Fauna of heathlands') (Webb, 1986; Kirby, 1992; John Day, pers. com.).

In contrast, an area of former forestry plantation at Arne has been actively restored to heathland. The site was a former dense plantation of forty to sixty-year-old pine trees. The trees were felled in the late 1980s to early 1990s and the timber was removed. The area was raked and pine litter was removed from the site. Bracken regrowth was sprayed with herbicide, and the site was left to regenerate from the seed bank. Regeneration was successful, and eleven to twelve years later the site is dominated by heather (Plate 8). The site has been lightly grazed in the last two years, and rhododendron and bracken will be cleared in the near future (John Day, pers. com.). On sites where pine has been felled, and the litter cleared down to the mineral soil, relatively large amounts of phosphorus are removed and the soil nutrient concentrations and pH are similar to those of heathland (Mitchell *et al.*, 2000). Forestry plantations on former heathland are therefore potential sites for the highly successful regeneration of heathland, although if the seed bank is no longer present it will be necessary to introduce seed (Webb *et al.*, 1995; Gimingham, 1992; Michael, 1993).

CONSERVATION OF THE SAND LIZARD (*LACERTA AGILIS*) IN THE UK

The sand lizard is one of only six reptiles native to Britain, and is one of the rarest (Species box 4.1). It is under threat in the UK, largely owing to loss, deterioration and fragmentation of suitable habitat. It is also declining elsewhere in northern Europe. Populations (excluding reintroduced colonies) are confined to three small areas of suitable habitat: the lowland heaths of Dorset and Hampshire, heathland in the Surrey Weald, and the coastal dunes of Merseyside and North Wales. Sand lizard populations have also become increasingly fragmented within the three distribution areas (Table 4.10). A genetic study by Beebee and Rowe (2001) confirmed that there have

Plate 4.6: Captive breeding of the sand lizard at Chester Zoo

Table 4.10: Summary Species Action Plan for the sand lizard

Current status	*Current factors causing loss or decline*	*Current action*	*Action plan objectives and targets*	*Proposed actions*
• In the UK natural populations have disappeared over much of the species' former range. Surviving colonies are mostly confined to heathland habitats (particularly the dry heaths of Dorset), the Merseyside sand dunes and one long-established introduced colony in the Inner Hebrides (Scotland). Sand lizards have recently been introduced to sites in the New Forest, the Weald and Wales. Recent estimates of sand lizard populations are approximately 200–500 adults in Merseyside, 1,000 in Surrey and 6,000 in Dorset • Populations are declining in Belgium, Denmark, northern France, northern Germany, Luxembourg, the Netherlands and Sweden • The species is listed under the Habitats Directive and the Bern Convention and protected under the Conservation Regulations (1994) and the Wildlife and Countryside Act (1981)	• Loss, deterioration and fragmentation of heathland and dune habitat as a result of a wide range of factors including development, forestry, mineral extraction, etc. • Scrub encroachment of dune and heathland habitats • Uncontrolled fires • Shortage of suitable breeding sand on heathland sites	• Populations have been successfully reintroduced to some heaths in south-east England, Dorset and Wales. An introduction to the Inner Hebrides has survived since 1970 • Research on the distribution, status and habitat resulted in a programme of habitat management, and a continued programme of translocation to former sites • The sand lizard is the subject of a species recovery programme, initiated in 1994 (English Nature and Herpetological Conservation Trust)	• Re-establish ten populations in suitable habitats within its former range • Maintain all breeding populations at current levels, and enhance where possible • Reverse the fragmentation of sites by habitat re-creation and management	• *Policy and legislation.* Encourage uptake of schemes for heathland management and restoration, remove limited areas of woodland to link fragmented populations; encourage appropriate dune management • *Site protection and management.* Review SSSI coverage of sand lizard sites in Wales; identify all sites with sand lizards to local authorities for identification in development plans; habitat re-creation to consolidate and expand range • *Species management and protection.* Ensure sand lizard needs are catered for in cutting, burning or grazing management programmes; maintain and enhance all breeding populations; consider 10 translocations to re-establish the species • *Advisory.* Ensure that relevant local authorities, landowners and managers of sand lizard sites are aware of the species' needs, legal status, etc., and that advice on management is available • *Future research and monitoring.* Investigate and refine methods of redressing habitat invasion by scrub; evaluate the genetic differences between the Merseyside, Weald and Dorset populations (achieved, Beebee and Rowe, 2001); encourage regular monitoring of known populations; incorporate information into a national database • *Communication and publicity.* Publicise the importance, rarity and conservation needs of the sand lizard, e.g. through interpretive materials, zoos, etc.

Sources: The UK Biodiversity Steering Group (1995); Inns (1999b); Beebee and Rowe (2001).

Table 4.11: Achievements of the sand lizard recovery programme.

Achievements	Between 1994 and 1997 site management for sand lizards was completed at 120 sites and initiated at twenty-six others, and fifteen habitat linkage schemes linked thirty populations. By 1997 a total of 972 (833 captive bred and 139 wild-caught) sand lizards were reintroduced to eleven sites (six heathland sites and five dune sites). All translocations appeared to be successful in 1997, and subsequent monitoring indicates that introduced populations have bred successfully. Initial sand lizard captive breeding stock of eighty-five adults was increased to 116 in 1997. During the programme seventeen previously undesignated sand lizard sites were given SSSI status. Most efficient sand exposure methods for egg incubation were found to be blading and turfing; south and south-west-facing exposures were favoured
Limitations	Continuing large housing developments. Heathland fires: e.g. fires in 1995 and 1996 caused on average 66% damage to thirty-one sand lizard sites and destroyed an estimated 85% of lizards in those populations. Inappropriate management. Lack of management for specific species. High cost of genetic studies: completed at a later date by Beebee and Rowe (2001)
Conclusions	By 1997 the programme had met its targets for site management and translocations, but the effectiveness of the programme was seriously reduced by continuing limitations such as development and uncontrolled fires. Enforced species and habitat policies and realistic funding are needed to achieve favourable sand lizard conservation status. Reintroduction of populations continues

Sources: Corbett and Moulton (1998); Inns (1999b); English Nature (2001a).

been substantial recent declines in the effective population size (the number of animals reproducing successfully) in Merseyside and Surrey, and indicated that the smallest and most isolated population (the Merseyside population) has the lowest genetic diversity. Estimates of genetic differentiation in sand lizards from the three regions suggest that the three populations are sufficiently distinct to be protected in their own right (Beebee and Rowe, 2001).

Because of its continuing decline, the species was included in English Nature's Species Recovery Programme (Plate 4.6). The Species Recovery Programme began in 1991, and has subsequently become the primary mechanism by which English Nature implements Species Action Plans (English Nature, 2001a, b). The sand lizard recovery programme was initiated in April 1994 (Table 4.10) (Corbett and Moulton, 1995, 1998). The achievements and limitations of the programme are summarised in Table 4.11. Despite its successes, the direct and indirect effects of urban encroachment continue to pose problems.

5

PRACTICAL WORK

•

By far the best way of understanding the ecology of a habitat is to investigate it for yourself. In order to gain maximum benefit from ecological studies it is important to plan in advance. Part of the planning is to be sure of the questions being asked or problems to be solved; the most elegant research tends to ask simple questions or looks at simple problems. Prior planning also includes identifying the methods of data analysis (see below). Equipment should be checked and the experimenter should be fully competent in its use before gathering data. A consistent approach, using the same techniques and only varying the factor to be analysed, is usually the best approach. Often a small pilot study will help the main investigation to run smoothly and will allow the methods to be refined.

It is important that permission is received from the owner(s) of land being used for the project, both to gain access and to carry out particular types of experiment. For example, it is against the law (Wildlife and Countryside Act 1981) for any unauthorised person (i.e. anyone who is not the owner, occupier or who has not been authorised by the owner or occupier of the land concerned) to uproot any wild plant. Other organisms are even more strictly protected; almost all birds and some other animals, together with several plant species, are fully protected under the law (see Jones, 1991, for further details). Care should be taken to comply with the law, to ensure that you do not damage or unduly disturb any plant or animal, and to design and implement any research project so as to leave the habitat as it was found.

Grassland and heathland provide opportunities for ecological research into population and community characteristics, ecosystem structure and function, the role of management and the effects of pollution (e.g. Brodie, 1985). The following projects may be investigated with relative ease. They are not intended to increase the pressure on sensitive grassland and heathland. Instead, they can be carried out in areas of accessible habitat such as amenity grassland, or in the laboratory. They are designed to ask simple questions and provide data to analyse with simple tests. As in most research, the results obtained may stimulate further work on supplementary questions.

EXPERIMENTAL DESIGN

Comparisons between situations in which only one factor varies are the easiest to interpret. For example, comparisons between several sites that differ in size but are similar in all other respects will allow an examination of the influence of size of site on whatever is being recorded. The experimental design must be considered in some detail. There are two major types of experimental design: observational and manipulative. A number of projects described below are observational. Here, a variable (the behaviour of animals, the percentage cover of different plant species, the number of animals of a particular species, etc.) is recorded under different circumstances (different sites, weather conditions, number of animals in a group). Analysis is often a matter of finding whether the variable measured differs in two or more circumstances (e.g. sites, times of day, etc.). If two variables (e.g. number of species and temperature) are measured, you may wish to examine the relationship (called correlation) between the two. Two variables may have a positive (as one increases, so does the other), negative (as one increases, the other decreases) or no relationship. However, with observational studies you will not be able to say definitively that a change in one variable causes the change in the other. This is because an observed increase in one variable (e.g. the number of a species) may be correlated with a measured rise in another factor (e.g. temperature) but may actually be due to changes in a third, unmeasured variable (e.g. the abundance of prey). Manipulative experiments, on the other hand, are designed such that one variable is altered by the experimenter (pH, temperature, fertilizer concentration) allowing a much greater emphasis on cause and effect when it comes to analysing the data (e.g. by regression). However, since these experiments are often conducted in fairly artificial conditions, they may have less relevance to real world situations than may be achieved with observational studies.

It is important to take several replicate samples, since data gathered from only a single sample may not be representative of the situation in general. In order to reduce bias, sampling should be systematic or random: instructions on sampling are given in each practical. It is essential that data gathered are recorded correctly to avoid later problems in analysis and interpretation. In order to help you collect data in an appropriate way, sample data recording sheets have been included where appropriate. Record your data in a hard-backed book rather than on loose sheets of paper, to reduce the risk of loss. Where data are to be analysed using a computer, it is useful to transcribe the data on to an appropriate spreadsheet or other data file as soon as possible, and to check the transcription carefully to identify and rectify any mistakes. The techniques that could be used in the analysis of such data are beyond the scope of this book, although methods of analysis are suggested in the project descriptions, and several texts which provide further details are listed in the 'further reading' section.

HEALTH AND SAFETY

Health and safety in field (and laboratory) work should be paramount. Do not engage in behaviour or activities that could harm yourself or others, and assess the risks and health and safety issues which are likely to be involved and protect against them. General safety issues include: wearing appropriate clothing for the time of year and terrain; using safe equipment (e.g. plastic tubes rather than glass); not working far from help (always work in groups of two or more); informing someone responsible of the details of planned fieldwork in advance (including location and duration) and 'signing off' with that person on return.

Soil and water contain organisms and compounds that are hazardous to health. Gloves should be worn for all fieldwork involving these materials, and when handling spiny, or otherwise hazardous, plants. Tetanus is a hazard for anyone working out of doors, especially those in contact with soil. Spores of the tetanus bacillus live in soil, and minor scratches (e.g. from bramble thorns) could provide a point of infection. Immunisation is the only safe protection and, being readily available, should be kept up to date. Weil's disease (or Leptospirosis) is caused by a bacteria carried by rodents (especially rats). Urine from infected animals contaminates fresh water and associated damp habitats such as river, stream and canal banks, and is more common in stagnant conditions and during warmer months. Infection is usually via cuts or grazes, or through the nose, mouth or eye membranes, and precautions should be taken to avoid contact between these areas and infected water. Cover cuts and grazes with waterproof dressings, use appropriate waterproof clothing, including strong gloves and footwear, and avoid eating, drinking or smoking near possible sources of infection. Lyme disease is another hazard for fieldworkers. This is transmitted by female ticks, which are especially likely to bite from early spring to late summer. To help avoid the disease, prevent ticks from biting by wearing appropriate clothing (e.g. long trousers), check for the presence of ticks (light-coloured clothing helps) and remove ticks as soon as possible if bitten. (Twist them slowly in an anticlockwise direction without pulling and seek medical help if mouth parts remain within the skin.)

In addition, it must be remembered that many field sites are dangerous places; watch out for obvious hazards such as sharp pieces of metal and broken glass, avoid traffic hazards when working on roadsides and be careful on uneven surfaces. The safety issues indicated here are not fully comprehensive. Before any fieldwork is undertaken you are advised to consult appropriate publications such as that produced by the Institute of Biology (Nichols, 1999).

WOODLICE AND NUTRIENT CYCLING

Ineffective decomposition and insufficient cycling of nutrients is a common cause of a decline in the vigour of restored vegetation and is most commonly seen in grassland (Harris *et al.*, 1996). Macroinvertebrates such as earthworms, millipedes and woodlice (isopods) have an important role in stimulating decomposition. Dead plant material in the litter layer is converted to faeces, which increases the surface area available for microbial attack (Hopkin, 1989). Absence of these invertebrates on derelict land or newly restored sites, or owing to metal pollution, may lead to ineffective plant decomposition and the accumulation of plant litter on the soil surface.

The aim of this experiment is to examine the role of woodlice in decomposition, nutrient cycling and facilitation of vegetation development. Briefly, the method is as follows:

1. Pots containing sand or limestone chippings should be evenly sown with a constant number or mass of seeds of a grass such as red fescue (obtained from a seed merchant) and placed in individual saucers.
2. A small amount of fertilizer should be applied to sustain initial plant growth. Once the grass is established (after about three weeks), it should be cut to a height of about 5 cm, and the cuttings should be oven dried at 70°C, weighed and returned to the appropriate pot.
3. Woodlice (for example, six individuals, three male and three female) should be added to half the pots; the other half should act as woodlice-free controls. Woodlice may be obtained from suppliers such as Philip Harris Science Education Equipment and Resources, Leicestershire, UK. Clear polythene sleeves or fine mesh covers should be

attached to the pots to prevent the escape of the woodlice and to create a humid microclimate. Fertiliser application should cease in all pots (including the controls) at the same time that woodlice are introduced.

4 About twenty replicates of the control and woodlice treatment should be used. The pots should be arranged randomly in a greenhouse or on a window sill, and rotated weekly to minimise the effects of local environmental variation. The pots should be watered from below as necessary, and should be monitored frequently to ensure that they do not become too dry or too wet.
5 The grass in each pot should be cut to a height level with the rim of the pot on a regular basis (for example, every fourteen days). Each time the clippings should be oven-dried at 70°C, weighed and returned to the appropriate pot.
6 At the end of the experiment (for example, after eight weeks from the introduction of woodlice) a number of variables can be recorded, including the mass of plant litter on the surface of each pot, the number of woodlice and mass of plants. The data may be analysed using the *t*-test or analysis of variance.
7 The results of the experiment should indicate that there is no significant difference between treatments in the mass of clippings before woodlice are introduced, but that subsequently the mass of clippings at each harvest, and the cumulative mass of clippings plus the final biomass of plants, are greater in the pots with woodlice than in pots without. The mass of accumulated litter is likely to be less in pots with woodlice than in pots without woodlice (Collins, 1997).

By stimulating decomposition of the clippings, the presence of woodlice may indirectly promote the growth of grass. This experiment could be repeated using different species of plant or different substrates, or substituting different detritivores for woodlice. For this experiment a small pilot study may be especially useful and will allow the methods to be refined.

METAL TOXICITY

The presence of heavy metals in substrates can influence the species composition of grassland communities, e.g. OV37 (Chapter 2, 'Lowland calcareous grasslands', and Chapter 3, 'Pollution'). The aim of this study is to examine the effect of heavy metals on pasture species, for instance cadmium (Cd) or zinc (Zn) on perennial rye grass or clover. This combination of species and metals has been selected because the sludge left behind after sewage treatment is commonly used as a low-cost nitrogen fertiliser on pasture, and sewage sludge is sometimes rich in cadmium and zinc (Newman, 2000).

To examine the effects of zinc and cadmium, plants can be grown from seed (purchased from seed merchants) under a range of experimental conditions. Seeds of perennial rye grass and clover should be evenly sown in sealed pots or seed trays containing peat-free compost. Sufficient trays should be sown to allow at least three replicates of each treatment. A range of metal solutions can be applied to determine the effect of different metals on different species. A constant number or mass of seeds should be sown in each tray. Some possible treatments are listed in Table 5.1.

Appropriate solutions should be applied to each tray (e.g. to 23 cm × 18 cm trays at a rate of 60 ml twice a week for six weeks). In addition, trays should be watered as necessary.

Table 5.1: Possible metal toxicity experimental treatments (mM)

Species	*Perennial rye grass*	*Perennial rye grass*	*White clover*	*White clover*
Control	Deionised water	Deionised water	Deionised water	Deionised water
Treatments	10mM Zn	1mM Cd	10mM Zn	1mM Cd
	20mM Zn	5mM Cd	20mM Zn	5mM Cd
	40mM Zn	10mM Cd	40mM Zn	10mM Cd
	60mM Zn	15mM Cd	60mM Zn	15mM Cd

Trays should be arranged randomly (e.g. randomised blocks, see Wheater and Cook, 2000) in a greenhouse or on a window sill, and rotated weekly to minimise the effects of local environmental variation. In this example, the growth of the two species should decline visibly as the concentration of metal increases. This can be quantified by recording biomass (dry weight). Legumes tend to be more sensitive than grasses to metals but metals are different in their effects (Bradshaw and Chadwick, 1980). The results of the experiment should indicate that clover is more sensitive to cadmium and zinc than perennial rye grass, and that cadmium is more toxic than zinc. (Note that higher concentrations of zinc than cadmium are used.) The data may be analysed using analysis of variance and regression.

Although the selection of species and metals in this example is based on a realistic scenario, the concentrations used are selected to demonstrate differences between species and metals. The experiment could be repeated using alternative species and different metal concentrations – for example, recommended threshold levels for soils where plants are grown (Petts *et al.*, 1997). In all cases, extreme care should be taken when working with toxic substances. Good laboratory practice should be observed at all times (e.g. COSHH risk assessment), protective clothing and gloves should be worn, and contaminated soil and plants should be carefully disposed of.

VEGETATION DESCRIPTION AND NVC ANALYSIS

Monitoring is an essential part of habitat management (Chapter 3, 'Opportunities'), and vegetation monitoring often forms the basis of habitat description. A number of detailed texts on vegetation description are available (for example, Kent and Coker, 1992; Bullock, 1996), and different approaches are appropriate, depending on the purpose of the study. The aim of this exercise is to collect data to identify vegetation types in terms of the National Vegetation Classification (Rodwell, 1991, 1992, 2000).

The most practical way to obtain quantitative information about the plants at a particular site is to take representative samples. A quadrat is a sample area of defined size and shape. If species are recorded from within a sufficiently large number of randomly placed quadrats, it is assumed that the area sampled within the quadrats will be representative of the whole site. The number of quadrats should be decided on the basis of the required precision of the results and the variability within the monitoring area. In general, vegetation with small plants and high species diversity requires small quadrats. The quadrat sizes most often used for grassland and dwarf shrub communities are 0.25–16 m^2 (Rodwell, 1991, 1992; Bullock, 1996). The way in which data are collected can influence the diagnosis of vegetation types using the NVC. For example, random sampling (e.g. using random co-ordinates) may produce a different result from sampling within selected relatively homogeneous stands of vegetation within the site (e.g. Manchester Airport case study, Chapter 4). This is because NVC communities are divisions of a continuum of vegetation types (Rodwell, 1992; Sanderson, 1998) (Chapter 2, 'Classification of vegetation types'). In addition, if quadrats are placed completely randomly, each sample will be independently positioned, which is important for statistical analysis purposes. If the layout is random within predefined areas, such as adjacent areas of relatively homogeneous vegetation, e.g. wet and dry heath (stratified random or randomised block layout), then different statistical methods are used to analyse the samples (Watt, 1993; Wheater and Cook, 2000).

The best time to identify plant species is in summer when they are typically most conspicuous (Plate 5.1), although the identification of dwarf shrubs is relatively straightforward throughout the year. Identification of grasses can present a challenge, as mowing or grazing often removes the characteristic flower heads. However, grasses can usually be identified by their vegetative characteristics (e.g. Hubbard, 1984). A number of plant identification guides are available, and some of these are listed in the 'Further reading' section. Frequency is a simple measure calculated as the percentage of quadrats in which a species is present, e.g. if five out of twenty quadrats contain the species, the species has a frequency of 25 per cent. Additionally, visual estimates of cover of each species within each quadrat can be made. These can be recorded as percentage cover of individual species or by using the Domin Scale (Rodwell, 1991, 1992) (Tables 5.2 and 5.3).

Information on species frequency and abundance (mean percentage cover or mean Domin value) can be used to assign a site to a particular NVC community using the TableFit computer program, version 1.0 (Hill, 1996). TableFit is a computer program for identifying vegetation types, which measures the degree of agreement between the sample under study and the community characteristics described in the NVC (e.g. Rodwell, 1991, 1992, 2000).

Plate 5.1: The dandelion genus (*Taraxacum*) comprises a large number of micro-species that are difficult to separate

Table 5.2: The Domin Scale for visual estimates of plant species cover

Cover %	*Recorded as Domin*
91–100	10
76–90	9
51–75	8
34–50	7
26–33	6
11–25	5
4–10	4
<4 with many individuals	3
<4 with several individuals	2
<4 with few individuals	1
1 individual, no measurable cover	+

Sources: Rodwell (1991); Bullock (1996)

Table 5.3: Recording sheet for vegetation description

Name:
Date:
Site location:
Site description:

	Quadrat 1		*Quadrat 2*		*Etc.*	*Quadrat X*		*Mean abundance (in X quadrats)*		*Frequency (ie. % of quadrats in which the species is present)*
Species	*% cover*	*Domin*	*% cover*	*Domin*	*Etc.*	*% cover*	*Domin*	*% cover*	*Domin*	
Perennial rye grass (*Lolium perenne*)					Etc.					
Crested dog's-tail (*Cynosurus cristatus*)					Etc.					
White clover (*Trifolium repens*)					Etc.					
Etc.					Etc.					

TableFit also identifies habitat types according to the EC CORINE system, which can be useful for international communication purposes.

The data can be entered in a number of formats depending on the type of data that you have been able to collect, including species composition and abundance, species composition only, or an incomplete species list only, as described by Hill (1996). Ideally, you should have a reasonably good species list, with some information on frequency and abundance, for example, Table 5.4.

A sample diagnosis using the data from Table 5.4 is shown in Table 5.5. The TableFit program identifies vegetation types by means of an index of goodness of fit (*G*) (Hill, 1996). The ratings of overall goodness of fit values (*G*) are outlined in Table 5.6. For this example, the community has been classified as MG6a (*Lolium perenne–Cynosurus cristatus*, typical sub-community), with a goodness of fit rating of 73, good (Table 5.5). Community description approaches, such as the NVC, are valuable because they allow identification of vegetation types, although if a community does not fit a particular NVC type it does not mean that the community is not ecologically valuable. NVC analysis can also be used to monitor changes in habitat extent and condition, or to compare sites. For instance, you

Table 5.4: Sample data from twenty quadrats for use in NVC analysis

Species	*Frequency (%)*	*Mean % cover*
Perennial rye grass *Lolium perenne*	80	55
Crested dog's-tail *Cynosurus cristatus*	65	5
White clover *Trifolium repens*	75	75
Yorkshire fog *Holcus lanatus*	70	10
Common mouse-ear *Cerastium fontanum*	60	25
Creeping bent *Agrostis stolonifera*	10	25
Creeping buttercup *Ranunculus repens*	20	35
Ribwort plantain *Plantago lanceolata*	40	10
Dandelion *Taraxacum* agg.	35	5
Daisy *Bellis perennis*	25	30

Notes: n = twenty quadrats.

Table 5.5: Sample NVC diagnosis using TableFit: analysis of top five possibilities

CORINE code	*NVC code*	*G (overall goodness of fit)*	*NVC community*
C38.111	MG6a	73	*Lolium perenne–Cynosurus cristatus.* Typical sub-community
C37.242	MG11a	66	*Festuca rubra–Agrostis stolonifera–Potentilla anserina. Lolium perenne* sub-community
C38.111	MG6	65	*Lolium perenne–Cynosurus cristatus*
C87.2	OV23c	61	*Lolium perenne–Dactylis glomerata. Plantago major–Trifolium repens* sub-community
C81	MG7a	55	*Lolium perenne* ley. *Lolium perenne–Trifolium repens* leys

Notes: The NVC type of the sample is MG6a with goodness of fit 73, good. NVC community: *Lolium perenne–Cynosurus cristatus.* Sub-community: typical. CORINE community: *Lolio–Cynosuretum*, C38.111.

Table 5.6: Ratings of overall goodness of fit values

Goodness of fit	*Rating*
From	
80–100	Very good
through	
60–69	Fair
to	
0–49	Very poor

Source: Hill (1996)

could compare a number of easily accessible areas of grassland to see how they differ in terms of NVC classification.

RESPONSES TO INCREASING LEVELS OF NITROGEN

Most ecosystems that are not under intensive agricultural management, including heathlands and unimproved grasslands, have low supplies of nitrogen (Chapter 2, 'Geology and soils'). These habitats are often of great conservation value, for example 65–80 per cent of the central European Red List species are restricted to ecosystems with the lowest nitrogen status (INDITE, 1994). There is already evidence of increasing nutrient levels and associated changes in species composition in lowland grasslands and heathlands throughout the UK (Haines-Young *et al.*, 2000) (Chapter 3, 'Problems'). A number of common grasses such as Yorkshire fog and perennial rye grass thrive on soils with high nitrogen. These species have a high potential growth rate and can respond vigorously to an increased supply of nitrogen. However, many species found in semi-natural communities such as sheep's fescue, devil's-bit scabious and nitrogen-fixing legumes such as common bird's-foot trefoil have much lower nitrogen demands. These species have a limited capacity for increased growth in response to a greater supply of nitrogen, and may be out-competed by more vigorous species (Grime *et al.*, 1988; INDITE, 1994; Hodgson *et al.*, 1995). An increase in the proportion of species with a competitor strategy might therefore indicate eutrophication (Crofts and Jefferson, 1999) (Chapter 3, 'Opportunities').

There are a number of approaches to assessing the response of species to resource availability and competition (Begon *et al.*, 1996), and the experiments suggested here are relatively simple. The effects of increased nitrogen inputs on grassland and heathland species can be investigated experimentally by growing plants in a range of conditions. For instance, different species could be grown under increasing levels of nitrogen fertiliser. Nitrogen levels could be selected to mimic levels of nitrogen fertiliser applied annually to improved grassland (e.g. 117 kg N ha^{-1} yr^{-1}, Little, 1998), or different levels of annual nitrogen deposition (e.g. 10 kg N ha^{-1} yr^{-1}, 15 kg N ha^{-1} yr^{-1}, 20 kg N ha^{-1} $yr^{-1}$and 30 kg N ha^{-1} yr^{-1}, INDITE, 1994). Higher levels could be used to represent several years of nitrogen deposition. These experiments could operate on a small scale by sowing sealed seed trays with appropriate species. For example, 1 ha is equivalent to 10,000 m^2, so an application rate of 2 g N m^{-2} yr^{-1} would be equivalent to an application of 20 kg N ha^{-1} yr^{-1}. The experiments could compare the responses of species that may coexist such as heather and wavy-hair grass, perennial rye grass and sheep's fescue, or Yorkshire fog and common bird's-foot trefoil (Rodwell, 1991, 1992) (Table 5.7).

Additions of different levels of nitrogen in the form of ammonium sulphate (Power *et al.*, 1995, 1998) or ammonium nitrate (Cawley, 2000) should be applied to each tray. In addition, trays should be watered as necessary. Trays should be arranged randomly (e.g. randomised blocks, Wheater and Cook, 2000) in a greenhouse or on a window sill, and rotated weekly to minimise the effects of local environmental variation. At the end of the experimental period, plants should be harvested, and the biomass (dry mass, oven-dried at 70°C) of each species should be recorded and plotted against nitrogen concentration. Data could be analysed using analysis of variance or regression.

In natural plant communities, species interactions such as competition influence responses to increases in nitrogen. Nitrogen levels can be manipulated to assess the responses of two competing species grown in a mixture, for example Yorkshire fog and common bird's-foot trefoil (Table 5.8). The proportion of each species can be assessed visually in each of the nitrogen treatments, and above-ground plant parts (shoots) can be harvested and separated according to species. The biomass (dry weight) of the above-ground parts of each species, and the relative proportion of each species, can be plotted against nitrogen treatment. The biomass data can be analysed using analysis of variance or regression. If proportions are statistically analysed, the data may need to be transformed before analysis (Wheater and Cook, 2000). Roots can also be harvested, but it will not be possible to separate the roots of the two species accurately. However, total biomass (roots and shoots of Species A and B combined) as a measure of productivity can be plotted against nitrogen treatment. In the field, other environmental factors may limit plant responses to nitrogen; for example, phosphate or water may be limiting (INDITE, 1994).

VEGETATION HEIGHT AND INVERTEBRATE DIVERSITY

The structure of grassland and heathland varies enormously and the composition, height

Table 5.7: Possible nitrogen addition experimental treatments (single species)

Species monoculture	*Control*	*Increasing level of nitrogen (kg N ha^{-1} y^{-1})*		
A	Deionised water	10	20	30
B	Deionised water	10	20	30

Table 5.8: Possible nitrogen addition experimental treatments (species mixture)

Species	*Control*	*Increasing level of nitrogen (kg N ha^{-1} y^{-1})*		
A and B (50:50 mix)	Deionised water	10	20	30

and structure of vegetation are particularly important to invertebrates (McLean, 1990; Key, 2000; Morris, 2000). For example, Morris (1971) found significant correlations between mean vegetation height and numbers of individuals and species diversity of a number of invertebrate taxa, and Haysom and Coulson (1998) found there was a significant progressive increase in Lepidoptera larval diversity with increase in the height of heather.

Structural complexity often reflects management intensity. For instance, the most intensively managed grasslands tend to be the most uniform (Curry, 1994), although unimproved grasslands that have been managed as hay meadows are usually less valuable for invertebrates than pasture because their vegetation structure is less varied and cutting management is less sympathetic to invertebrates (Kirby, 1992; Morris 2000) (Chapter 3, 'Mowing and cutting'). The aim of this project is to investigate the effect of grassland sward height on invertebrate species richness or the species diversity and number of individuals. Several grassland sites should be selected that differ in vegetation height but are as similar as possible in terms of vegetation type, topography, aspect, etc. It is best to carry out this exercise in spring or summer on days when the vegetation is dry, although tussocks of grass can support large numbers of invertebrates during winter (Ausden, 1996). At each site the height of the vegetation should be recorded. Invertebrates may be sampled by systematically searching through the vegetation, using randomly placed quadrats (direct searching), using sweep nets for a set period of time, and or suction sampling for a standard length of time (see Wheater and Cook, 2003, for further details). All these methods have some limitations – for example, sweep netting is not effective in vegetation less than 15 cm high, whereas suction sampling is effective only in vegetation less than 15 cm high (Ausden, 1996) – but the same techniques should be used at each sampling site. Invertebrates can be removed from the net or vegetation using a pooter (a device that allows specimens to be sucked into a collecting tube), or by hand. Animals should be sorted, identified, using appropriate identification guides such as Tilling (1987) and Chinery (1986, 1993), and counted. All unharmed animals should be returned to the collection sites. The number of individuals in each species (or group), and species diversity at each site, can be presented graphically, and the relationship between vegetation height and the number of individuals or species richness can be determined by correlation.

INSECTS ON NETTLES, DOCKS AND THISTLES

Common stinging nettles, docks and thistles such as creeping thistle (Plate 5.2) are common grassland species that provide a number of opportunities for experimental or observational work. For instance, a number of insects are associated with thistles, docks and nettles, and the invertebrate fauna of nettles growing in the sun or shade could be compared, or the insect community of thistle heads could be compared with that of stems and leaves. The effect of nettle or thistle patch size on invertebrate species richness or the number of individuals could also be assessed. (Try to select patches that are similar in all other respects.) For more detailed information on studies involving insects on thistles, nettles and docks refer to Redfern (1983), Davis (1991) and Salt and Whittaker (1998).

Plate 5.2: A number of insects are associated with creeping thistle (*Matt Brierley*)

SEED BANK ASSESSMENT

The seed bank is the population of viable dormant seeds that accumulates in the soil. Assessment of the seed bank is an important management tool, as the buried seed bank can be a valuable source of propagules for the re-establishment of vegetation (Putwain and Rae, 1988; Shaw, 1996; Gilbert and Anderson, 1998) (Chapter 3, 'Vegetation establishment'). Some species have transient seed banks, in which all the seeds germinate or die within one year of dispersal, others have persistent seed banks, in which a proportion of seed survives for more than a year (Fenner, 1985; Hutchings, 1997). A number of methods are available for determining the viable fraction of the seed bank, but the easiest way is to provide optimal conditions to stimulate seed germination (Hutchings, 1986). Seed bank germination trials allow the range and abundance of species present to be determined, and can be used to compare the seed bank under different vegetation types, for example to compare heathland with grassland on former heathland, or improved grassland with unimproved grassland.

To conduct a seed bank trial, the vegetation should be sampled from a number of randomly placed quadrats, and samples of soil should be collected from within each quadrat, beneath any undecomposed litter or from the bare soil or substrate surface. A bulb planter can be used to extract soil samples. Samples should be collected to a standard depth, for example the top 3 cm or the top 10 cm. If 10 cm deep plugs of soil are removed, they can be divided into sections according to depth, e.g. 1–3 cm, 3–6 cm and 6–10 cm. All samples should be placed in labelled sealed plastic bags for transport. The samples should be spread thinly on to sterile peat-free compost in seed trays labelled with the date, site of collection and sample identification. Trays should be watered, placed in a greenhouse or on a window sill and covered with polythene to prevent drying out. Trays should be arranged randomly, rotated weekly to minimise the effects of local environmental variation, and checked on a regular basis to ensure that they do not become too wet or too dry. Different species often have different germination requirements, e.g. in terms of light, temperature or abrasion (Grime *et al.*, 1988; Rees, 1997), and environmental conditions are likely to influence the outcome of your germination trials. Emerging seedlings can be removed once they are identifiable, or alternatively they can be carefully transplanted to pots and grown on until they can be identified. The soil substrate can be periodically disturbed or turned over to encourage more seeds to germinate. All seedlings should be counted over the experimental period (for example, three months). At the end of the study period, the range of species present in the seed bank and their relative abundance can be determined. If soil cores were sectioned according to depth, the proportion of seeds germinating from each layer can be established. For example, about 90 per cent of viable heather seed is contained in the top 5 cm of soil (Putwain and Rae, 1988). The seed bank figures can be compared with the range and relative abundance of species in the vegetation. For instance, in species that have persistent seed banks, dormant seeds may outnumber growing plants (Hutchings, 1997). Seed bank trials can give a useful indication of the range of species in the soil, but germination conditions are likely to be more hostile in the field, and species establishment will also be affected by factors such as competition and herbivory.

OTHER OPTIONS

If none of these options appeals to you, think of a group of organisms that you are interested in, or a question that you would like to address, e.g. rabbit grazing preferences, habitat use by birds, foraging by bumblebees, or the effects of nutrient addition on invertebrate herbivory. Search the literature for examples of that kind of work, and use them as a guide to design a simple project. It is advisable to seek advice before you begin and to start with a pilot study.

GLOSSARY

Acid grassland Grassland on nutrient-poor, generally free-draining soils with pH 4–6 (calcifugous grassland).
Allele Alternative forms of a gene.
Anaerobic Lacking in oxygen.
Apical meristem Meristem or growing point (region of active cell division) at the tip of stems or roots.
ASSI Area of Special Scientific Interest (Table 3.7).
Bioaccumulation Increasing concentration of compounds (often toxins) in organisms at successive trophic levels, sometimes leading to lethal amounts at higher levels.
Biodiversity Action Plan (BAP) Action plan launched to address the needs of species and habitat types of most concern to biodiversity conservation.
Calcareous grassland Grassland on shallow, lime-rich soils, typically of pH 7.0–8.4 (calcicolous grassland).
Calcicolous grassland Grassland on shallow, lime-rich soils, typically of pH 7.0–8.4 (calcareous grassland). Calcicolous: lime loving.
Calcifugous grassland Grassland on nutrient-poor, generally free-draining soils with pH 4–6 (acid grassland). Calcifugous: lime hating.
CAP Common Agricultural Policy of the European Union.
Carnivorous plants Plants that consume animals to obtain energy and nutrients.
Climax community Final plant community at the end of a succession.
Clonal plants Plants growing and propagating by self-replication of genetically identical units (ramets, tillers, rhizome fragments, etc.) that can survive and function on their own if separated.
Community Collection of species found in a common environment or habitat.
Competitor Species suited to low-stress, low-disturbance conditions (Table 2.2).
Continental climate Large annual temperature range, less rainfall than oceanic climate.
Coprophilous Dung-feeding.
Creative conservation Creating simplified habitats using a few common species (Table 3.14).
Crepuscular Active at dawn and dusk.
Critical level Refers to threshold *gaseous* concentrations of pollutants: the concentrations in the atmosphere above which direct adverse effects on receptors such as plants, ecosystems or materials may occur according to present knowledge (INDITE, 1994).
Critical load Refers only to the *deposition* of pollutants: a quantitative estimate of exposure to one or more pollutants below which significant harmful effects on sensitive elements of the environment do not occur according to present knowledge (INDITE, 1994).

CSS Countryside Stewardship Scheme (Table 3.7).

Culm grassland Rush pasture communities.

Decomposition The breakdown of complex organic molecules into simple inorganic constituents.

Detritivory Consumption of dead organic matter.

Diapause A state of arrested development or growth, often related to the seasons, and usually only applied to insects.

Dormancy State of almost complete absence of metabolism despite the organism or organ being alive.

Dry heaths Heaths on dry, freely draining soils.

Ecological succession Natural directional change in the species composition and structure of a community.

Ecotype A population of a widespread species which is adapted to local conditions.

Ecotypic differentiation Adaptive genetic differences between populations of a species.

Effective population size The breeding population size of a population which takes account of the fact that not all adults breed.

EIA Environmental Impact Assessment.

Elytra Tough forewings of a beetle.

Endangered *See* Red Data Book category 1.

Enhancement Operations that are carried out to improve the quality of existing habitats (Table 3.14).

Ephemerals Species that germinate, set seed and die in a short period within a year.

Ericaceous Belonging to the family Ericaceae (the heather family).

ESA Environmentally Sensitive Area (Table 3.7).

Eutrophication Nutrient enrichment.

Evapotranspiration Water evaporated from the soil surface or transpired by plants.

Gametophyte Phase of the life cycle of plants with alternation of generations during which gametes are produced (*see* Sporophyte).

Genetic drift Random changes in allele frequencies.

Gley soils Soils characterised by permanent or temporary waterlogging.

Grazing pressure A measure of the amount of vegetation that a number of grazing animals will remove during the time that they are grazing.

Habitat Action Plan *See* Biodiversity Action Plan.

Habitat creation Establishment of new habitats, often mimicking semi-natural ones (Table 3.14).

Habitats Directive European law which provides for the creation of a network of protected areas to be known as 'Natura 2000' (Table 3.7).

Herbaceous dicotyledons (herbs) Members of one of the two classes of flowering plants, distinguished by having seedlings with usually a pair of seedling leaves (cotyledons), and commonly with leaves with net venation (see monocotyledons).

Humid heath Damp/moist heath intermediate between wet and dry heaths, often on soils with impeded drainage.

Hyphae Fungal filaments, components of a mycelium.

Improved grassland Agriculturally improved, i.e. sown or created by modification of unimproved grassland by the use of drainage, fertilisers, herbicides, etc. Species-poor (Table 2.5).

Inbreeding depression A loss of vigour among offspring when closely related individuals mate.

Indicator species Species whose presence can indicate habitat age or changes in the habitat such as pollution.

Instar The form of an insect between moults.

Intercalary meristem Meristem or growing point (region of active cell division) situated between regions of permanent tissue, e.g. at the base of nodes and leaves.

IUCN World Conservation Union.

Larva Pre-adult form that turns into an adult by a rather radical metamorphosis (compare with nymphal stages).

LNR Local Nature Reserve (Table 3.7).

Maritime Situated near the sea; influenced by the sea.

Meristem Localized region of active cell division in plants (growing point), from which permanent tissues are derived.

Mesotrophic grassland Grassland on moderately fertile to nutrient-rich mineral soils of pH 4.5–6.5 (neutral grassland).

Metapopulation A population that exists as a series of sub-populations, linked by limited migration between them.

Minimum viable population The smallest population size that can be predicted to persist in the future.

Mitigation To appease (Table 3.14).

Monocotyledons Members of one of the two classes of flowering plants, distinguished by having seedlings with usually a single seedling leaf (cotyledon), and commonly with parallel leaf veins (*see* Herbaceous dicotyledons).

Mycelium Mass of hyphae constituting the body of a fungus.

Mycorrhizas Intimate symbiotic associations of fungi and plant roots.

Myrmecochrous plants Plants whose seed dispersal is aided by ants.

Nationally Notable (Scarce) Taxa which do not fall into Red Data Book categories but which are uncommon in Britain. Where it is not possible to determine which of the Notable categories (A or B) is most appropriate for scarce species, these species are assigned to an undivided category.

Nationally Notable (Scarce) category A Taxa which do not fall into Red Data Book categories, but which are uncommon in Britain and are thought to occur in thirty or fewer 10 × 10 km squares, or within seven or fewer vice-counties.

Nationally Notable (Scarce) category B Taxa which do not fall into Red Data Book categories but which are uncommon in Britain and are thought to occur in between thirty-one and one hundred 10 × 10 km squares, or between eight and twenty vice-counties.

Nationally rare *See* Red Data Book category 3.

Nationally scarce Taxa occur in between sixteen and one hundred 10 × 10 km squares (term combines Nationally Notable (Scarce) categories A and B).

Neutral grassland Grassland on moderately fertile to nutrient-rich mineral soils of pH 4.5–6.5 (mesotrophic grassland).

Nitrogen fixation Symbioses between plants and nitrogen-fixing bacteria.

NNR National Nature Reserve (Table 3.7).

NVC National vegetation classification.

Nymphal stages Young stages of insects which have gradual metamorphosis, e.g. wings develop gradually by successive moults (compare with larva).

Oceanic climate Small annual temperature range and precipitation linked with cyclonic activity.

Oligotrophic Nutrient-poor.

Oviparous Egg-laying.

Phytophagous Feeding on plant material.

Poaching The break-up of ground into wet muddy patches, usually by animals.

Podsol (podzol) Soils with a characteristic bleached, ash-grey layer and hard pan layers, formed in acid, coarse textured, well-drained materials (Plate 2.1).

Population Group of organisms of the same species, occupying a common area at the same time.

Ramet Individual root and shoot units of clonal plants that can potentially survive and function on their own if separated from the parent plant.

Rare *See* Red Data Book category 3.

Red Data Book category 1 – Endangered Taxa in danger of extinction and whose survival is unlikely if the causal factors continue to operate (Kirby, 1994).

Red Data Book category 2 – Vulnerable Taxa believed likely to move into the Endangered category in the near future if the causal factors continue to operate.

Red Data Book category 3 – Rare Taxa with small populations that are not at present Endangered or Vulnerable, but are at risk. Usually unlikely to exist in more than fifteen 10 × 10 km squares.

Red Data Book category I – Indeterminate Taxa considered to be Endangered, Vulnerable or Rare, but for which there is not enough information to say which of the three categories is appropriate.

Red Data Book category K – Insuffieciently Known Taxa suspected of falling into categories 1–3, but about which there is insufficient information to be certain.

Rendzina soils Shallow soils, with an organic layer overlying calcareous bedrock.

Restoration A term with a number of meanings, broadly to restore semi-natural vegetation, or produce something fairly close to it (Table 3.14).

Rhizome An underground stem, which helps a plant to survive unfavourable periods (storage organ) and to reproduce vegetatively (bears dormant buds).

Ruderal Species suited to low-stress, high-disturbance conditions (Table 2.2).

SAC Special Area of Conservation (Table 3.7).

Saprophyte Organism which carries out extracellular digestion of dead organic matter and absorbs the products.

Scarce *See* Nationally scarce.

Scutellum Large, often triangular, plate extending back from the top of the second or the third thoracic segments.

Seed bank The population of viable dormant seeds that accumulates on and in soil.

Semi-improved grassland A transition category, between agriculturally improved and unimproved grassland (Table 2.5).

Semi-natural vegetation Vegetation composed of native species created over centuries of traditional, low-input, low-output management, e.g. types of heathland and unimproved grassland.

Semiochemicals Signalling chemicals which can alter behaviour.

Sexual reproduction Production of offspring by the fusion of two haploid (i.e. with only one set of chromosomes) cells, typically from two different individuals.

Soil transfer Spreading of soil from a donor site on to a recipient site (Table 3.16).

Species Action Plan *See* Biodiversity Action Plan.

Sporophore Specialist fungal structure (fruiting body), which liberates spores.

Sporophyte Phase of the life cycle of plants with alternation of generations during which spores are produced (*see* Gametophyte).

SSSI Site of Special Scientific Interest (Table 3.7).

Stolon Horizontal stems that can root at the nodes (the positions where leaves arise).

Stress tolerator Species suited to high-stress, low-disturbance conditions (Table 2.2).

Sward A stretch of turf or grassy vegetation.

Symbiosis Individuals of different species living together in close association.

Taxon General term for a taxonomc group, whatever its rank, e.g. family, genus, species. Plural: taxa.

Tiller A sideshoot of a grass, arising at ground level.

Trophic level Position in a food chain.

Turf The surface layer of grassland or heathland consisting of soil containing a dense growth of plants and their roots.

Turf translocation Transplantation of intact turf from a donor to a recipient site (Table 3.16).

Unimproved grassland Not reseeded, nor treated with chemical fertiliser or pesticides, and any agricultural treatments, such as the application of low levels of farmyard manure, insufficient to modify sward composition. Species-rich (Table 2.5).

Vegetative reproduction Asexual production of new plants.

Viviparous Give birth to live young.

Vulnerable *See* Red Data Book category 2.

Water potential A measure of plant water status.

WCMC World Conservation Monitoring Centre.

Wet grassland Grassland types occurring in areas with a high water table (e.g. ill-drained permanent pasture) or subject to periodic flooding (e.g. inundation grassland).

Wet heaths Heaths on soils which are waterlogged for varying periods of the year.

SPECIES LIST

•

This section draws the species mentioned in the book together into their appropriate taxonomic groupings. Organisms (except viruses) are first divided into one of five kingdoms (Table 1). There are several more subdivisions: the major levels generally used here are in Table 2. The final division is into species. The naming of organisms in a standard way (following international convention) enables everyone to be certain which species is under consideration. Every species is identified by a unique scientific name consisting of a binomial term (two words, the first the genus and the second the species). Sometimes, especially where there are several species which are difficult to separate, a different term (e.g. agg. or sect.) is placed after the genus, to indicate that several species are involved. In other cases, the name may include an x, indicating that the organism (usually a plant) is a cross between two different species. A third term may be used to designate a sub-species or a variety of the species. Note that species and sub-species names are Latinised and should be shown in italics, with the generic name beginning with a capital letter.

Table 1: Summary of the five kingdoms

Kingdom	*Organisms*	*Characteristics*
Prokaryotae (Monera)	Bacteria and Cyanobacteria	Single-celled, prokaryotic (lack a membrane-bounded nucleus)
Protoctista	Nucleated algae (including seaweeds), protozoa and slime moulds	Those eukaryotes (i.e. possess a membrane-bounded nucleus) which are not fungi, plants or animals. Often single-celled, mainly aquatic (including in damp environments and the tissues of other species), often autotrophic
Fungi	Fungi and lichens	Eukaryotic, mainly multicellular, develop directly from spores with no embryological development, heterotrophic and often saprotrophic (feed on non-living organic matter)
Plantae	Plants, including mosses, ferns, liverworts, conifers and flowering plants	Eukaryotic, multicellular, develop from an embryo (multicellular young organisms supported by maternal tissue), often photoautotrophic (autotrophs obtaining energy from sunlight)
Animalia	Invertebrate and vertebrate animals	Eukaryotic, multicellular, develop from a blastula (hollow ball of cells), heterotrophic

Source: Margulis and Schwartz (1988)

Table 6.2: Taxonomic levels used in the species list.

KINGDOM
PHYLUM
CLASS
Order
Family
Species Authority [Common name]

The species name is followed by the name of the person who originally described it (the authority). Where the authority is well known (e.g. if they have named many species), an abbreviation may be used (e.g. *L.* for *Linnaeus*, *F.* for *Fabricius*). An authority in parentheses indicates that the original species name has been altered (e.g. if the organism has been placed in a different genus). In the absence of parentheses around the authority, the original name is still in use today.

The names used in this book follow a variety of sources, depending on the taxonomic group, including: Littlewood and Bray (2001) for flatworms; Roberts (1985–1987) for spiders; Kloet and Hinks (1964–1975) for insects; Cramp (1977–1994) for birds; Corbet and Harris (1991) for mammals; Ainsworth *et al.* (1973) for fungi; Duncan (1970) for lichens; Hill *et al.* (1991–1994) for mosses and Stace (1997) for vascular plants. Vascular plant names used in NVC community titles that are based on an earlier source (Clapham *et al.* 1987) are also included where appropriate.

PROKARYOTAE
NITROGEN-FIXING AEROBIC BACTERIA
EUBACTERIA
Gram-negative aerobic rods and cocci
Rhizobiaceae
Rhizobium Frank species
FUNGI
BASIDIOMYCOTA (Mushrooms and Toadstools)
HYMENOMYCETES
Agaricales
Agaricaceae
Agaricus campestris L. ex Fr. syn. [Field Mushroom]
Hygrophoraceae
Hygrocybe spadicea (Scop.) [Date-coloured Wax Cap]
PYRENOMYCETES
Sphaerialies
Xylariaceae
Poronia punctata (L. ex Fr.) [Nail Fungus]
DISCOMYCETES
Helotiales
Microglossum olivaceum (Pen. ex Fr.) [Earth-tongue]
MYCOPHYCOPHYTA (Lichens)
Cladoniaceae
Cladonia species (Hill)
Cladonia peziziformis (With.)
Cladonia mediterranea Durign and Abbays
Lecanoraceae
Squamarina lentigera (Web.) [Scaly Breck-lichen]
Usneaceae
Cornicularia aculeata (Schreb.)
Physciaceae
Buellia asterella Poelt and Sulzer [Starry Breck-lichen]
PLANTAE (Plants)
BRYOPHYTA (Mosses and Liverworts)
MUSCI (Mosses)
Sphagnales
Sphagnaceae
Sphagnum capillifolium (Ehrh.)
Sphagnum compactum DC.

Dicranales
Dicranaceae
Dicranum scoparium Hedw.
Hypnobryales
Hypnaceae
Hypnum cupressiforme Hedw.
Bryales
Bryaceae
Pohlia nutans Hedw.
FILICINOPHYTA (Ferns)
LYCOPSIDA
Lycopodiaceae (Clubmosses)
Lycopodiella inundata (L.) [Marsh Clubmoss]
PTEROPSIDA
Dennstaedtiaceae (Bracken family)
Pteridium aquilinum (L.) [Bracken]
CONIFEROPHYTA (Conifers)
PINOPSIDA
Pinaceae (Pine family)
Pinus sylvestris L. [Scots Pine]
ANGIOSPERMOPHYTA (Flowering Plants)
MAGNOLIOPSIDA (Dicotyledons)
Nymphaeales
Ceratophyllaceae (Hornwort family)
Ranunculales
Ranunculaceae (Buttercup family)
Ranunculus repens L. [Creeping Buttercup]
Ranunculus acris L. [Meadow Buttercup]
Ranunculus bulbosus L. [Bulbous buttercup]
Caltha palustris L. [Marsh Marigold]
Trollius europaeus L. [Globeflower]
Pulsatilla vulgaris Mill. [Pasqueflower]
Urticales
Urticaceae (Nettle family)
Urtica dioica L. [Common Nettle]
Fagales
Fagaceae (Beech family)
Quercus spp. L. [Oak]
Betulaceae (Birch family)
Betula spp. Roth [Birch]
Corylus avellana L. [Hazel]
Caryophyllales
Caryophyllaceae (Pink family)
Dianthus armeria L. [Deptford Pink]
Dianthus deltoides L. [Maiden Pink]
Cerastium pumilum Curtis [Dwarf Mouse-ear]
Cerastium fontanum Baumg. [Common Mouse-ear]
Scleranthus perennis L. [Perennial Knawel]
Arenaria serpyllifolia L. [Thyme-leaved Sandwort]
Minuartia verna (L.) [Spring Sandwort]
Silene conica L. [Sand Catchfly]
Silene otites (L.) [Spanish Catchfly]
Polygonales
Polygonaceae (Knotweed family)
Rumex L. species [Docks]
Rumex acetosella L. [Sheep's Sorrel]
Rumex obtusifolius L. [Broad-leaved Dock]
Rumex acetosa L. [Common Sorrell]
Rumex crispus L. [Curled Dock]
Plumbaginales
Plumbaginaceae (Thrift family)
Armeria maritima Willd. [Thrift]
Theales
Clusiaceae (St John's-wort family)
Hypericum perforatum L. [Perforate St John's-wort]
Malvales
Nepenthales
Droseraceae (Sundew family)
Drosera rotundifolia L. [Round-leaved Sundew]
Drosera anglica Huds. [Great Sundew]
Drosera intermedia Hayne [Oblong-leaved Sundew]
Violales
Cistaceae (Rock-rose family)
Helianthemum nummularium (L.) [Common Rock-rose]
Helianthemum apenninum (L.) [White Rock-rose]
Helianthemum oelandicum (L.) [Hoary Rock-rose]
Violaceae (Violet family)
Viola L. species [Violets]
Viola lutea Huds. [Mountain Pansy]
Salicales
Capperales
Brassicaceae (Cabbage family)
Cochlearia officinalis ssp. *scotica* (Druce) [Scottish Scurvygrass]
Thlaspi caerulescens J. and C. Presl. [Alpine Penny-cress]
Alyssum alyssoides (l.) [Small Alison]
Resedaceae (Mignonette family)
Ericales
Empetraceae (Crowberry family)
Empetrum nigrum L. [Crowberry]
Ericaceae (Heather family)
Rhododendron ponticum L. [Rhododendron]
Calluna vulgaris (L.) [Heather]

Erica cinerea L. [Bell Heather]
Erica tetralix L. [Cross-leaved Heath]
Erica ciliaris L. [Dorset Heath]
Erica vagans L. [Cornish Heath]
Vaccinium myrtillus L. [Bilberry]
Vaccinium vitis-idaea L. [Cowberry]

Primulales

Primulaceae (Primrose family)

Primula veris L. [Cowslip]
Primula scotica Hook [Scottish Primrose]
Glaux maritima L. [Sea-milkwort]

Rosales

Crassulaceae (Stonecrop family)

Crassula tillaea Lest.-Garl. [Mossy Stonecrop]

Rosaceae (Rose family)

Filipendula ulmaria (L.) [Meadowsweet]
Sanguisorba minor Scop. [Salad Burnet]
Sanguisorba officinalis L. [Great Burnet]
Alchemilla acutiloba Opiz. [Lady's Mantle]
Alchemilla monticola Opiz [Lady's Mantle]
Alchemilla subcrenata Buser [Lady's Mantle]
Potentilla erecta (L.) Raeusch [Tormentil]
Potentilla anserina L. [Silverweed]
Crataegus monogyna Jacq. [Hawthorn]

Fabales

Fabaceae (Pea family)

Lathyrus L. species [Pea species]
Lotus L. species [Bird's-foot trefoils]
Lotus corniculatus L. [Common Bird's-foot trefoil]
Hippocrepis comosa L. [Horseshoe Vetch]
Anthyllis vulneraria L. [Kidney Vetch]
Genista tinctoria L. [Dyer's Greenweed]
Trifolium pratense L. [Red Clover]
Trifolium repens L. [White Clover]
Trifolium suffocatum L. [Suffocated Clover]
Trifolium ochroleucon Huds [Sulphur Clover]
Medicago sativa ssp. *falcata* (L.) [Sickle Medick]
Ulex minor Roth [Dwarf Gorse]
Ulex gallii Planch [Western Gorse]
Ulex europaeus L. [Gorse]

Myrtales

Onagraceae (Willowherb family)

Oenothera biennis L. [Common Evening-primrose]

Geraniales

Geraniaceae (Crane's-bill family)

Erodium cicutarium (L.) [Common Stork's-bill]
Geranium sanguineum L. [Bloody Crane's-bill]
Geranium molle L. [Dove's-foot Crane's-bill]
Geranium pratense L. [Meadow Crane's-bill]
Geranium sylvaticum L. [Wood Crane's-bill]

Apiales

Araliaceae (Ivy family)

Hedera helix L. [Ivy]

Apiaceae (Carrot family)

Trinia glauca (L.) [Honewort]
Anthriscus sylvestris L. [Cow Parsley]
Daucus carota ssp. *gummifer* (Syme) Hook [Sea Carrot]
Daucus carota L. [Wild Carrot]
Silanum silaus (L.) [Pepper-saxifrage]
Meum athamanticum Jacq. [Spignel]
Apium repens (Jacq.) [Creeping Marshwort]
Oenanthe silaifolia M. Bieb. [Narrow-leaved Water-dropwort]

Gentianales

Gentianaceae (Gentian family)

Centaurium erythraea Rafn. [Common Centaury]
Blackstonia perfoliata (L.) [Yellow-wort]
Gentiana pneumonanthe L. [Marsh Gentian]
Gentianella anglica (Pugsley) E.F. Warb. [Early Gentian]

Solanales

Cuscutaceae (Dodder family)

Cuscuta epithymum (L.) [Dodder]

Convolvulaceae (Bindweed family)

Convolvulus arvesnsis L. [Field Bindweed]

Polemoniaceae (Jacob's-ladder family)

Polemonium caeruleum L. [Jacob's Ladder]

Lamiales

Lamiaceae (Deadnettle family)

Punella vulgaris L. [Selfheal]
Thymus serpyllum L. [Breckland Thyme]
Thymus polytrichus A. Kern. Ex Borbas [Wild Thyme]
Thymus pulegiodes L. [Large Thyme]
Mentha pulegium L. [Pennyroyal]

Plantaginales

Plantaginaceae (Plantain family)

Plantago L. species [Plantains]
Plantago lanceolata L. [Ribwort Plantain]
Plantago maritima L. [Sea Plantain]

Scrophulariales

Oleraceae (Ash family)

Fraxinus excelsior L. [Ash]

Scrophulariaceae (Figwort family)
Verbascum nigrum L. [Dark Mullein]
Veronica spicata L. [Spiked Speedwell]
Euphrasia officinalis L. [Eyebrights]
Euphrasia L. species [Eyebrights]
Euphrasia vigursii Davey [Eyebright]
Rhinanthus minor L. [Yellow-rattle]
Rhinanthus angustifolius C. C. Gmel. [Greater Yellow-rattle]

Campanulales

Campanulaceae (Bellflower family)
Campanula glomerata L. [Clustered Bellflower]
Campanula rotundifolia L. [Harebell]

Rubiales

Rubiaceae (Bedstraw family)
Galium saxatile L. [Heath Bedstraw]
Galium verum L. [Lady's Bedstraw]
Galium sterneri Ehrend. [Limestone Bedstraw]
Asperula cynanchica L. [Squinancywort]

Dipsacales

Caprifoliaceae (Honeysuckle family)
Sambucus nigra L. [Elder]
Viburnum lanata L. [Wayfaring–tree]

Dipsacaceae (Teasel family)
Knautia arvensis (L.) [Field Scabious]
Scabiosa columbaria L. [Small Sacbious]
Succisa pratensis Moench [Devil's-bit Scabious]
Dipsacus fullonum L. [Wild Teasel]

Asterales

Asteraceae (Daisy family)
Cirsium Mill. Species [Thistles]
Cirsium arvense (L.) [Creeping Thistle]
Cirsium vulgare (Savi) [Spear Thistle]
Cirsium acaule (L.) [Dwarf Thistle]
Cirsium tuberosum (L.) [Tuberous Thistle]
Cirsium palustre (L.) [Marsh Thistle]
Carlina vulgaris L. [Carline Thistle]
Centaurea L. species [Knapweeds]
Centaurea nigra L. [Common Knapweed]
Centaurea scabiosa L. [Greater Knapweed]
Hypochaeris radicata L. [Cat's-ear]
Hypochaeris glabra L. [Smooth Cat's-ear]
Leontodon autumnalis L. [Autumn Hawkbit]
Taraxacum Wigg. sect. [Dandelions]
Pilosella officinarum F. W. Schulz and Sch. Bip. [Mouse-ear-hawkweed]
Hieracium L. sect. [Hawkweeds]
Bellis perennis L. [Daisy]
Achillea millefolium L. [Yarrow]
Leucanthemum vulgare Lam. [Oxeye Daisy]
Senecio jacobaea L. [Common Ragwort]
Senecio aquaticus Hill [Marsh Ragwort]
Serratula tinctoria L. [Saw-wort]
Tripleurospermum maritimum (L.) [Sea Mayweed]

LILIIDAE (Monocotyledons)

Juncales

Juncaceae (Rush family)
Juncus inflexus L. [Hard Rush]
Juncus articulatus L. [Jointed Rush]
Juncus effusus L. [Soft Rush]
Luzula campestris (L.) [Field Wood-rush]

Cyperales

Cyperaceae (Sedge family)
Eriophorum angustifolium Honck. [Common Cottongrass]
Isolepis setacea (L.) [Bristle Club-rush]
Schoenus nigricans L. [Black Bog-rush]
Carex flacca Schreb. [Glaucous Sedge]
Carex panicea L. [Carnation Sedge]
Carex nigra (L.) [Common Sedge]
Carex filiformis L. [Downy-fruited Sedge]
Carex hirta L. [Hairy Sedge]
Carex arenaria L. [Sand Sedge]
Carex muricata ssp. *muricata* L. [Prickly Sedge]
Trichophorum cespitosum (L.) [Deergrass]

Poaceae (Grass family)
Festuca L. species [Fescues]
Festuca ovina L. [Sheep's-fescue]
Festuca rubra L. agg. [Red Fescue]
Festuca longifolia Thuill. [Blue Fescue]
Festuca arundinacea Schreb. [Tall Fescue]
Lolium L. species [Rye-grasses]
Lolium perenne L. [Perennial Rye-grass]
Lolium multiflorum Lam. [Italian Rye-grass]
Cynosurus cristatus L. [Crested Dog's-tail]
Briza media L. [Quaking-grass]
Sesleria caerulea (L.) [Blue Moor Grass]
Poa trivialis L. [Rough Meadow-grass]
Dactylis glomerata L. [Cock's-foot]
Agrostis curtisii Kerguelen [Bristle Bent]
Arrhenatherum elatius (L.) [False Oat-grass]
Nardus stricta L. [Mat Grass]
Holcus lanatus L. [Yorkshire fog]
Alopercus geniculatus L. [Marsh Foxtail]
Alopercus pratensis L. [Meadow Foxtail]
Agrostis capillaris L. [Common Bent]
Agrostis stolonifera L. [Creeping Bent]
Ammophila arenaria (L.) [Marram]

Helictotrichon pratense (L.) [Meadow Oat Grass]
Helictotrichon pubescens (Huds.) [Downy Oat Grass]
Brachypodium pinnatum (L.) [Tor Grass]
Molinia caerulea (L.) [Purple Moor Grass]
Aira caryophyllea L. [Silver-hair Grass]
Aira praecox L. [Early hair Grass]
Anthoxanthum odoratum L. [Sweet Vernal Grass]
Pleum pratense L. [Timothy]
Bromus ercta (Huds.) [Upright Brome]
Deschampsia flexuosa (L.) [Wavy-hair Grass]
Deschampsia cespitosa L. [Tufted hair Grass]
Leersia oryyzoides (L.) [Cut-grass]

Liliales
Liliaceae (Lily family)
Fritillaria meleagris L. [Snake's-head Fritillary]
Allium schoenoprasum L. [Chives]
Muscari neglectum Guss. Ex Ten. [Grape Hyacinth]
Scilla verna Huds. [Spring Squill]
Ornithogalum L. species [Star-of-Bethlehems]
Asparagus officinalis ssp. *prostratus* (Dumort) [Wild Asparagus]

Orchidales
Orchidaceae (Orchid family)
Ophrys apifera Huds. [Bee-orchid]
Ophrys sphegodes Mill. [Early Spider-orchid]
Dactylorhiza praetermissa (Druce) [Southern Marsh-orchid]
Orchis simia Lam. [Monkey Orchid]
Orchis morio L. [Green-winged Orchid]
Spiranthes romanzoffiana Cham. [Irish Lady's-tresses]
Cypripedium calceolus L. [Lady's-slipper]
Platanthera chlorantha (Custer) [Greater Butterfly-orchid]

ANIMALIA (animals)
PLATYHELMINTHES
Seriata
Geoplanidae
Arthurdendyus triangulates (Dendy) [New Zealand Flatworm]
NEMATODA (Nematode worms or Roundworms)
MOLLUSCA
GASTROPODA
Stylommatophora
Helicidae
Cepeaea nemoralis (Linne) [Brown lip Banded Snail]
Cepeaea hortensis (Muller) [White lip Banded Snail]

ANNELIDA (Segmented Worms)
OLIGOCHAETA
Haplotaxida
Lumbricidae (Earthworms)
Aporrectodea longa (Ude)
Lumbricus terrestris L.

ARTHROPODA (Arthropods)
MALACOSTRACA
Isopoda (Woodlice)
ARACHNIDA
Araneae (Spiders)
Micryphantidae
Tapinocyba pygmaea (Menge)
Uloboridae
Uloborus walckenaerius Latreille
Eresidae
Eresus cinnaberinus Olivier [Ladybird Spider]
Thomisidae
Thomisus onustus Walckenaer [Crab Spider]
Misumena vatia Clerck [Crab Spider]

INSECTA (Insects)
Collembola (Springtails)
Odonata (Dragonflies and Damselflies)
Aeshnidae
Aeshna species (Fabricius) [Hawker Dragonflies]
Coenagrionidae
Coenagrion mercuriale Charpentier [Southern Damselfly]
Ceriagrion tenellum (de Villers, C.J.) [Small Red Damselfly]
Libellulidae
Leucorrhinia dubia (Van der Linden) [White-faced Darter]
Orthetrum coerulescens (Fabricius) [Keeled Skimmer]

Orthoptera (Grasshoppers and Crickets)
Gryllidae (Crickets)
Gryllus campestris L. [Field Cricket]
Tettigoniidae (Bush Crickets)
Concephalus discolour (Thunberg) [Long-winged Cone-head]
Metrioptera roeselii (Hagenbach) Roesel's Bush Cricket]
Decticus verrucivorus (L.) [Wart-biter Cricket]
Gryllotalpidae (Mole Crickets)
Gryllotalpa gryllotalpa (L.) [Mole Cricket]
Acrididae (Grasshoppers)

Chorthippus brunneus (Thunberg) [Common Field Grasshopper]
Chorthippus vagans (Eversmann) [Heath Grasshopper]
Gomphocerippus rufus (L.) [Rufous Grasshopper]
Omocestus rufipes (Zetterstedt) [Woodland Grasshopper]

Hemiptera (True Bugs)
Heteroptera
Homoptera
Auchenorrhyncha
Cercopidae (Froghoppers)
Cicadellidae (Leafhoppers)

Diptera (Two-winged or True Flies)
Tipulidae (Craneflies)
Syrphidae (Hoverflies)
Doros conopseus Fabricius
Chrysotoxum octomaculatum Curtis
Asilidae (Robber Flies)
Asilus cracroniformis L. [Hornet Robber-fly]
Bombyliidae (Bee Flies)
Bombylius minor L. [Heath Bee-fly]
Thyridanthrax fenestratus (Fallen) [Mottled Bee-fly]
Otitidae
Dorycera graminum (Fabricius) [Picture-winged Fly]

Coleoptera (Beetles)
Carabidae (Ground Beetles and Tiger Beetles)
Amara famelica Zimmermann
Amara strenua Zimmermann
Anisodactylus nemoravagus (Duftschmid)
Anisodactylus poeciloides (Stephens)
Harpalus dimidiatus (Rossi)
Harpalus parallelus (Sharp)
Harpalus cordatus (Duftschmid)
Harpalus punctatulus (Duftschmid)
Pterostichus kugelanni (Panzer)
Panagaeus crux-major L. [Crucifix Ground-beetle]
Cicindela species L. [Tiger Beetles]
Cicindela campestris L. [Green Tiger Beetle]
Cicindela sylvatica L. [Heath Tiger Beetle]
Clavigeridae
Claviger testaceus Preyssler [Guest Beetle]
Geotrupidae
Geotrupes species Latreille [Large Dung Beetles]
Geotrupes pyrenaeus Charpentier [Dor Beetle]
Typhaeus typhoeus (L.) [Minotaur Beetle]
Scarabaeidae (Scarabs and Chafers)
Aphodius species Illiger [Dung beetles]
Melolontha melolontha L. [Cockchafer]
Melyridae
Malachius aeneus (L.) [False Soldier Beetle]
Chrysomelidae (Leaf Beetles)
Lochmaea suturalis Thompson, C.G. [Heather Beetle]
Cryptocephalus primarius Harold
Psylliodes sophiae Heikertinger [Flea Beetle]
Curculionidae (Weevils)
Cathormiocerus britannicus Blair [Broad-nosed Weevil]
Protapion ryei Blackburn

Hymenoptera (Bees, Wasps, Ants and Allies)
Formicidae (Ants)
Myrmica sabuleti Meinert
Lasius flavus (F.) [Yellow Meadow Ant]
Formica pratensis Retzius [Black-backed Meadow Ant]
Formica exsecta Nylander [Narrow-headed Ant]
Formica rufibarbis Fabricius [Red Barbed Ant]
Anergates atratulus (Schenck) [Dark Guest Ant]
Vespidae (Wasps)
Philanthus triangulum Fabricius [Bee-killer Digger wasp]
Cerceris quinquefasciata (Rossi) [Solitary Wasp]
Cerceris quadricincta (Panzer) [Solitary Wasp]
Evagetes pectinipes L. [Spider Wasp]
Homonotus sanguinolentus Fabricius [Spider Wasp]
Chrysis fulgida L. [Ruby-tailed Wasp]
Pseudepipona herrichii Saussure [Purbeck Mason Wasp]
Apidae (Bees)
Colletes floralis Eversmann [Northern Colletes]
Andrena gravida Imhoff [Banded Mining Bee]
Andrena lathyri Altken [Mining Bee]
Andrena nigroaena (Kirby, W.) [Solitary Bee]
Bombus ruderatus (Fabricius) [Large Garden Bumble Bee]
Bombus hortorum (L.) [Garden Bumble Bee]
Bombus terrestris (L.) [Bumble Bee]
Bombus distinguendus Morawitz [Great Yellow Bumble Bee]
Bombus subterraneus (L.) [Short-haired Bumble Bee]
Bombus sylvarum (L.) [Shrill Carder Bee]
Bombus humilis (Illiger) [Carder Bumble Bee]
Nomada errans Lepeletier [Cuckoo Bee]
Nomada armata Herrich-Schaeffer [Cuckoo Bee]

Psithyrus barbutellus (Kirby, W.) [Cuckoo Bee]
Apis mellifera L. [Honey Bee]
Lepidoptera (Butterflies and Moths)
Nymphalidae
Boloria euphrosyne L. [Pearl-bordered Fritillary]
Mesoacidalia aglaja (L.) [Dark Green Fritillary]
Argynnis adippe (L.) [High Brown Fritillary]
Eurodryas aurinia Rottemburg [Marsh Fritillary]
Mellicta athalia Rottemburg [Heath Fritillary]
Nemeobiidae
Hamearis lucina (L.) [Duke of Burgundy]
Satyridae (Brown Butterflies)
Maniola jurtina L. [Meadow Brown]
Lasiommata megera L. [Wall Brown]
Pyronia tithonus (L.) [Gatekeeper]
Hipparchia semele (L.) [Grayling]
Melanargia galathea (L.) [Marbled White]
Coenonympha pamphilus (L.) [Small Heath]
Aphantopus hyperantus (L.) [Ringlet]
Lycaenidae (Blues, Coppers and Hairstreaks)
Lycaena phlaeas L. [Small Copper]
Cupido minimus (Fuessly) [Small Blue]
Plebejus argus L. [Silver-studded Blue]
Polyommatus icarus Rott. [Common Blue]
Lysandra coridon Poda [Chalkhill Blue]
Lysandra bellargus Rott. [Adonis Blue]
Maculinea arion L. [Large Blue]
Aricia agestis (Schiffermueller) [Brown Argus]
Callophrys rubi L. [Green Hairstreak]
Hesperiidae (Skippers)
Pyrgus malvae (L.) [Grizzled Skipper]
Thymelicus flavus (Poda) [Small Skipper]
Ochlodes venatus Turati [Large Skipper]
Erynnis tages (L.) [Dingy Skipper]
Thymelicus aceton Rott. [Lulworth Skipper]
Hesperia comma (L.) [Silver-spotted Skipper]
Thymelicus lineola (Ochsenheimer) [Essex Skipper]
Zygaenidae
Zygaena viciae Briggs [New Forest Burnet]
Geometridae
Lycia zonaria britannica Harrison [Belted Beauty]
Scotopteryx bipunctaria Prout [Chalk Carpet]
Cyclophora pendularia Clerck [Dingy Mocha]
Idaea dilutaria Hubner [Silky Wave]
Aspitates gilvaria Denis and Schiffermuller [Straw Belle]
Sphingidae (Hawkmoths)
Hemaris tityus (L.) [Narrow-bordered Bee Hawkmoth]
Coleophoridae
Coleophora tricolor Walsingham [Basil-thyme Case-bearer]
Selidosemidae
Siona lineata (Scopoli) [Black-veined Moth]
Melanchrinae
Mythimna turca (L.) [Double Line]
Caradrinidae
Athetis pallustris Hubner [Marsh Moth]
Acosmetia caliginosa Huebner [Reddish Buff]
Lithophaninae
Polia bombycina Hufnagel [Pale Shining Brown]
Micropterigidae
Coscinia cribraria L. [Speckled Footman]
Noctuidae
Cucullia lychnitis Rambur [Striped lychnitis]
Heliophobus reticulata marginosa Howarth [Bordered Gothic]
Tyta luctuosa Schiffermueller [Four Spotted]
Noctua L. species [Yellow Underwings]
Noctua orbona Hufnagel [Lunar Yellow Underwing]
CHORDATA
AMPHIBIA (Amphibians)
Caudata
Salamandridae (Newts and Salamanders)
Triturus cristatus (Laurenti) [Great-crested Newt]
Triturus helveticus (Razoumowski) [Palmate Newt]
Triturus vulgaris (L.) [Smooth Newt]
Salientia
Bufonidae (Toads)
Bufo bufo (L.) [Common Toad]
Bufo calamita Laurenti [Natterjack Toad]
Ranidae (Frogs)
Rana temporaria L. [Common Frog]
REPTILIA (Reptiles)
Squamata
Anguidae
Anguis fragilis L. [Slow-worm]
Lacertidae
Lacerta agilis L. [Sand Lizard]
Lacerta vivipara Jacquin [Common Lizard]
Colubridae
Natrix natrix (L.) [Grass Snake]
Coronella austriaca Laurenti [Smooth Snake]

Viperidae
Vipera berus L. [Adder]
AVES (Birds)
Podicipediformes
Podicipedidae (Grebes)
Anseriformes
Anatidae (Wildfowl)
Cygnus columbianus (Ord) [Bewick's Swan]
Cygnus cygnus (L.) [Whooper Swan]
Branta leucopsis (Bechstein) [Barnacle Goose]
Anas acuta L. [Pintail]
Accipitriformes
Accipitridae (Hawks and Allies)
Falconiformes
Accipitridae (Kites, Buzzards, Eagles and Allies)
Buteo buteo (L.) [Buzzard]
Circus cyaneus (L.) [Hen Harrier]
Circus pygarus (L.) [Montagu's Harrier]
Falconidae (Falcons and Allies)
Falco tinnunculus L. [Kestrel]
Falco columbarius L. [Merlin]
Galliformes
Phasianidae (Partridges, Pheasants and Allies)
Perdix perdix (L.) [Grey Partridge]
Coturnix coturnix (L.) [Quail]
Gruiformes
Rallidae (Rails)
Crex crex (L.) [Corncrake]
Charadiiformes (Plovers and Allies)
Charadriidae (Plovers and Lapwings)
Vanellus vanellus (L.) [Lapwing]
Charadrius morinellus L. [Dotterel]
Pluvialis apricaria L. [Golden Plover]
Burhinidae (Stone Curlews and Stone Plovers)
Burhinus oedicnemus L. [Stone Curlew]
Scolopacidae (Sandpipers and Allies)
Gallinago gallinago (L.) [Snipe]
Tringa totanus (L.) [Redshank]
Limosa limosa (L.) [Black-tailed Godwit]
Numenius arquata (L.) [Curlew]
Philomachus pugnax (L.) [Ruff]
Laridae (Gulls)
Larus L. species [Gulls]
Columbiformes
Columbidae (Pigeons)
Cuculiformes
Cuculidae (Cuckoos)
Stringiformes
Tytonidae (Barn Owls and Allies)
Tyto alba (Scopoli) [Barn Owl]
Strigidae (Brown Owls and Allies)
Asio flammeus (Pontoppidan) [Short-eared Owl]
Caprimulgiformes
Caprimulgidae (Nightjars)
Caprimulgus europaeus L. [Nightjar]
Apodiformes
Apodidae (Swifts)
Coraciiformes
Alcedinidae (Kingfishers)
Piciformes
Picidae (Woodpeckers and Allies)
Picus viridis (L.) [Green Woodpecker]
Passeriformes (Perching Birds)
Alaudidae (Larks)
Alauda arvensis L. [Skylark]
Lullula arborea (L.) [Woodlark]
Hirundinidae (Swallows)
Motacillidae (Pipits and Wagtails)
Anthus pratensis (L.) [Meadow Pipit]
Bombycillidae (Waxwings and Hypocolicuses)
Troglodytidae (Wrens)
Prunellidae (Accentors)
Turdidae (Chats and Thrushes)
Turdus pilaris L. [Fieldfare]
Turdus iliacus L. [Redwing]
Turdus torquatus L. [Ring Ouzel]
Oenanthe oenanthe (L.) [Wheatear]
Saxicola rubetra (L.) [Whinchat]
Sylviidae (Warblers and Allies)
Sylvia communis Latham [Whitethroat]
Sylvia undata (Boddaert) [Dartford Warbler]
Laniidae (Shrikes)
Lanius collurio L. [Red-backed Shrike]
Corvidae (Crows and Allies)
Pyrrhocorax pyrrhocorax (L.) [Chough]
Corvus frugilegus L. [Rook]
Corvus species L. [Crows]
Fringillidae (Finches)
Carduelis cannabina (L.) [Linnet]
Emberizidae (Buntings)
Emberiza citrinella L. [Yellow Hammer]
Miliaria calandra (L.) [Corn Bunting]
MAMMALIA (Mammals)
Insectivora (Insectivores)
Erinaceida
Erinaceus europaeus L. [Hedgehog]
Talpidae
Talpa europaea L. [Mole]
Soricidae
Sorex araneus L. [Common Shrew]
Sorex minutus (L.) [Pygmy Shrew]

Chiroptera (Bats)
  Rhinolophidae
    Rhinolophus ferrumequinum (Schreber) [Greater Horseshoe Bat]
Lagomorpha (Lagomorphs)
  Leporidae (Rabbits and Hares)
    Lepus europaeus (Pallas) [Brown Hare]
    Oryctolagus cuniculus (L.) [Rabbit]
Rodentia (Rodents)
  Muridae (Voles, Rats and Mice)
    Clethrionomys glareolus (Schreber) [Bank Vole]
    Microtus agrestis (L.) [Field Vole]
    Apodemus sylvaticus (L.) [Wood Mouse]
    Micromys minutus (Pallas) [Harvest Mouse]
Carnivora (Terrestrial Carnivores)
  Mustelidae (Weasels and Allies)
    Mustela nivalis (L.) [Weasel]
    Mustela ermina (L.) [Stoat]
    Meles meles (L.) [Badger]
Artiodactyla
  Cervidae (Deer)
    Cervus nippon Temminck [Sika Deer]
    Capreolus capreolus (L.) [Roe Deer]

FURTHER READING

The list below gives some of the major texts which assist in the study of the habitats discussed in this book. They are listed under sections dealing with practical techniques (including the identification of plants and animals), practical conservation, journals and internet resources. Useful contact addresses are also included.

PRACTICAL TECHNIQUES

Several texts help with the design of experiments and methods of approaching a particular ecological problem:

Chalmers N. and Parker P. (1989) *The Open University Project Guide* (second edition). Field Studies Occasional Publications 9. Milton Keynes: Open University Press.
Gilbertson D.D., Kent M. and Pyatt F.B. (1985) *Practical Ecology for Geography and Biology*, London: Unwin Hyman.
Wheater, C.P. and Cook, P.A. (2003) *Studying Invertebrates*. Naturalists' Handbooks 28, Slough: Richmond Publishing.
Williams G. (1987) *Techniques and Field Work in Ecology*, London: Bell and Hyman.

Data analysis

It is important to incorporate statistical techniques into experimental design, since a poorly designed experiment can be difficult to interpret. There are a number of reference texts, but most are quite heavy going. The following are a few of the more user-friendly student texts:

Chalmers, N. and Parker, P. (1989) *The Open University Project Guide* (second edition). Field Studies Occasional Publications 9. Milton Keynes: Open University Press.
Fowler, J. and Cohen, L. (1990) *Practical Statistics for Field Biology*, Milton Keynes: Open University Press.
Watt, T.A. (1993) *Introductory Statistics for Biology Students*, London: Chapman and Hall.
Wheater, C.P. and Cook, P.A. (2000) *Using Statistics to Understand the Environment*, London: Routledge.

Identification

There are two types of identification guide. Some are descriptive and usually contain colour illustrations. Care should be taken when using descriptive guides, since it is easy to confuse superficially similar species. Better are those texts that incorporate keys (where organisms are sequentially separated out using diagnostic characters).

Field guides are accessible descriptive guides either to specific groups of plants or animals, or to particular habitats. Collins and Country Life publish mainly descriptive guides, while Warne keys to wildflowers, birds and trees provide alternatives which contain identification keys. Field Studies Council AIDGAP keys are user-friendly and cover a wide range of plant and animal groups, especially invertebrates. Naturalists' Handbooks (Richmond Publishing Company) are other accessible keys to either specific groups of invertebrate animals (mainly insects), or to the occupants of particular habitats. Other identification texts exist which are not part of a series (e.g. Skinner, 1984; Marshall and Haes, 1988).

More specialist keys are aimed towards the professional and, although they may be difficult for the beginner to use, they are usually more complete than the examples given above. Specialist keys include the Linnean Society Synopses of the British Fauna covering a large number of invertebrate groups (e.g. earthworms, harvestmen, woodlice, millipedes) and the Royal Entomological Society of London handbooks for the identification of British insects. Other texts such as Stace (1997) for identifying plants do not form part of a series.

PRACTICAL CONSERVATION

Andrews, J. and Rebane, M. (1994) *Farming and Wildlife. A Practical Management Handbook*, Sandy: RSPB.

Baines, C. and Smart, J. (1991) *A Guide to Habitat Creation* (second edition). Ecology Handbook 2. London: London Ecology Unit.

Crofts, A. and Jefferson, R.G. (eds) (1999) *The Lowland Grassland Management Handbook*, Peterborough: English Nature/Wildlife Trusts.

Gilbert, O.L. and Anderson, P. (1998) *Habitat Creation and Repair*, Oxford: Oxford University Press.

Kirby, P. (2001) *Habitat Management for Invertebrates. A Practical Handbook* (second edition), Sandy: RSPB.

Landlife and the Urban Wildlife Partnership (2000) *Creative Conservation. Guidelines and Principles*, Nottingham: Urban Wildlife Partnership.

Lane, A. (1992) *Practical Conservation. Grasslands, Heaths and Moors*, London: Hodder and Stoughton.

Parker, D.M. (1995) *Habitat Creation. A Critical Guide*. English Nature Science, 21, Peterborough: English Nature.

Sutherland, W.J. and Hill, D.A. (eds) (1995) *Managing Habitats for Conservation*, Cambridge: Cambridge University Press.

Tait, J., Lane, A. and Carr, S. (1988) *Practical Conservation. Site Assessment and Management Planning*, London: Hodder and Stoughton.

Treweek, J., Jose, P. and Benstead, P. (eds) (1997) *The Wet Grassland Guide: Managing Floodplain and Coastal Wet Grassland for Wildlife*, Sandy: RSPB.

JOURNALS

Several journals cover ecology, management and conservation issues including those related to grasslands and heathlands, for example:

Biological Conservation
British Wildlife
Environmental Management
Journal of Applied Ecology
Journal of Environmental Management
Restoration Ecology

Local natural history journals give details of surveys and sites of local importance, for example:

Essex Naturalist
Lancashire Wildlife Journal
The London Naturalist
The Naturalist (North of England)
North West Naturalist
Proceedings of the Bristol Naturalists' Society
Sorby Record (covering Sheffield)

INTERNET RESOURCES

The following list is not exhaustive, but it gives an introduction to some interesting web sites. Please note that internet addresses may have changed since publication and it may be necessary to use a search engine to find the information that you require.

Butterfly Conservation http://www.butterfly-conservation.org/main.html

Centre for Ecology and Hydrology (Land Cover Map 2000 information) http://www.ceh.ac.uk/data/lcm/index.htm

Countryside Survey 2000 http://www.cs2000.org.uk/

DEFRA Summary of UK Food and Farming Statistics http://www.defra.gov.uk/esg/summary.htm

DEFRA Wildlife and Countryside http://www.defra.gov.uk/wildlife-countryside/index.htm

Food and Agriculture Organization of the United Nations statistics databases (FAOSTAT) http://apps.fao.org/

JNCC (Government's wildlife adviser on behalf of English Nature, Scottish Natural Heritage and the Countryside Council for Wales) http://www.jncc.gov.uk/content.htm

Landlife http://www.landlife.org.uk

Peak District National Park Authority http://www.peakdistrict.org/Pages/homepage.htm

Plantlife – The Wild-Plant Conservation Charity http://www.plantlife.org.uk/html/homepage.htm

RSNC (Royal Society for Nature Conservation) and the Wildlife Trusts partnership http://www.quiet-storm.net/wildlifetrusts/mainframe.php

RSPB (Royal Society for the Protection of Birds) http://www.rspb.org.uk/main.html

The British Herpetological Society (Reptiles and Amphibians) http://www.number19.free-online.co.uk/society.html

The Mammal Society http://www.abdn.ac.uk/mammal/

The UK Climate Impacts Programme http://www.ukcip.org.uk

UK Biodiversity Action Plan information http://www.ukbap.org.uk

USEFUL INFORMATION

Copies of the TableFit program and documentation can be obtained from: TableFit, CEH Monks Wood, Abbots Ripton, Huntingdon, Cambridgeshire PE28 2LS, UK. Woodlice may be obtained from: Philip Harris Science Education Equipment and Resources, Novara House, Excelsior Road, Ashby Park, Ashby de la Zouch, Leicestershire, LE65 1NG. www.philipharris.co.uk/education

REFERENCES

•

Adamec, L. (1997) Mineral nutrition in carnivorous plants: a review. *Botanical Review*, 63: 273–29.

Aerts, R. and Berendse, F. (1988) The effects of increased nutrient availability on vegetation dynamics in wet heathlands. *Vegetatio*, 76: 63–69.

Ainsworth, G.C., Sparrow, F.K. and Sussman, A.S. (1973) *The Fungi. An Advanced Treatise*. Volumes 4A–4B. *A Taxonomic Review with Keys*, London: Academic Press.

Aitchinson, J. and Medcalf, K. (1994) *Common Land and Conservation. Biological Surveys in England and Wales – a Synthesis*. English Nature Research Report 77, Peterborough: English Nature.

Akeroyd, J. (1994) *Seeds of Destruction? Non-native Wild Flower Seed and British Floral Biodiversity*, London: Plantlife.

Akinola, M.O., Thompson, K. and Buckland, S.M. (1998) Soil seed bank of an upland calcareous grassland after six years of climate and management manipulations. *Journal of Applied Ecology*, 35: 544–552.

Allen, M.F. (ed.) (1992) *Mycorrhizal Functioning. An Integrative Plant–Fungal Process*, London: Chapman and Hall.

Alonso, I. (2001) *Review of the Value of Published Lowland Heathland Re-creation Plans in Progressing the BAP Objectives*. English Nature Research Report 409, Peterborough: English Nature.

Alpert, P. and Stuefer, J.F. (1997) Division of labour in clonal plants. In: H. de Kroon and J. van Groenendael (eds) *The Ecology and Evolution of Clonal Plants*, pp. 137–154. Leiden: Backhuys.

Anderson, P. (1994) *Outline Landscape and Habitat Management Plan Associated with the Proposed Second Runway*. (Document MA1160), Manchester: Manchester Airport plc.

Andrews, J. and Rebane, M. (1994). *Farming and Wildlife. A Practical Management Handbook*, Sandy: RSPB.

Angold, P.G. (1997) The impact of a road upon adjacent heathland vegetation: effects on plant species composition. *Journal of Applied Ecology*, 34: 409–417.

Anon. (2000): http://www.hmso.gov.uk/acts/acts2000/20000037.htm

Archibold, O.W. (1995) *Ecology of World Vegetation*, London: Chapman and Hall.

Arnold, E.N. and Burton, J.A. (1978) *A Field Guide to the Reptiles and Amphibians of Britain and Europe*, London: Collins.

Ash, H.J., Gemmell, R.P. and Bradshaw, A.D. (1994) The introduction of native plant species on industrial waste heaps: a test of immigration and other factors affecting primary succession. *Journal of Applied Ecology*, 31: 74–84.

Ashbee, P., Bell, M. and Proudfoot, E. (1989) *Wilsford Shaft. Excavations 1960–62*. English Heritage Archaeological Report 11, London: Historic Buildings and Monuments Commission for England.

Ashmore, M. (1997) Plants and pollution. In: M.J. Crawley (ed.) *Plant Ecology* (second edition), Oxford: Blackwell.

Atkinson, R.P.D., MacDonald, D.W. and Johnson, P.J. (1994) The status of the European mole *Talpa europaea* L. as an agricultural pest and its management. *Mammal Review*, 24: 73–90.

Ausden, M. (1996) Invertebrates. In: W.J. Sutherland (ed.) *Ecological Census Techniques*, pp. 139–177. Cambridge: Cambridge University Press.

Ausden, M. (2001) The effects of flooding of grassland on food supply for breeding waders. *British Wildlife*, 12: 179–187.

Ausden, M. and Treweek, J. (1995) Grasslands. In W.J. Sutherland and D.A. Hill (eds) *Managing*

Habitats for Conservation, pp. 197–229. Cambridge: Cambridge University Press.

Ayasse, M., Schiestl, F.P., Paulus, H.F., Lofstedt, C., Hansson, B., Ibarra, F. and Francke, W. (2000) Evolution of reproductive strategies in the sexually deceptive orchid *Ophrys sphegodes*: how does flower-specific variation of odor signals influence reproductive success? *Evolution*, 54: 1996–2006.

Baines, C. (1995) Urban areas. In W.J. Sutherland and D.A. Hill (eds) *Managing Habitats for Conservation*, pp. 362–380. Cambridge: Cambridge University Press.

Baker, A.J.M. (1987) Metal tolerance. *New Phytologist*, 106: 93–111.

Bakker, J.P. and Berendse, F. (1999) Constraints in the restoration of ecological diversity in grassland and heathland communities. *Trends in Ecology and Evolution*, 14: 63–68.

Baldock, D., Bishop, K., Mitchell, K. and Phillips, A. (1996) *Growing Greener. Sustainable Agriculture in the UK*, London: WWF-UK and CPRE.

Bannister, P. (1986) Water relations and stress. In: P.D. Moore and S.B. Chapman (eds) *Methods in Plant Ecology* (second edition), pp. 73–143. Oxford: Blackwell.

Bardgett, R.D. and McAlister, E. (1999) The measurement of soil fungal: bacterial biomass ratios as an indicator of ecosystem self-regulation in temperate meadow grasslands. *Biology and Fertility of Soils*, 29: 282–290.

Baxter, D. and Farmer, A. (1993) *The Control of* Brachypodium pinnatum *in Chalk Grasslands. Influence of Management and Nutrients.* English Nature Research Report 100, Peterborough: English Nature.

Beebee, T.J.C. (1987) Eutrophication of heathland ponds at a site in southern England: causes and effects, with particular reference to the amphibia. *Biological Conservation*, 42: 39–52.

Beebee, T.J.C. (2001) British wildlife and human numbers: the ultimate conservation issue? *British Wildlife*, 13: 1–8.

Beebee, T.J.C. and Rowe, G. (2001) A genetic assessment of British populations of the sand lizard (*Lacerta agilis*). *Herpetological Journal*, 11: 23–27.

Begon, M., Harper, J.L. and Townsend, C.R. (1996) *Ecology. Individuals, Populations and Communities* (third edition) Oxford: Blackwell Science.

Bell, J.R., Wheater, C.P. and Cullen, W.R. (2001) The implications of grassland and heathland management for the conservation of spider communities: a review. *Journal of Zoology, London*, 255: 377–387.

Bell, M. and Walker, M.J.C. (1992) *Late Quaternary Environmental Change. Physical and Human Perspectives*, Harlow: Longman.

Benstead, P.J., Jose, P.V., Joyce, C.B. and Wade, P.M. (1999) *European Wet Grassland. Guidelines for Management and Restoration*, Sandy: RSPB.

Bignal, E., Jones, G. and McCracken, D. (2001) Future directions in agricultural policy and nature conservation. *British Wildlife*, 13: 16–20.

Biodiversity Information Group (2000) Species of Conservation Concern List (SoCC). http://www.ukbap.org.uk/SoCC.htm

Blackstock, T.H., Stevens, J.P., Howe, E.A. and Stevens, D.P. (1995) Changes in the extent and fragmentation of heathland and other semi-natural habitats between 1920–22 and 1987–88 in the Llyn Peninsula, Wales, UK. *Biological Conservation*, 72: 33–44.

Blake, S. and Foster, G. N. (1998) The influence of grassland management on body size in Carabidae (ground beetles) and its bearing on the conservation of wading birds. In: C.B. Joyce and P.M. Wade (eds) *European Wet Grasslands*, pp. 163–169. Chichester: Wiley.

Blakemore, R.J. (2000) Ecology of earthworms under the 'Haughley experiment' of organic and conventional management regimes. *Biological Agriculture and Horticulture*, 18: 141–159.

Blomqvist, M.M., Olff, H., Blaauw, M.B., Bongers, T. and van der Putten, W.H. (2000) Interactions between above- and below-ground biota: importance for small-scale vegetation mosaics in a grassland ecosystem. *Oikos*, 90: 582–598.

Bobbink, R., and Roelofs, J.G.M. (1995) Nitrogen critical loads for natural and semi-natural ecosystems: the empirical approach. *Water, Air and Soil Pollution*, 85: 2413–2418.

Bobbink, R. and Willems, J.H. (1987) Increasing dominance of *Brachypodium pinnatum* (L.) Beauv. in chalk grasslands: a threat to a species-rich ecosystem. *Biological Conservation*, 40: 301–314.

Bobbink, R., Hornung, M. and Roelofs, J.G.M. (1998) The effects of airborne nitrogen pollutants on species diversity in natural and semi-natural European vegetation. *Journal of Ecology*, 86: 717–738.

Bonkowski, M., Griffiths, B.S. and Ritz, K. (2000) Food preferences of earthworms for soil fungi. *Pedobiologia*, 44: 666–676.

Bradley, R., Burt, A.J. and Read, D.J. (1982) The biology of mycorrhiza in the Ericaceae VIII, The role of mycorrhizal infection in heavy metal resistance. *New Phytologist*, 91: 197–209.

Bradshaw, A.D. (1996) Underlying principles of restoration. *Canadian Journal of Fisheries and Aquatic Sciences*, 53: 3–9.

Bradshaw, A.D. and Chadwick, M.J. (1980). *The Restoration of Land. The Ecology and Reclamation of Derelict and Degraded Land*, Berkeley CA: University of California Press.

Brodie, J. (1985) *Grassland Studies*. Practical Ecology series, London: Allen and Unwin.

Brown, A. (1992) *The UK Environment*, London: HMSO.

Brown, M. and Farmer, A. (1996) *Excess Sulphur and Nitrogen Deposition in England's Natural Areas*. English Nature Research Report 201, Peterborough: English Nature.

Brown, V.K (1990) *Grasshoppers*. Naturalists Handbooks 2 (revised edition), Slough: Richmond Publishing.

Brown, V.K and Gibson, C.W.D. (1993) Re-creation of species-rich calcicolous grassland communities. In: R.J. Haggar and S. Peel (eds) *Grassland Management and Nature Conservation*. pp. 125–136. Occasional Symposium 28, British Grassland Society, Aberystwyth: British Grassland Society.

Brown, V.K., Gibson, C.W.D. and Sterling, P.H. (1990) The mechanisms controlling insect diversity in calcareous grasslands. In: S.H. Hillier, D.W.H. Walton and D.A. Wells (eds) *Calcareous Grasslands. Ecology and Management*, pp. 79–88. Huntingdon: Bluntisham Books.

Buckingham, H., Chapman, J. and Newman, R. (1999) The future for hay meadows in the Peak District National Park. *British Wildlife*, 10: 311–318.

Buckland, S.M., Grime, J.P., Hodgson, J.G. and Thompson, K. (1997) A comparison of plant responses to the extreme drought of 1995 in northern England. *Journal of Ecology*, 85: 875–882.

Bullock, J.M. (1996) Plants. In W.J. Sutherland (ed.) *Ecological Census Techniques*, pp. 111–138. Cambridge: Cambridge University Press.

Bullock, J.M. (1998) Community translocation in Britain: setting objectives and measuring consequences. *Biological Conservation*, 84: 199–214.

Bullock, J.M. and Pakeman, R.J. (1996) Grazing lowland heath in England: management methods and their effects on heathland vegetation. *Biological Conservation*, 79: 1–13.

Bullock, J.M. and Webb, N.R. (1995) A landscape approach to heathland restoration. In: K.M. Urbanska and K. Grodzinska (eds) *Restoration Ecology in Europe*, pp. 71–91. Zurich: Geobotanical Institute SFIT.

Bunce, R.G.H., Smart, S.M., van de Poll, H.M., Watkins, J.W. and Scott, W.A. (1999) *Measuring Change in British Vegetation*, London: HMSO.

Burke, M.J.W. and Grime, J.P. (1996) An experimental study of plant community invasibility. *Ecology*, 77: 776–790.

Burton, J.F. (2001) The responses of European insects to climate change. *British Wildlife*, 12: 188–198.

BUTT (Butterflies Under Threat Team) (1986) *The Management of Chalk Grassland for Butterflies*. Focus on Nature Conservation 17, Peterborough: Nature Conservancy Council.

Byfield, A. and Pearman, D. (1996) *Dorset's Disappearing Heathland Flora. Changes in the Distribution of Dorset's Rarer Heathland Species 1931 to 1993*, London, Sandy: Plantlife/RSPB.

Byrne, S. (1990) *Habitat Transplantation in England. A Review of the Extent and Nature of the Practice and the Techniques Employed*. England Field Unit Project 104, Peterborough: NCC.

Cain, M.L., Pacala, S.W., Silander, J.A. and Fortin, M.-J. (1995) Neighborhood models of clonal growth in the white clover *Trifolium repens*. *American Naturalist*, 145: 888–917.

Cameron, R.A.D. (1976) *British Land Snails. Mollusca: Gastropoda. Keys and Notes for the Identification of the Species*. Synopses of the British Fauna 6, London: Academic Press.

Carr, S. and Lane, A. (1993) *Practical Conservation. Urban Habitats*, London: Hodder and Stoughton.

Carroll, J.A., Caporn, S.J.M., Morecroft, M. and Lee, J.A. (1997) *Natural Vegetation Responses to Atmospheric Nitrogen Deposition. Critical Levels and Loads of Nitrogen for Vegetation growing on Contrasting Native Soils. Final Report, May 1994–May 1997.* Department of the Environment Research Contract reference EPG 1/3/11, Sheffield: University of Sheffield.

Carroll, J.A., Johnson, D., Morecroft, M., Taylor, A., Caporn, S.J.M. and Lee, J.A. (2000) The effect of long-term nitrogen additions on the bryophyte cover of upland acidic grasslands. *Journal of Bryology*, 22: 83–89.

Carvell, C. (2002) Habitat use and conservation of bumblebees (*Bombus* spp.) under different grassland management regimes. *Biological Conservation*, 103: 33–49.

Cawley, L.E. (2000) Pollutant Nitrogen and Drought Tolerance in Heathland Plants. Ph.D. thesis, Manchester Metropolitan University.

Chapman, J.L. and Reiss, M.J. (1992) *Ecology. Principles and Applications*, Cambridge: Cambridge University Press.

Chinery, M. (1986) *Collins Guide to the Insects of Britain and Western Europe*, London: Collins.

Chinery, M. (1993) *Collins Field Guide. Insects of Britain and Northern Europe* (third edition), London: Collins.

Chittka, L. (2001) Camouflage of predatory crab spiders on flowers and the colour perception of bees (Aranida: Thomisidae/Hymenoptera: Apidae). *Entomologia Generalis*, 25: 181–187.

Clapham, A.R., Tutin, T.G. and Moore, D.M. (1987) *Flora of the British Isles* (third edition), Cambridge: Cambridge University Press.

Coghlan, A. (1997) Lamb's liver with cadmium garnish. *New Scientist*, 22 March 1997: 4.

Collins, A.J. (1997) The Facilitation of Vegetation Establishment (*Festuca rubra*) by Primary Decomposer Invertebrates (*Porcellio scaber*). Unpublished B.Sc. (Hons) Project, Manchester Metropolitan University.

Corbet, G.B. and Harris, S. (1991) *The Handbook of British Mammals* (third edition), Oxford: Blackwell.

Corbett, K.F. and Moulton, N.R. (1995) *Sand Lizard Species Recovery Programme. First Year (1994–95) Report.* English Nature Research Report 134, Peterborough: English Nature.

Corbett, K.F. and Moulton, N.R. (1998) *Sand Lizard Species Recovery Programme Project (1994–97). Final Report.* English Nature Research Report 288, Peterborough: English Nature.

Countryside Council for Wales (1995) *European Habitats Directive*, Bangor: Countryside Council for Wales.

Countryside Council for Wales (1997) *Habitats Series. Lowland and Coastal Heath*, Bangor: Countryside Council for Wales.

Cowie, R.H. and Jones, J.S. (1998) Gene frequency changes in *Cepaea* snails on the Marlborough Downs over twenty-five years. *Biological Journal of the Linnean Society*, 65: 233–255.

Cowley, M., Thomas, C., Thomas, J. and Warren, M. (2000) Assessing butterflies' status and decline. *British Wildlife*, 11: 243–249.

Cox, J. (1999) The nature conservation importance of dung. *British Wildlife*, 11: 28–36.

Cramp, S. (1977–1994) *Handbook of the Birds of Europe, the Middle East and North Africa. The Birds of the Western Palearctic*, 1–8, Oxford: Oxford University Press.

Cramp, S. (1985) *Handbook of the Birds of Europe, the Middle East and North Africa. The Birds of the Western Palearctic*, 4, *Terns to Woodpeckers*, Oxford: Oxford University Press.

Cramp, S. (1992) *Handbook of the Birds of Europe, the Middle East and North Africa. The Birds of the Western Palearctic*, 6, *Warblers*, Oxford: Oxford University Press.

Critchley, C.N.R., Chambers, B.J., Fowbert, J.A., Sanderson, R.A., Bhogal, A. and Rose, S.C. (2002) Association between lowland grassland plant communities and soil properties. *Biological Conservation*, 105: 199–215.

Crofts, A. and Jefferson, R.G. (1999) (eds) *The Lowland Grassland Management Handbook*, Peterborough: English Nature/Wildlife Trusts.

Cullen, W.R. and Wheater, C.P. (1993) The flora and invertebrate fauna of a relocated grassland at Thirslington Plantation, County Durham, England. *Restoration Ecology*, 1: 130–137.

Curry, J.P. (1994) *Grassland Invertebrates. Ecology, Influence on Soil Fertility and Effects on Plant Growth*, London: Chapman and Hall.

Dargie, T.C. (1993) *The Distribution of Lowland Wet Grassland in England.* English Nature Research Report 49, Peterborough: English Nature.

Davies, D.M., Graves, J.D., Elias, C.O. and Williams, P.J. (1997) The impact of *Rhinanthus* spp. on sward productivity and composition: implications for the restoration of species-rich grasslands. *Biological Conservation*, 82: 87–93.

Davis, A.J., Jenkinson, L.S., Lawton, J.H., Shorrocks, B. and Wood, S. (1998) Making mistakes when predicting shifts in species range in response to global warming. *Nature*, 391: 783–786.

Davis, B.N.K. (1991) *Insects on Nettles.* Naturalists' Handbooks 1 (revised edition), Slough: Richmond Publishing.

Davy, A.J. (1980) Biological flora of the British Isles. *Deschampsia caespitosa* (L.) Beauv. *Journal of Ecology*, 68: 1075–1096.

DEFRA (2001a) DEFRA Economics Consultation Papers. Agenda 2000 CAP Reform. Strategy for Agriculture. Annex 2, *Current Situation and Future Prospects in the Dairy Sector* and Annex 3, *Current Situation and Future Prospects in the Cattle and Sheep Sector.* http://www.defra.gov.uk/esg/economics/r_consult.htm

DEFRA (2001b) Summary of UK Food and Farming. http://www.defra.gov.uk/esg and http://www.defra.gov.uk/esg/m_overview.htm

DEFRA (2001c) *European Community Directive 92/43/EEC on the Conservation of Natural Habitats and of Wild Fauna and Flora. First Report by the United Kingdom under Article 17 on Implementation of the Directive from June 1994 to December 2000*, Bristol: DEFRA.

DEFRA (2001d) htttp://www.defra.gov.uk

DEFRA (2002a) http://www.defra.gov.uk/wildlife-countryside/issues/common/index.htm

DEFRA (2002b) *Guidelines. Environmental Impact Assessment for Use of Uncultivated Land or Semi-natural Areas for Intensive Agricultural Purposes*, London: DEFRA.

de Kroon, H. and Bobbink, R. (1997) Clonal plant dominance under elevated nitrogen deposition, with special reference to *Brachypodium pinnatum* in chalk grassland. In: H. de Kroon and J. van Groenendael (eds) *The Ecology and Evolution of Clonal Plants*, pp. 359–379. Leiden: Backhuys.

de Rooij-van der Goes, P.C.E.M., Peters, B.A.M. and van der Putten, W.H. (1998) Vertical migration of nematodes and soil-borne fungi to developing roots of *Ammophila arenaria* (L.) link after sand accretion. *Applied Soil Ecology*, 10: 1–10.

Devon County Council (2000) *Devon Biodiversity Action Plan. Nightjar.* http://www.devon.gov.uk/biodiversity/nightjar.html

Diaz, A. (2000) Can plant palatability trials be used to predict the effect of rabbit grazing on the flora of ex-arable land? *Agriculture, Ecosystems and Environment*, 78: 249–259.

Dimbleby, G.W. (1984) Anthropogenic changes from Neolithic through medieval times. *New Phytologist*, 98: 57–72.

Dix, N.J. and Webster, J. (1995) *Fungal Ecology*, London: Chapman and Hall.

DOE (1995) *The Habitats Directive. How it will Apply in Great Britain*, London: DOE.

DOE (1996a) *Indicators of Sustainable Development for the United Kingdom*, London: HMSO.

DOE (1996b) *Reclamation of Damaged Land for Nature Conservation*, London: HMSO.

Dolman, P.M. and Land, R. (1995) Lowland heathland. In W.J. Sutherland and DA. Hill (eds) *Managing Habitats for Conservation*, pp. 267–291. Cambridge: Cambridge University Press.

Donald, P.F. and Aebisher, N.J. (eds), assisted by Bratton, J.H., Davies, S.M. and Grice, P.V. (1997) *The Ecology and Conservation of Corn Buntings* Miliaria calandra, Peterborough: JNCC.

Donnison, L.M., Griffith, G.S., Hedger. J., Hobbs, P.J. and Bardgett, R.D. (2000) Management influences soil microbial communities and their function in botanically diverse haymeadows of northern England and Wales. *Soil Biology and Biochemistry*, 32: 253–263.

Drake, C.M. and Denman, D.J. (1993). *A Survey of the Invertebrates of Five Dunes in Northumberland.* English Nature Research Report 46, Peterborough, English Nature.

Drake, M. (1998) The important habitats and characteristic rare invertebrates of lowland wet grasslands in England. In: C.B. Joyce and P.M. Wade (eds) *European Wet Grasslands*, pp. 137–149. Chichester: Wiley.

Dryden, R. (1997) *Habitat Restoration Project: Fact Sheets and Bibliographies.* English Nature Research Report 260, Peterborough: English Nature.

Duffey, E., Morris, M.G., Sheail, J., Ward, L.K., Wells, D.A. and Wells, T.C.E. (1974) *Grassland*

Ecology and Wildlife Management, London: Chapman and Hall.
Duncan, U.K. (1970) *Introduction to the British Lichens*, Arbroath: Buncle.
Edwards, C.A. and Bohlen, P.J. (1996) *Biology and Ecology of Earthworms* (third edition), London: Chapman and Hall.
Edwards, G.R., Crawley, M.J. and Heard, M.S. (1999) Factors influencing molehill distribution in grassland: implications for controlling the damage caused by molehills. *Journal of Applied Ecology*, 36: 434–442.
Edwards, R. (ed.) (1997) *Provisional Atlas of the Aculeate Hymenoptera of Britain and Ireland* 1, Huntingdon: Institute of Terrestrial Ecology.
Ellison, A. M. and Gotelli, N.J. (2001) Evolutionary ecology of carnivorous plants. *Trends in Ecology and Evolution*, 16: 623–629.
English Nature (2000) Ambitious plans: tomorrow's heathland heritage. *English Nature Magazine*, 52: 5–7.
English Nature (2001a) Species recovery special. *English Nature Magazine*, 58: 6–11.
English Nature (2001b) Species Recovery Programme. www.english-nature.org.uk/science/srp
Ennos, R. and Sheffield, E. (2000) *Plant Life*, Oxford: Blackwell.
FAOSTAT (2001) http://apps.fao.org
Fenner, M. (1985) *Seed Ecology*, London: Chapman and Hall.
Fielding, A.H. and Haworth, P.F. (1999) *Upland Habitats*, London: Routledge.
Firbank, L.G., Smart, S.M., van de Poll, H.M., Bunce, R.G.H., Hill, M.O., Howard, D.C., Watkins, J.W. and Stark, G.J. (2000) *Causes of Change in British Vegetation*, London: HMSO.
Fitter, A. (1997) Nutrient acquisition. In: M.J. Crawley (ed.) *Plant Ecology* (second edition), pp. 51–72. Oxford: Blackwell.
Fitter, A.H. and Hay, R.K.M. (1987) *Environmental Physiology of Plants* (second edition), London: Academic Press.
Fitter, R., Fitter, A. and Blamey, M. (1985) *The Wild Flowers of Britain and Northern Europe* (second edition), London: Collins.
Fitter, R., Fitter A., and Farrer, A. (1984) *Collins Pocket Guide. Grasses, Sedges, Rushes and Ferns*, London, Collins.
Flora Locale (1999) *Planting with Wildlife in Mind. An Overview of Issues Concerning the Sources and Use of Native Plants: a Flora Locale Guidance Note*. Floralocale@naturebureau.co.uk.
Flora Locale (2000) *Planting with Wildlife in Mind. The Supply and Use of Native Flora for Projects in the Town or Countryside: a Cross-cutting Action Plan for UK Biodiversity*. Floralocale@naturebureau.co.uk.
Forsythe, T.G. (1987) *Common Ground Beetles*. Naturalists' Handbooks 8, Slough: Richmond Publishing.
Fox, L.R., Ribeiro, P., Brown, V.K., Masters, G.J. and Clarke, I.P. (1999) Direct and indirect effects of climate change on St John's wort, *Hypericum perforatum* L. (Hypericaceae). *Oecologia*, 120: 113–122.
Frid, C.L.J. and Evans, P.R. (1995) Coastal habitats. In W.J. Sutherland and D.A. Hill (eds) *Managing Habitats for Conservation*, pp. 59–83. Cambridge: Cambridge University Press.
Fuller, R.M. (1987) The changing extent and conservation interest of lowland grasslands in England and Wales: a review of grassland surveys 1930–84. *Biological Conservation*, 40: 281–300.
Garbisu, C. and Alkorta, I. (2001) Phytoextraction: a cost-effective plant-based technology for the removal of metals from the environment. *Bioresource Technology*, 77: 229–236.
Gasc, J.-P., Cabela, A., Crnobrnja-Isailovic, J., Dolman, D., Grossenbacher, K., Haffner, P., Lesure, J., Martens, H., Martinez Rica, J.P., Maurin, H., Olivera, M.E., Sofianidon, T.S., Veith, M. and Zuiderwijk, A. (1997) *Atlas of Amphibians and Reptiles in Europe*, Paris: Societas Europaea Herpetologica, Musée national d'histoire naturelle.
Gibson, C.W.D. (1996) *The Effects of Horse Grazing on Species-rich Grasslands*. English Nature Research Reports 164, Peterborough: English Nature.
Gibson, C.W.D. (1997) *The Effects of Horse Grazing on Species-rich Grasslands*. English Nature Research Reports 210, Peterborough: English Nature.
Gibson, C.W.D. (1998) *Brownfield: Red data. The Values Artificial Habitats have for Uncommon Invertebrates*. English Nature Research Report 273, Peterborough: English Nature.
Gibson, C.W.D. and Brown, V.K. (1992) Grazing and vegetation change: deflected or modified

succession? *Journal of Applied Ecology*, 29: 120–131.
Gibson, C.W.D., Watt, T.A. and Brown, V.K. (1987) The use of sheep grazing to recreate species-rich grassland from abandoned arable land. *Biological Conservation*, 42: 165–183.
Gilbert, O.L. and Anderson, P. (1998) *Habitat Creation and Repair*. Oxford: Oxford University Press.
Gimingham, C.H. (1972) *Ecology of Heathlands*, London: Chapman and Hall.
Gimingham, C.H. (1975) *An Introduction to Heathland Ecology*, Edinburgh: Oliver and Boyd.
Gimingham, C.H. (1992) *The Lowland Heathland Management Handbook*. English Nature Science 8, Peterborough: English Nature.
Glaves, D.J. (1998) Environmental monitoring of grassland management in the Somerset Levels and Moors Environmentally Sensitive Area, England. In: C.B. Joyce and P.M. Wade (eds) *European Wet Grasslands*, pp. 73–94. Chichester: Wiley.
Godwin, H. (1975) *History of the British Flora* (second edition), London: Cambridge University Press.
Good, J.E.G., Wallace, H.L., Stevens, P.A. and Radford, G.L. (1999) Translocation of herb-rich grassland from a site in Wales prior to opencast coal extraction. *Restoration Ecology*, 7: 336–347.
Goulson, D. and Stout, J.C. (2001) Homing ability of the bumblebee *Bombus terrestris* (Hymenoptera: Apidae). *Apidologie*, 32: 105–111.
Graves, J. and Reavey, D. (1996) *Global Environmental Change. Plants, Animals and Communities*, Harlow: Longman.
Grayston, S.J., Griffith, G.S., Mawdsley, J.L., Campbell, C.D. and Bardgett, R.D. (2001) Accounting for variability in soil microbial communities of temperate upland grassland ecosystems. *Soil Biology and Biochemistry*, 33: 533–551.
Green, R.E. and Stowe, T.J. (1993) The decline of the corncrake *Crex crex* in Britain and Ireland in relation to habitat change. *Journal of Applied Ecology*, 30: 689–695.
Grime, J.P., Hodgson, J.G. and Hunt, R. (1988) *Comparative Plant Ecology. A Functional Approach to Common British Plants*, London: Unwin Hyman.
Grime, J.P., Brown, V.K., Thompson, K., Masters, G.J., Hillier, S., Clarke, I.P., Askew. A.P., Corker, D. and Kielty, J.P. (2000) The response of two contrasting limestone grasslands to simulated climate change. *Science*, 289: 762–765.
Grubb, P.J., Green, H.E. and Merrifield, R.C.J. (1969) The ecology of chalk heath: its relevance to the calcicole-calcifuge and soil acidification problems. *Journal of Ecology*, 57: 175–212.
Gurnell, A.M. and Gregory, K.J. (1995) Interactions between semi-natural vegetation and hydrogeomorphological processes. *Geomorphology*, 13: 49–69.
Hadley, G. and Pegg, G.F. (1989) Host–fungus relationships in orchid mycorrhizal systems. In: H.W. Pritchard (ed.) *Modern Methods in Orchid Conservation. The Role of Physiology, Ecology and Management*, Cambridge: Cambridge University Press.
Haines-Young, R.H., Barr, C.J., Black, H.I.J., Briggs, D.J., Bunce, R.G.H., Clarke, R.T., Cooper, A., Dawson, F.H., Firbank, L.G., Fuller, R.M., Furse, M.T., Gillespie, M.K., Hill, R., Hornung, M., Howard, D.C., McCann, T., Morecroft, M.D., Petit, S., Sier, A.R.J., Smart, S.M., Smtih, G.M., Stott, A.P., Stuart, R.C. and Watkins, J.W. (2000) *Accounting for Nature. Assessing Habitats in the UK Countryside*, London: Department of the Environment, Transport and the Regions (DETR).
Hanley, N., Whitby, M. and Simpson, I. (1999) Assessing the success of agri-environmental policy in the UK. *Land Use Policy*, 16: 67–80.
Harris, J.A., Birch, P. and Palmer, J. (1996) *Land Restoration and Reclamation. Principles and Practice*, Harlow: Longman.
Harrison, P.A., Berry, P.M. and Dawson, T.P (eds) (2001) *Climate Change and Nature Conservation in Britain and Ireland. Modelling Natural Resource Responses to Climate Change (the MONARCH project)*, Oxford: UKCIP.
Hartley, S.E. and Amos, L. (1999) Competitive interactions between *Nardus stricta* L. and *Calluna vulgaris* (L.) Hull: the effect of fertiliser and defoliation on above- and below-ground performance. *Journal of Ecology*, 87: 330–340.
Haskins, L.E. (1978) The Vegetational History of South East Dorset, Ph.D. thesis, University of Southampton.
Haskins, L.E. (2000) Heathlands in an urban

setting: effects of urban development on heathlands of south-east Dorset. *British Wildlife*, 11: 229–237.

Haysom, K.A. and Coulson, J.C. (1998) The Lepidoptera fauna associated with *Calluna vulgaris*: effects of plant architecture on abundance and diversity. *Ecological Entomology*, 23: 377–385.

Hester, A.J., Miles, J. and Gimingham, C.H. (1991) Succession from heather moorland to birch woodland 2, Growth and competition between *Vaccinium myrtillus*, *Deschampsia flexuosa* and *Agrostis cappilaris*. *Journal of Ecology*, 79: 317–328.

Hill, J.K., Thomas, C.D. and Huntley, B. (1999) Climate and habitat availability determine twentieth century changes in a butterfly's range margin. *Proceedings of the Royal Society of London Biological Sciences*, 266: 1197–1206.

Hill, M.O. (1996) *TABLEFIT Version 1.0 for Identification of Vegetation Types*, Huntingdon: Institute of Terrestrial Ecology.

Hill, M.O., Preston, C.D. and Smith, A.J.E. (1991–1994) *Atlas of the Bryophytes of Britain and Ireland* Volumes I–III, Colchester: Harley Books.

Hillier, S.H., Walton, D.W.H. and Wells, D.A. (1990) *Calcareous Grasslands. Ecology and Management*, Huntingdon: Bluntisham Books.

Hindmarch, C. and Pienkowski, M. (2000) *Land Management: the Hidden Costs*, Oxford: Blackwell Science/British Ecological Society.

Hodgson, J.G., Colasanti, R. and Sutton, F. (1995) *Monitoring Grasslands* I, English Nature Research Report 156, Peterborough: English Nature.

Hopkin, S.P. (1989) *Ecophysiology of Metals in Terrestrial Invertebrates*, London: Elsevier.

Hopkins, A., Pywell, R.F., Peel, S., Johnson, R.H. and Bowling, P.J. (1999) Enhancement of botanical diversity of permanent grassland and impact on hay production in Environmentally Sensitive Areas in the UK. *Grass and Forage Science*, 54: 163–173.

House, S.M. and Spellerberg, I.F. (1983) Ecology and conservation of the sand lizard (*Lacerta agilis* L.) habitat in southern England. *Journal of Applied Ecology*, 20, 417–437.

Hubbard, C.E. (1984) *Grasses. A Guide to their Structure, Identification, Uses and Distribution in the British Isles* (third edition), London: Penguin Books.

Hudson, R., Tucker, G.M. and Fuller, R.J. (1994) Lapwing *Vanellus vanellus* populations in relation to agricultural changes: a review. In: G.M. Tucker, S.M. Davies and R.J. Fuller (eds) *The Ecology and Conservation of Lapwings* Vanellus vanellus, pp. 1–33. UK Nature Conservation 9, Peterborough: Joint Nature Conservation Committee.

Hulme, M. and Jenkins, G. (1998) *Climate change Scenarios for the United Kingdom. Scientific Report.* UK Climate Impacts Programme Technical 1, Norwich: Climatic Research Unit.

Humphries, C.J., Press, J.R. and Sutton, D.A. (1981) *Trees of Britain and Europe*, London: Hamlyn.

Hunter, B.A., Johnson, M.S., Thompson, D.J. (1987a) Ecotoxicology of copper and cadmium in a contaminated grassland ecosystem I, Soil and vegetation contamination. *Journal of Applied Ecology*, 24: 573–586.

Hunter, B.A., Johnson, M.S., Thompson, D.J. (1987b) Ecotoxicology of copper and cadmium in a contaminated grassland ecosystem II, Invertebrates. *Journal of Applied Ecology*, 24: 587–599.

Hunter, B.A., Johnson, M.S., Thompson, D.J. (1987c) Ecotoxicology of copper and cadmium in a contaminated grassland ecosystem III, Small mammals. *Journal of Applied Ecology*, 24: 601–614.

Hurst, A. and John, E. (1999) The biotic and abiotic changes associated with *Brachypodium pinnatum* dominance in chalk grassland in south-east England. *Biological Conservation*, 88: 75–84.

Hutchings, M.J. (1986) Plant population biology. In: P.D. Moore and S.B. Chapman (eds) *Methods in Plant Ecology* (second edition), pp. 377–435. Oxford: Blackwell.

Hutchings, M.J. (1987) The population biology of the early spider orchid, *Ophrys sphegodes* Mill I, A demographic study from 1975 to 1984. *Journal of Ecology*, 75: 711–727.

Hutchings, M.J. (1989) Population biology and conservation of *Ophrys sphegodes*. In: H.W. Pritchard (ed.) *Modern Methods in Orchid Conservation. The Role of Physiology, Ecology and Management*, Cambridge: Cambridge University Press.

Hutchings, M.J. (1997) The structure of plant

populations. In: M.J. Crawley (ed.) *Plant Ecology* (second edition), pp. 325–358. Oxford: Blackwell.

Hutchings, M.J. and Booth, K.D. (1996a) Studies on the feasibility of re-creating chalk grassland vegetation on ex-arable land I, The potential roles of the seed bank and the seed rain. *Journal of Applied Ecology*, 33: 1171–1181.

Hutchings, M.J. and Booth, K.D. (1996b) Studies on the feasibility of re-creating chalk grassland vegetation on ex-arable land II, Germination and early survivorship of seedlings under different management regimes. *Journal of Applied Ecology*, 33: 1182–1190.

Hutchings, M.R. and Harris, S. (1996) *The Current Status of the Brown Hare* (Lepus europaeus) *in Britain*, Peterborough: Joint Nature Conservation Committee.

INDITE (1994) *Impacts of Nitrogen Deposition on Terrestrial Ecosystems. Report of the United Kingdom Review Group on Impacts of Atmospheric Nitrogen*, London: DOE.

Ingold, C.T. (1984) *The Biology of Fungi* (fifth edition), London: Hutchinson.

Inns, H. (1999a) Wildlife reports: reptiles and amphibians. *British Wildlife*, 10: 349.

Inns, H. (1999b) Wildlife reports: reptiles and amphibians. *British Wildlife*, 11: 131.

ITE (1997) *Ecology and Twyford Down*, Furzebrook: Institute of Terrestrial Ecology.

Jefferson, R.G. and Grice, P.V. (1998) The conservation of lowland wet grassland in England. In: C.B. Joyce and P.M. Wade (eds) *European Wet Grasslands*, pp. 31–48. Chichester: Wiley.

Jefferson, R.G. and Robertson, H.J. (1996) *Lowland Grassland. Wildlife Value and Conservation Status.* English Nature Research Report 169, Peterborough: English Nature.

Jefferson, R.G., Gibson, C.W.D., Leach, S.J., Pulteney, C.M., Wolton, R. and Robertson, H.J. (1999) *Grassland Habitat Translocation. The Case of Brocks Farm, Devon.* English Nature Research Report 304, Peterborough: English Nature.

Jeffrey, D.W. and Pigott, C.D. (1973) The response of grasslands on sugar-limestone in Teesdale to application of phosphorus and nitrogen. *Journal of Ecology*, 61: 85–92.

JNCC (1998) *Nature Conservation in the UK. An Introduction to the Statutory Framework*, Peterborough: JNCC.

JNCC (2001) *Biological Translocation. A Conservation Policy for Britain.* Consultation Draft, Peterborough: JNCC.

Jones, A. (1991) British wildlife and the law: a review of the species protection provisions of the Wildlife and Countryside Act 1981. *British Wildlife*, 2: 345–358.

Jones, A. (2001) Comment: we plough the fields, but what do we scatter? A look at the science and practice of grassland restoration. *British Wildlife*, 12: 229–235.

Jones, A.T., Hayes, M.J. and Hamilton, N.R.S. (2001a) The effect of provenance on the performance of *Crataegus monogyna* in hedges. *Journal of Applied Ecology*, 38: 952–962.

Jones, H.D., Santoro, G., Boag, B. and Neilson, R. (2001b) The diversity of earthworms in 200 Scottish fields and the possible effect of New Zealand land flatworms (*Arthurdendyus triangulatus*) on earthworm populations. *Annals of Applied Biology*, 139: 75–92.

Joyce, C.B and Wade, P.M. (1998) *European Wet Grasslands. Biodiversity, Management and Restoration*, Chichester: Wiley.

Kampf, H. (2000) The role of large grazing animals in nature conservation – a Dutch perspective. *British Wildlife*, 12:37–46.

Kent, M. and Coker, P. (1992) *Vegetation Description and Analysis*, London: Belhaven Press.

Kerridge, E. (1953) The sheepfold in Wiltshire and the floating of the watermeadows. *Economic History Review*, second series, 6: 282–289.

Key, R. (2000) Bare ground and the conservation of invertebrates. *British Wildlife*, 11: 183–191.

Keymer, R.J. and Leach, S.J. (1990) Calcareous grassland – a limited resource in Britain. In: S.H. Hillier, D.W.H. Walton and D.A. Wells (eds) *Calcareous Grasslands. Ecology and Management*, pp. 11–17. Huntingdon: Bluntisham Books.

King, C. (1989) *The Natural History of Weasels and Stoats*, Bromley: Christopher Helm.

Kirby, P. (1992) *Habitat Management for Invertebrates. A Practical Handbook* (first edition), Sandy: RSPB.

Kirby, P. (1994) *Habitat Fragmentation. Species at Risk. Invertebrate Group Identification.* English Nature 89, Peterborough: English Nature.

Kirby, P. (2001) *Habitat Management for Invertebrates. A Practical Handbook* (second edition), Sandy: RSPB.

Kloet, G.S. and Hincks, W.D. (1964–1975) *A Check List of British Insects* (second edition), London: Royal Entomological Society of London.

Kovarova, M., Dostal, P. and Herben, T. (2001) Vegetation of ant hills in a mountain grassland: effects of mound history and of dominant ant species. *Plant Ecology*, 156: 215–227.

Landlife and Urban Wildlife Partnership (2000) *Creative Conservation. Guidelines and Principles*, Nottingham: Urban Wildlife Partnership.

Landlife (2001) *Landlife Wildflowers. Seed and Plant Catalogue 2001–2003*, Liverpool: Landlife.

Landlife (2002) *Break New Ground. A Landlife Project funded by the DEFRA Environment Action Programme, First Quarterly Report, April–June 2002*, Liverpool: Landlife.

Lane, A. (1992) *Practical Conservation. Grasslands, Heaths and Moors*, London: Hodder and Stoughton.

Lee, J.A. and Caporn, S.J.M. (1998) Ecological effects of atmospheric reactive nitrogen deposition on semi-natural terrestrial ecosystems. *New Phytologist*, 139: 127–134.

Leishman, M.R., Masters, G.J., Clarke, I.P and Brown, V.K. (2000) Seed bank dynamics: the role of fungal pathogens and climate change. *Functional Ecology*, 14: 293–299.

Little, W. (1998) *Environmental Effects of Agriculture. Final Report*, London: DETR.

Littlewood, D.T.J. and Bray, R.A. (2001) *Interrelationships of the Platyhelminthes.* Systematics Association Special Volume Series 60, London: Taylor and Francis.

Lowe, J.J. and Walker, M.J.C. (1997) *Reconstructing Quaternary Environments* (second edition), Harlow: Addison Wesley Longman.

Lumaret, J.P., Galante, E., Lumbreras, C., Mena, J., Bertrand, M., Bernal, J.L., Cooper, J.F., Kadiri, N. and Crowe, D. (1993) Field effects of ivermectin residues on dung beetles. *Journal of Applied Ecology*, 30: 428–436.

Lynn, D.E. and Waldren, S. (2001) Variation in life history characteristics between clones of *Ranunculus repens* grown in experimental garden conditions. *Weed Research*, 41: 421–432.

Macdonald, D. and Barrett, P. (1993) *Collins Field Guide to Mammals of Britain and Europe*, London: HarperCollins.

MacGillivray, C.W., Grime, J.P., Band, S.R., Booth, R.E., Campbell, B., Hendry, G.A.F., Hillier, S.H., Hodgson, J.G., Hunt, R., Jalili, A., Mackey, J.M.L., Mowforth, M.A., Neal, A.M., Reader, R., Rorison, I.H., Spencer, R.E., Thompson, K. and Thorpe, P.C. (1995) Testing predictions of the resistance and resilience of vegetation subjected to extreme events. *Functional Ecology*, 9: 640–649.

Madsen, T., Stille, B. and Shine, R. (1996). Inbreeding depression in an isolated population of adders *Vipera berus. Biological Conservation*, 75: 113–118.

Madsen, T., Olsson, M., Wittzell, H., Stille, B., Gullberg, A., Shine, R., Andersson, S. and Tegelstrom, H. (2000) Population size and genetic diversity in sand lizards (*Lacerta agilis*) and adders (*Vipera berus*). *Biological Conservation*, 94: 257–262.

Mammal Society (2001) *British Mammal Fact Sheets*, London: Mammal Society.

Manchester Airport (1997) *Landscape and Habitat Management Plan*, Manchester: Manchester Airport plc (confidential).

Manchester Airport (2001) *Manchester Airport Second Runway. Ecological Monitoring of Grassland Sites, 2000*, Manchester: Manchester Airport plc (confidential)

Manchester, S., Treweek, J., Mountford, O., Pywell, R. and Sparks, T. (1998) Restoration of a target wet grassland community on ex-arable land. In: C.B. Joyce and P.M. Wade (eds) *European Wet Grasslands*, pp. 277–294. Chichester: Wiley.

Marchant, J.H., Hudson, R., Carter, S.P. and Whittington, P.A. (1990) *Population Trends in British Breeding Birds*, Tring: British Trust for Ornithology.

Margulis, L. and Schwartz, K.V. (1988) *Five Kingdoms. An Illustrated Guide to the Phyla of Life on Earth*, New York: Freeman.

Marrs, R.H. (1993) An assessment of change in *Calluna* heathlands in Breckland, eastern England, between 1983 and 1991. *Biological Conservation*, 65: 133–139.

Marrs, R.H., Johnson, S.W. and Le Duc, M.G. (1998a) Control of bracken and restoration of heathland VII, The response of bracken rhizomes to eighteen years of continued bracken control or six years of control followed by recovery. *Journal of Applied Ecology*, 35: 748–757.

Marrs, R.H., Snow, C.S.R., Owen, K.M. and Evans, C.E. (1998b) Heathland and acid grassland creation on arable soils at Minsmere: identification of potential problems and a test of cropping to impoverish soils. *Biological Conservation*, 85: 69–82.

Marshall, C. and Price, E.A.C. (1997) Sectoriality and its implications for physiological integration. In: H. de Kroon and J. van Groenendael (eds) *The Ecology and Evolution of Clonal Plants*, pp. 79–107. Leiden: Backhuys.

Marshall. J.A. and Haes, E.C.M. (1988) *The Grasshoppers and Allied Insects of the British Isles*, Essex: Harley Books.

Masters, G.J., Brown, V.K., Clarke, I.P., Whittaker, J.B. and Hollier, J.A. (1998) Direct and indirect effects of climate change on insect herbivores: Auchenorrhyncha (Homoptera). *Ecological Entomology*, 23: 45–52.

McClure, R. (2001) Planning for wildlife in new housing estates. *British Wildlife*, 12: 389–393.

McDonald, R.A., Webbon, C. and Harris, S. (2000) The diet of stoats (*Mustela ermina*) and weasels (*Mustela nivalis*) in Great Britain. *Journal of Zoology*, 252: 363–371.

McLean, I.F.G. (1990) The fauna of calcareous grasslands. In: S.H. Hillier, D.W.H. Walton and D.A. Wells (eds) *Calcareous Grasslands. Ecology and Management*, pp. 41–46. Huntingdon: Bluntisham Books.

Meikle, A., Paterson, S., Finch, R.P., Marshall, G. and Waterhouse, A. (1999) Genetic characterisation of heather (*Calluna vulgaris* (L.) Hull) subject to different management regimes across Great Britain. *Molecular Ecology*, 8: 2037–2047.

Merryweather, J. (2001) Meet the Glomales – the ecology of mycorrhiza. *British Wildlife* 13: 86–93.

Michael, N. (1993) *The Lowland Heathland Management Booklet.* English Nature Science 11, Peterborough: English Nature.

Michael, N. (1994) *A Brief Review of the Extent, Nature and Costs of Lowland Heathland Management in England.* English Nature Research Report 101, Peterborough: English Nature.

Michael, N. (1996) *Lowland Heathland. Wildlife Value and Conservation Status.* English Nature Research Report 188, Peterborough: English Nature.

Mitchell, R.J., Auld, M.H.D., Hughes, J.M. and Marrs, R.H. (2000) Estimates of nutrient removal during heathland restoration on successional sites in Dorset, southern England. *Biological Conservation*, 95: 233–246.

Mitchell, R.J., Marrs, R.H., Le Duc, M.G. and Auld, M.H.D. (1999) A study of the restoration of heathland on successional sites: changes in vegetation and soil chemical properties. *Journal of Applied Ecology*, 36: 770–783.

Mitchley, J. and Malloch, J.C. (1991) *Sea Cliff Management Handbook for Great Britain*, Lancaster: University of Lancaster and Joint Nature Conservation Committee in association with the National Trust.

Moore, I. (1966) *Grass and Grasslands*, London: Collins.

Morris, A., Burges, D., Fuller, R.J., Evans, A.D. and Smith, K.W. (1994) The status and distribution of nightjars *Caprimulgus europaeus* in Britain in 1992. A report to the British Trust for Ornithology. *Bird Study*, 41: 181–191.

Morris, M.G. (1971) The management of grassland for the conservation of invertebrate animals. In: E. Duffey and A.S. Watt (eds) *The Scientific Management of Animal and Plant Communities for Conservation*, pp. 527–552. Oxford: Blackwell.

Morris, M.G. (2000) The effects of structure and its dynamics on the ecology and conservation of arthropods in British grasslands. *Biological Conservation*, 95: 129–142.

Mortimer, S.R., Turner, A.J., Brown, V.K., Fuller, R.J., Good, J.E.G., Bell, S.A., Stevens, P.A., Norris, D., Bayfield, N. and Ward, L.K. (2000) *The Nature Conservation Value of Scrub in Britain.* JNCC 308, Peterborough: JNCC.

Mountford, J.O., Lakhani, K.H. and Kirkham, F.W. (1993) Experimental assessment of the effects of nitrogen addition under hay-cutting and aftermath grazing on the vegetation of meadows on a Somerset peat moor. *Journal of Applied Ecology*, 30: 321–332.

Mountford, J.O, Cooper, J.M., Roy, D.B. and Warman, E.A. (1999) *Targeting Areas for the Restoration and Re-creation of Coastal Floodplain Grazing Marsh.* English Nature Research Report 332, Peterborough: English Nature.

Multimap (2001) http://www.multimap.com/

Nature Conservancy Council (1990) *Handbook for Phase 1 Habitat Survey. A Technique for*

Environmental Audit, Peterborough: Nature Conservancy Council.

Newman, E.I. (2000) *Applied Ecology and Environmental Management* (second edition), Oxford: Blackwell.

Newsham, K.K., Fitter, A.H. and Watkinson, A.H. (1994) Root pathogenic and arbuscular mycorrhizal fungi determine fecundity of asymptomatic plants in the field. *Journal of Ecology*, 82: 805–814.

Nichols, D. (1999) *Safety in Biological Fieldwork. Guidance Notes for Codes of Practice* (fifth edition), London: Institute of Biology.

Nicholson, A.M. and Spellerberg, I.F. (1989) Activity and home range of the lizard *Lacerta agilis* L. *Herpetological Journal*, 1: 362–365.

Oates, M. (1999) Sea cliff slopes and combes – their management for nature conservation. *British Wildlife*, 10: 394–402.

Oates, M. (2000) The Duke of Burgundy – conserving the intractable. *British Wildlife*, 11: 250–257.

Osborne, J.L., Clark, S.J., Morris, R.J., Williams, I.H., Riley, J.R., Smith, A.D., Reynolds, D.R. and Edwards, A.S. (1999) A landscape-scale study of bumblebee foraging range and constancy, using harmonic radar. *Journal of Applied Ecology*, 36: 519–533.

Owen, K.M. and Marrs, R.H. (2000) Creation of heathland on former arable land at Minsmere, Suffolk, UK: the effects of soil acidification on the establishment of *Calluna* and ruderal species. *Biological Conservation*, 93: 9–18.

Pakeman, R.J. and Marrs, R.H. (1992) The conservation value of bracken *Pteridium aquilinum* (L.) Kuhn-dominated communities in the UK, and an assessment of the ecological impact of bracken expansion or its removal. *Biological Conservation*, 62: 101–114.

Parmesan, C., Ryrholm, N., Stefanescu, C., Hill, J.K., Thomas, C.D., Descimon, H., Huntley, B., Kaila, L., Kullberg, J., Tammaru, T., Tennent, W.J., Thomas, J.A. and Warren, M. (1999) Poleward shifts in geographical ranges of butterfly species associated with regional warming. *Nature*, 399: 579–583.

Peach, W.J., Lovett, L.J., Wotton, S.R. and Jeffs, C. (2001) Countryside stewardship delivers cirl buntings (*Emberiza cirlus*) in Devon, UK. *Biological Conservation*, 101: 361–373.

Petts, J., Cairney, T. and Smith, M. (1997) *Risk-based Contaminated Land Investigation and Assessment*, Chichester: Wiley.

Phillips, M. and Huggett, D. (2001) From passive to positive – the Countryside Act 2000 and British Wildlife. *British Wildlife*, 12: 237–243.

Phillips, R. (1980) *Grasses, Ferns, Mosses and Lichens of Great Britain and Ireland*, London: Pan Books.

Phillips, R. (1981) *Mushrooms and other Fungi of Great Britain and Europe*, London: Pan Books.

Pitcairn, C.E.R., Fowler, D. and Grace, J. (1995) Deposition of fixed atmospheric nitrogen and foliar nitrogen content of bryophytes and *Calluna vulgaris* (L.) Hull. *Environmental Pollution*, 88: 193–205.

Planning Inspectorate (2002) http://www.planning-inspectorate.gov.uk

PCFFF (Policy Commission on the Future of Farming and Food) (2002) *Farming and Food: a Sustainable Future*, London: Stationery Office.

Pollard, E. and Yates, T.J. (1993) *Monitoring Butterflies for Ecology and Conservation*, London: Chapman and Hall.

Power, S.A., Ashmore, M.R. and Cousins, D.A. (1998) Impacts of experimentally enhanced nitrogen deposition on a British lowland heath. *Environmental Pollution*, 102: 27–34.

Power, S.A., Ashmore, M.R., Cousins, D.A. and Ainsworth, N. (1995) Long term effects of enhanced nitrogen deposition on a lowland dry heath in southern Britain. *Water, Air and Soil Pollution*, 85: 1701–1706.

Prestt, I. (1971) An ecological study of the viper *Vipera berus* in southern Britain. *Journal of Zoology, London*, 164: 373–418.

Primack, R.B. (1998) *Essentials of Conservation Biology* (second edition), Sunderland, MA: Sinauer Associates.

Prys-Jones, O.E. and Corbet, S.A. (1987) *Bumblebees*. Naturalists' Handbooks 6, Slough: Richmond Publishing.

Putwain, P.D. and Rae, P.A.S. (1988) *Heathland Restoration. A Handbook of Techniques*, Southampton: British Gas plc and Environmental Advisory Unit, University of Liverpool.

Pywell, R.F., Webb, N.R. and Putwain, P.D. (1996) Harvested heather shoots as a resource for heathland restoration. *Biological Conservation*, 75: 247–254.

Pywell, R.F., Bullock, J.M., Hopkins, A., Walker, K.J., Sparks, T.H., Burke, M.J.W. and Peel, S. (2002) Resoration of species-rich grassland on arable land: assessing the limiting process using a multi-site experiment. *Journal of Applied Ecology*, 39: 294–309.

Rackham, O. (1986) *The History of the Countryside*, London: Phoenix.

Ransome, R.D. (1997) *The Management of Greater Horseshoe Bat Feeding Areas to Enhance Population Levels*. English Nature Research Report 241, Species Recovery Programme, Peterborough: English Nature.

Ransome, R.D. (2000) *Monitoring Diets and Population Changes of Greater Horseshoe Bats in Gloucestershire and Somerset.* English Nature Research Report 341, Species Recovery Programme, Peterborough: English Nature.

Raskin, I. and Ensley, B. (eds) (2000) *Phytoremediation of Toxic Metals. Using Plants to Clean up the Environment*. New York: Wiley.

Ratcliffe, D.A. (1984) Post-medieval and recent changes in British vegetation: the culmination of human influence. *New Phytologist*, 98: 73–100.

Read, H.J. and Frater, M. (1999) *Woodland Habitats*, London: Routledge.

Rebbeck, M., Corrick, R., Eaglestone, B. and Stainton, C. (2001) Recognition of individual nightjars *Caprimulgus europaeus* from their song. *Ibis*, 143: 468–475.

Redfern, M. (1983) *Insects and Thistles*. Naturalists' Handbooks 4, Cambridge: Cambridge University Press.

Rees, M. (1997) Seed dormancy. In: M.J. Crawley (ed.) *Plant Ecology* (second edition), pp. 214–238. Oxford: Blackwell.

Roberts, G.M. and Hutson, A.M. (1993) *Greater Horseshoe Bat* (Rhinolophus ferrumequinum), London: Bat Conservation Trust.

Roberts, M.J. (1985) *The Spiders of Great Britain and Ireland* I, Atypidae to Theridiosomidae, Colchester: Harley Books.

Roberts, M.J. (1985–1987) *The Spiders of Great Britain and Ireland* I–III, Colchester: Harley Books.

Roberts, M.J. (1995) *Spiders of Britain and Northern Europe*, London: HarperCollins.

Roberts, N. (1989) *The Holocene: an Environmental History*, Oxford: Blackwell.

Robertson, H.J. and Jefferson, R.G. (2000) *Monitoring the Condition of Lowland Grassland SSSIs* I, *English Nature's Rapid Assessment Method*. English Nature Research Report 315, Peterborough: English Nature.

Robertson, H.J., Crowle, A. and Hinton, G. (2001) *Interim Assessment of the Effects of the Foot and Mouth Disease Outbreak on England's Biodiversity*. English Nature Research Reports 430, Peterborough: English Nature.

Robinson, M.F., Webber, M. and Stebbings, R.E. (2000) *Dispersal and Foraging Behaviour of Greater Horseshoe Bats, Brixham, Devon*. English Nature Research Reports 344, Peterborough: English Nature.

Robson, G., Percival, S.M. and Brown, A.F. (1994) *The Breeding Ecology of Curlew: a Pilot Study*. English Nature Research Report 127, Peterborough: English Nature.

Rodwell, J.S. (ed.) (1991) *British Plant Communities* II, *Mires and Heaths*, Cambridge: Cambridge University Press.

Rodwell, J.S. (ed.) (1992) *British Plant Communities* III, *Grasslands and Montane Communities*, Cambridge: Cambridge University Press.

Rodwell, J.S. (ed.) (2000) British Plant Communities V, *Maritime Communities and Vegetation of Open Habitats*, Cambridge: Cambridge University Press.

Rorison, I.H. (1986) The response of plants to acid soils. *Experientia*, 42: 357–362.

Rorison, I.H. (1990) Soils, mineral nutrition and climate. In: S.H. Hillier, D.W.H. Walton and D.A. Wells (eds) *Calcareous Grasslands. Ecology and Management*, pp. 21–28. Huntingdon, Bluntisham Books.

Rose, F. (2002) Living in interesting times: climate change spells an uncertain future for our nation's wild plants. *Plantlife*, spring 2002: 10–11.

Rose, R.J., Webb, N.R., Clarke, R.T. and Traynor, C.H. (2000) Changes on the heathlands in Dorset, England, between 1987 and 1996. *Biological Conservation*, 93: 117–125.

Rossiter, S.J., Jones, G., Ransome, R.D. and Barratt, E.M. (2000) Parentage, reproductive success and breeding behaviour in the greater horseshoe bat (*Rhinolophus ferrumequinum*). *Proceedings of the Royal Society of London* B *Biological Sciences*, 267: 545–551.

Sala, O.E., Chapin, F.S. III, Armesto, J.J., Berlow,

E., Bloomfield, J., Dirzo, R., Huber-Sanwald, E., Huenneke, L.F., Jackson, R.B., Kinzig, A., Leemans, R., Lodge, D.M., Mooney, H.A., Oesterheld, M., Poff, N.L., Sykes, M.T., Walker, B.H., Walker, M. and Wall, D.H. (2000). Global biodiversity scenarios for the year 2100. *Science*, 287: 1770–1774.

Salt, D.E., Blaylock, M., Kumar, N.P.B.A., Dushenkov, V., Ensley, B.D., Chet, I. and Raskin, I. (1995) Phytoremediation: a novel strategy for the removal of toxic metals from the environment using plants. *Biotechnology*, 13: 468–474.

Salt, D.T. and Whittaker, J.B. (1998) *Insects on Dock Plants*. Naturalists' Handbooks 26, Slough: Richmond Publishing.

Sanderson, N.A (1998) *A Review of the Extent, Conservation Interest and Management of Lowland Acid Grassland in Britain* I–II, English Nature Research Report 259, Peterborough: English Nature.

Schiestl, F.P and Ayasse, M. (2001) Post-pollination emission of a repellent compound in a sexually deceptive orchid: a new mechanism for maximising reproductive success? *Oecologia*, 126: 531–534.

Schiestl, F.P., Ayasse, M., Paulus, H.F., Lofstedt, C., Hansson, B.S., Ibarra, F., Franke W. (2000) Sex pheromone mimicry in the early spider orchid (*Ophrys sphegodes*): patterns of hydrocarbons as the key mechanism for pollination by sexual deception. *Journal of Comparative Physiology* A *Sensory Neural and Behavioural Physiology*, 186: 567–574.

Schlapfer, F. and Fischer, M. (1998) An isozyme study of clone diversity and relative importance of sexual and vegetative recruitment in the grass *Brachypodium pinnatum*. *Ecography*, 21: 351–360.

Scottish Natural Heritage (1996) *Soils. Scotland's Living Landscapes*, Perth: Scottish Natural Heritage.

Scottish Natural Heritage (2001) *Managing Grasslands for Wildlife on Scottish Farms*, Perth: Scottish Natural Heritage.

Shaw, P. (1994) Orchid woods and floating islands – the ecology of fly ash. *British Wildlife*, 5: 149–157.

Shaw, P.J.A. (1996) Role of seedbank substrates in the revegetation of fly ash and gypsum in the United Kingdom. *Restoration Ecology*, 4: 61–70.

Shaw, P. (2000) *The Acid Tests? Studies of the Ecological Effects of Atmospheric Pollutants*, Swindon: Innogy.

Simmons, E. (1999) Restoration of landfill sites for ecological diversity. *Waste Management and Research*, 17: 511–519.

Simpson, N.A. (1993) *A Summary Review of Information on the Autecology and Control of Six Grassland Weed Species*. English Nature 44, Peterborough: English Nature.

Simpson, N.A. and Jefferson, R.G. (1996) *Use of Farmyard manure on Semi-natural (Meadow) Grassland*. English Nature Research Report 150, Peterborough: English Nature.

Skinner, B. (1984) *Moths of the British Isles*, Harmondsworth: Viking.

Skinner, G.J. and Allen, G.W. (1996) *Ants*. Naturalists' Handbooks 24, Slough: Richmond Publishing.

Small, R.W., Poulter, C., Jeffreys, D.A. and Bacon, J.C. (1999) *Towards Sustainable Grazing for Biodiversity. An Analysis of Conservation Grazing Projects and their Constraints*. English Nature Research Report 316, Peterborough: English Nature.

Smith, C.J. (1980) *Ecology of the English Chalk*, London: Academic Press.

Smith, P.H. (2000) Classic wildlife sites: the Sefton Coast sand dunes, Merseyside. *British Wildlife*, 12: 28–36.

Snow, C.S.R. and Marrs, R.H. (1997) Restoration of *Calluna* heathland on a bracken *Pteridium*-infested site in north-west England. *Biological Conservation*, 81: 35–42.

Spellerberg, I.F. (1991) Biogeographical basis of conservation. In: I.F. Spellerberg, F.B. Goldsmith and M.G. Morris (eds) *The Scientific Management of Temperate Communities for Conservation*, Oxford: Blackwell.

Spurgeon, D.J. and Hopkin, S.P. (1999) Seasonal variation in the abundance, biomass and biodiversity of earthworms in soils contaminated with metal emissions from a primary smelting works. *Journal of Applied Ecology*, 36: 173–183.

Stace, C. (1997) *New Flora of the British Isles* (second edition), Cambridge: Cambridge University Press.

Sternberg, M., Brown, V.K., Masters, G.J. and Clarke, I.P. (1999) Plant community dynamics in a calcareous grassland under climate change manipulations. *Plant Ecology*, 143: 29–37.

Strandberg, M. and Johansson, M. (1999) Uptake of nutrients in *Calluna vulgaris* seed plants grown with and without mycorrhiza. *Forest Ecology and Management*, 114: 129–135.

Strohm, E. (2000) Factors affecting body size and fat content in a digger wasp. *Oecologia*, 123: 184–191.

Strohm, E. and Linsenmair, K.E. (1994) Leaving the cradle: how beewolves (*Philanthus triangulum* F.) obtain the necessary spatial information for emergence. *Zoology: Analysis of Complex Systems*, 98: 137–146.

Strohm, E. and Linsenmair, K.E. (1997) Female size affects provisioning and sex allocation in a digger wasp. *Animal Behaviour*, 54: 23–34.

Strohm, E. and Linsenmair, K.E. (2001) Females of the European beewolf preserve their honeybee prey against competing fungi. *Ecological Entomology*, 26: 198–203.

Sumpton, K.J. and Flowerdew, J.R. (1985) The ecological effects of the decline in rabbits (*Oryctolagus cuniculus* L.) due to myxomatosis. *Mammal Review*, 15: 151–186.

Sutherland, W.J. (1996). *Ecological Census Techniques*, Cambridge: Cambridge University Press.

Svensson, L. and Grant, P. J. (1999) *Bird Guide*, London: HarperCollins.

Tallowin, J.R.B. (1997) *The Agricultural Productivity of Lowland Semi-natural Grassland. A Review.* English Nature Research Report 233, Peterborough: English Nature.

Tallowin, J.R.B. and Smith, R.E.N. (1994) *The Effects of Inorganic Fertilizers in Flower-rich Hay Meadows on the Somerset Levels.* English Nature Research Report 87, Peterborough: English Nature.

Tallowin, J.R.B. and Smith, R.E.N. (2001) Restoration of a *Cirsio-Molinietum* fen meadow on an agriculturally improved pasture. *Restoration Ecology*, 9: 167–178.

Telfer, M.G. and Hassall, M. (1999) Ecotypic differentiation in the grasshopper *Chorthippus brunneus*: life history varies in relation to climate. *Oecologia*, 121: 245–254.

Thomas, C.D., Bodsworth, E.J., Wilson, R.J., Simmons, A.D., Davies, Z.G., Musche, M. and Conradt, L. (2001) Ecological and evolutionary processes at expanding range margins. *Nature*, 411: 577–581.

Thomas, J.A. (1990) The conservation of Adonis Blue and Lulworth Skipper butterflies – two sides of the same coin. In: S.H. Hillier, D.W.H. Walton and D.A. Wells (eds) *Calcareous Grasslands. Ecology and Management*, pp. 112–117. Huntingdon: Bluntisham Books.

Thomas, J.A. (1999) The large blue butterfly – a decade of progress. *British Wildlife*, 11: 22–27.

Thomas, J.A. and Elmes, G.W. (2001) Food-plant niche selection rather than the presence of ant nests explains oviposition patterns in the myrmecophilous butterfly genus *Maculinea. Proceedings of the Royal Society of London* B *Biological Sciences*, 268: 471–477.

Thomas, J.A., Elmes, G.W. and Wardlaw, J.C. (1998) Polymorphic growth in larvae of the butterfly *Maculinea rebeli*, a social parasite of *Myrmica* ant colonies. *Proceedings of the Royal Society of London Series* B *Biological Sciences*, 265: 1895–1901.

Thompson, D.B.A., Hester, A.J. and Usher, M.B. (1995) *Heaths and Moorland. Cultural Landscapes*, Edinburgh: HMSO.

Thompson, H.V. (1994) The rabbit in Britain. In: H.V. Thompson and C.M. King (eds) *The European Rabbit. The History and Biology of a Successful Colonizer*, pp. 64–107. Oxford: Oxford University Press.

Thompson, S., Larcom, A. and Lee, J.T. (1999) Restoring and enhancing rare and threatened habitats under agri-environment agreements: a case study of the Chiltern Hills Area of Outstanding Natural Beauty, UK. *Land Use Policy*, 16: 93–105.

Tilling, S.M. (1987) *A Key to the Major Groups of British Terrestrial Invertebrates*, Shrewsbury: Field Studies Council.

Tilman, D. (1997) Mechanisms of plant competition. In: M.J. Crawley (ed.) *Plant Ecology* (second edition), pp. 239–261. Oxford: Blackwell.

Tilman, D., Dodd, M.E., Silvertown, J., Poulton, P.R., Johnston, A.E. and Crawley, M.J. (1994) The Park Grass Experiment: insights from the most long-term ecological study. In: R.A. Leigh and A.E. Johnston (eds) *Long-term Experiments in Agricultural and Ecological Sciences*, Wallingford: CAB International.

Treweek, J. (1999) *Ecological Impact Assessment*, Oxford: Blackwell.

Treweek, J., Jose, P. and Benstead, P. (eds) (1997) *The Wet Grassland Guide. Managing Floodplain and Coastal Wet Grassland for Wildlife*, Sandy: RSPB.

Trist, P.J.O. (1981) *Fritillaria meleagris* L.: its survival and habitats in Suffolk, England. *Biological Conservation*, 20: 5–14.

Tubbs, C.R. (1986) *The New Forest*, London: Collins.

Tubbs, C.R. (1991) Grazing the lowland heaths. *British Wildlife*, 2: 276–289.

Tubbs, C.R. (2001) *The New Forest* (second edition), Lyndhurst: New Forest Ninth Centenary Trust.

UK BAP (2001) http://www.ukbap.org.uk

UK Biodiversity Group (1998a) *UK Biodiversity Group Tranche 2 Action Plans* I, *Vertebrates and Vascular Plants*, Peterborough: English Nature.

UK Biodiversity Group (1998b) *UK Biodiversity Group Tranche 2 Action Plans* II, *Terrestrial and Freshwater Habitats*, Peterborough: English Nature.

UK Biodiversity Group (1999a) *UK Biodiversity Group Tranche 2 Action Plans* III, *Plants and Fungi*, Peterborough: English Nature.

UK Biodiversity Group (1999b) *UK Biodiversity Group Tranche 2 Action Plans* IV, *Invertebrates*, Peterborough: English Nature.

UK Biodiversity Group (1999c) *UK Biodiversity Group Tranche 2 Action Plans* V, *Maritime Species and Habitats*, Peterborough: English Nature.

UK Biodiversity Group (1999d) *UK Biodiversity Group Tranche 2 Action Plans* VI, *Terrestrial and Freshwater Species and Habitats*, Peterborough: English Nature.

UK Biodiversity Steering Group (1995) *Biodiversity. The UK Steering Group Report* II, *Action Plans*, London: HMSO.

van den Berg, L.J.L., Bullock, J.M., Clarke, R.T., Langston, R.H.W. and Rose, R. (2001). Territory selection by the Dartford warbler (*Sylvia undata*) in Dorset, England: the role of vegetation type, habitat fragmentation and population size. *Biological Conservation*, 101: 217–228.

van der Heijden, M.G.A., Klironomos, J.N., Ursic, M., Moutoglis, P., Streitwolf-Engel, R., Boller, T., Wiemken, A. and Sanders, I.R. (1998) Mycorrhizal fungal diversity determines plant biodiversity, ecosystem variability and productivity. *Nature*, 396: 69–72.

van der Hoek, D. and Braakhekke, W. (1998) Restoration of soil chemical conditions of fen-meadow plant communities by water management in the Netherlands. In: C.B. Joyce and P.M. Wade (eds) *European Wet Grasslands*, pp. 265–275. Chichester: Wiley.

Veitch, N., Webb, N.R. and Wyatt, B.K. (1995) The application of geographic information systems and remotely sensed data to the conservation of heathland fragments. *Biological Conservation*, 72: 91–97.

Venus, C. (1997) *Conservation and the Farm Business*. English Nature Research Report 255. Peterborough: English Nature.

Vogelei, A. and Greissl, R. (1989) Survival strategies of the crab spider *Thomisus onustus* Walckenaer 1806 (Chelicerata, Arachnida, Thomisidae). *Oecologia*, 80: 513–515.

Waite, S. (1994) Field evidence of plastic growth response to habitat heterogeneity in the clonal herb *Ranunculus repens*. *Ecological Research*, 9: 311–316.

Wakeham-Dawson, A., Szoszkiewicz, K., Stern, K. and Aebisher, N.J. (1998) Breeding skylarks *Alauda arvensis* on Environmentally Sensitive Area arable reversion grass in southern England: survey-based and experimental determination of density. *Journal of Applied Ecology*, 35: 635–648.

Walker, C.H., Hopkin, S.P., Sibly, R.M. and Peakall, D.B. (1996) *Principles of Ecotoxicology*, London: Taylor and Francis.

Warren, J.M (2000) The role of white clover in the loss of diversity in grassland habitat restoration. *Restoration Ecology*, 8: 318–323.

Warren, M.S. (1994) The UK status and suspected metapopulation structure of a threatened European butterfly, the marsh fritillary *Eurodryas aurinia*. *Biological Conservation*, 67: 239–249.

Warren, M., Clarke, S. and Currie, F. (2001) The Coppice for Butterflies challenge: a targeted grant scheme for threatened species. *British Wildlife*, 13: 21–28.

Watson, D., Hack, V. and Fasham, M. (2000) *Wildlife Management and Habitat Creation on Landfill Sites. A Manual of Best Practice*, Crake Holme: Ecoscope Applied Ecologists.

Watt, A.S. (1981a) A comparison of grazed and ungrazed grassland A in East Anglian breckland. *Journal of Ecology*, 69: 499–508.

Watt, A.S. (1981b) Further observations on the effects of excluding rabbits from grassland A in East Anglian breckland: the pattern of change and factors affecting it (1936–73). *Journal of Ecology*, 69: 509–536.

Watt, T.A. (1993) *Introductory Statistics for Biology Students*, London: Chapman and Hall.

Webb, N.R. (1986) *Heathlands*, London: Collins.

Webb, N.R. (1990) Changes on the heathlands of Dorset, England, between 1978 and 1987. *Biological Conservation*, 51: 273–286.

Webb, N.R. (1998) The traditional management of European heathlands. *Journal of Applied Ecology*, 35: 987–990.

Webb, N.R. and Haskins, L.E. (1980) An ecological survey of heathland in the Poole Basin, Dorset, England, in 1978. *Biological Conservation*, 17: 281–296.

Webb, N.R., Veitch, N. and Pywell, R.F. (1995) Increasing the extent of heathland in southern England. In: K.M. Urbanska and K. Grodzinska (eds) *Restoration Ecology in Europe*, pp. 93–111. Zurich: Geobotanical Institute SFIT.

Wheater, C.P. (1999) *Urban Habitats*, London: Routledge.

Wheater, C.P. and Cook, P.A. (2000) *Using Statistics to Understand the Environment*, London: Routledge.

Wheater, C.P. and Cook, P.A. (2003) *Studying Invertebrates*. Naturalists' Handbooks 28, Slough: Richmond Publishing.

Wheater, C.P. and Cullen, W.R. (1997) The flora and invertebrate fauna of abandoned limestone quarries in Derbyshire, United Kingdom. *Restoration Ecology*, 5: 77–84.

White, R.E. (1997) *Principles and Practice of Soil Science. The Soil as a Natural Resource* (third edition), Oxford: Blackwell.

Wickman, P.-O. and Jansson, P. (1997) An estimate of female mate searching costs in the lekking butterfly *Coenonympha pamphilus*. *Behavioral Ecology and Sociobiology*, 40: 321–328.

Willems, J.H. (1990) Calcareous grassland in continental Europe. In: S.H. Hillier, D.W.H. Walton and D.A. Wells (eds) *Calcareous Grasslands. Ecology and Management*, pp. 3–10. Huntingdon: Bluntisham Books.

Willems, J.H. (2001) Problems, approaches and results in restoration of Dutch calcareous grassland during the last thirty years. *Restoration Ecology*, 9: 147–154.

Williams, P.H. (1989) *Bumblebees and their Decline in Britain*, Ilford: Association of Beekeepers.

Wilson, A.M., Vickery, J.A. and Browne, S.J. (2001) Numbers and distribution of northern lapwings *Vanellus vanellus* breeding in England and Wales in 1998. *Bird Study*, 48: 2–17.

Winter, M., Evans, N. and Gaskell, P. (1998) *The CAP Beef Regime in England and its Impact on Nature Conservation*. English Nature Research Report 265. Peterborough: English Nature.

Winter, M., Mills, J. and Wragg, A. (2000) *Practical Delivery of Farm Conservation Management in England*. English Nature Research Report 393, Peterborough: English Nature.

Wolters, V., Silver, W.L., Bignell, D.E., Coleman, D.C., Lavelle, P., van der Putten, W.H., de Ruiter P., Rusek, J., Wall, D.H., Wardle, D.A., Brussaard, L., Dangerfield, J.M., Brown, V.K., Giller, K.E., Hooper, D.U., Sala, O., Tiedje, J. and van Veen, J.A. (2000) Effects of global changes on above- and below-ground biodiversity in terrestrial ecosystems: implications for ecosystem functioning. *BioScience*, 50: 1089–1098.

Wood, M. (1995) *Environmental Soil Biology* (second edition), London: Blackie.

SUBJECT INDEX

•

SPECIES INDEX

•

The common name is followed by the *scientific name*. Where sources used in this book have used alternative scientific names, these are included in brackets. For further information, refer to the species list.